# PERCHLORATE IN THE ENVIRONMENT

# ENVIRONMENTAL SCIENCE RESEARCH



*Recent Volumes in this Series*

Volume 46—PRODUCED WATER: Technological/Environmental Issues and Solutions
Edited by James P. Ray and F. Rainer Engelhardt

Volume 47—GLOBAL ENERGY STRATEGIES: Living with Restricted Greenhouse Gas Emissions
Edited by James C. White

Volume 48—GLOBAL ATMOSPHERIC–BIOSPHERIC CHEMISTRY
Ronald G. Prinn

Volume 49—BIODEGRADATION OF NITROAROMATIC COMPOUNDS
Edited by Jim C. Spain

Volume 50—BIOMONITORS AND BIOMARKERS AS INDICATORS OF ENVIRONMENTAL CHANGE: A Handbook
Edited by Frank M. Butterworth, Lynda D. Corkum, and Judith Guzmán-Rincón

Volume 51—CHEMISTRY FOR THE PROTECTION OF THE ENVIRONMENT 2
Edited by Lucjan Pawłowski, William J. Lacy, Christopher G. Uchrin, and Marzenna R. Dudzińska

Volume 52—PRODUCED WATER 2: Environmental Issues and Mitigation Technologies
Edited by Mark Reed and Ståle Johnsen

Volume 53—EVALUATING CLIMATE CHANGE ACTION PLANS: National Actions for International Commitment
Edited by James C. White

Volume 54—BIOTECHNOLOGY IN THE SUSTAINABLE ENVIRONMENT
Edited by Gary S. Sayler, John Sanseverino, and Kimberly L. Davis

Volume 55—CHEMISTRY FOR THE PROTECTION OF THE ENVIRONMENT 3
Edited by Lucjan Pawłowski, Marjorie A. Gonzales, Marzenna R. Dudzińska, and William J. Lacy

Volume 56—BIOMONITORS AND BIOMARKERS AS INDICATORS OF ENVIRONMENTAL CHANGE 2
Edited by Frank M. Butterworth, Amara Gunatilaka, and Maria Eugenia Gonsebatt

Volume 57—PERCHLORATE IN THE ENVIRONMENT
Edited by Edward Todd Urbansky

A Continuation Order Plan is available for this series. A continuation order will bring delivery of each new volume immediately upon publication. Volumes are billed only upon actual shipment. For further information please contact the publisher.

# PERCHLORATE IN THE ENVIRONMENT

**Edited by**

**Edward Todd Urbansky**

United States Environmental Protection Agency
Cincinnati, Ohio

*This volume is not a publication of the EPA and should not be construed to reflect agency policy.*

Kluwer Academic / Plenum Publishers
New York, Boston, Dordrecht, London, Moscow

For my devoted and loving parents
Edward James and Carol Ann Urbansky

# Preface

Why and how did this volume come to be? The chapters in this book are derived primarily from papers presented at an American Chemical Society symposium in the Division of Environmental Chemistry held during the 218th national meeting in August 1999 in New Orleans, Louisiana. A few presentations did not make it as chapters into the book and a few chapters were added to round out the coverage and make it a more nearly complete work.

My own involvement with perchlorate began two months after I joined EPA in 1997 when Steve Pia at EPA–Las Vegas called to ask me about detecting perchlorate at trace concentrations in drinking water. As an inorganic chemist who spent his graduate school years soaking in 0.1–8.5 M sodium perchlorate solutions, I was immediately intrigued—and hooked. And thus began the path down the road of *Perchlorate in the Environment.*

The authors are among the best investigators in the field of environmental perchlorate research in the United States and Canada. Their careers span industry, universities, trade organizations, and government laboratories. Their work ranges from pure and applied chemistry to engineering to biology to hydrology and more. They define the state of the science and I daresay even its very frontier. I am especially pleased to have papers from such distinguished authors as Bruce Logan, Jim Espenson, John Coates, and Joe Earley. Each paper represents a significant and timely contribution to a knowledge base in real time that we have come to know as *Perchlorate in the Environment.* In addition, we have the benefit of first hand word from industry insiders, such as Raman Venkatesh, Evan Cox, and Mark Greene, whose companies have active large scale projects going on right now.

Unfortunately, some of the presentations made at the ACS meeting did not make it into the book for one reason or another. Harding Lawson Associates has active biological reactors in EPA's Region 9, and John Catts had hoped to contribute. I would encourage readers to check out the extended abstract he wrote for the meeting and to contact him for more information. In addition, there is ongoing work at EPA's National Exposure Research Laboratory (Athens, GA) that did not make it into this book, most of it dealing with interactions of plants with perchlorate. Besides work in North America, research in this area is taking place in other countries. By the time this book is available, Markus Forstmeier, shall have completed his thesis in Berlin, Germany. Markus has studied anaerobic perchlorate remediation in up-flow fluidized bed reactors. Without any question, we will continue to see publication of many papers in this rapidly developing field.

Although this book is mostly by research scientists, I have challenged the authors to make the concepts and lessons accessible for a wide variety of audiences—regulators, policy makers, utility plant operators, biologists, hydrogeologists, chemists, toxicologists, engineers, students, or teachers—anyone who might reasonably be called an interested party. The authors should be weary from my overshadowing demand to make the volume comprehensible to anyone with a bachelor's degree in a science field. If I have succeeded, this book is more than the sum of its parts.

Portions of this book will undoubtedly be out of date by the time it is published. However, I have strived to speed the process up as much as possible. Four months elapsed between the symposium and the time this book reached the publisher. From symposium to publication will be about 9 months. This has been no small feat as anyone who has edited a camera-ready book will attest. The primary objective was to get the information out. I hope it will be a useful reference to everyone whose work and life is touched by perchlorate—a remarkable chemical species.

*Perchlorate in the Environment* represents a monumental effort by many people—directly or indirectly. First, I must recognize my parents, to whom this volume is dedicated. Second, I note my maternal grandparents, Helen and Bill Seman. Third, I would be remiss if I failed to name the two dogs, Annie and Taffy—who have faithfully and patiently sat by while I have spent many hours working on this book—and cat Maggie who is no longer with us. Recognition also goes to my colleagues on the Interagency Perchlorate Steering Committee who have helped and encouraged me along the way, especially Annie Jarabek, Peter Grevatt, Mike Osinski, Dave Tsui, Kevin Mayer, and Dan Rogers.

There are many people who have touched my life over the years, from elementary school teachers to university professors to friends. Among those teachers, professors, and friends who guided this path: Mary Eaton, Elizabeth and Delores Peach, Mary Hodge, Mary Ann Boylan, Mary D'Alessio, Grace Haley, Pam Byers, Tresalyn Gentile, Bob Prehoda, Cliff Jaszcar, Susan Dalla Betta, Monica Sinai, Helen Wilson, Martha McKee, Debbie Morinello, Terry Phillips, Bonnie Harshbarger, Richard Smith, Ted Shlanta, John Salamone, Frank Blankenbuehler, Dave Vandermer, David R. Spahr, Linda Hickey, Don Brancolini, (Baldwin-Whitehall School District, Pittsburgh, Pennsylvania); Glen Rodgers, Ann Sheffield, Doug Smith, Ed Walsh, Richard Bivens, Paul Deutsch, Charles Cable, Jim Lombardi, Tony LoBello, Michael Barry, Joel Smith, Sally Hair, Brian Reid, Christine Nebiolo (Allegheny College, Meadville, Pennsylvania); Fu-Mei and Fu-Tyan Lin, Dick Devine (Calgon Corporation, Pittsburgh, Pennsylvania); Thomas Bein, Bill Robinson, Dick Walton, Bartow Culp, Harry Pardue, and especially Dale Margerum (Purdue University, West Lafayette, Indiana).

There are those, without whose intellectual, logistical, or editorial support, this volume would not have been possible: co-presiders Mike Schock and James Clark, Ruth Hathaway and Allan Ford of the ACS Environmental Division, and Susan Safren at Kluwer/Plenum. Of course, the majority of credit goes to the authors themselves who have written the very words that line the pages. I have come to know many as friends over the past few years. Special mention goes to Matt Magnuson, who is responsible for the success of our ESI-MS work. Our library's reference and ILL staff have been outstanding: Anna Hood, Jennifer Heffron, Kathy Connick, and Jennie Thomas. And I must not forget support and encouragement from EPA-NRMRL management: Tim Oppelt, Bob Clark, Bob Thurnau, Hugh McKinnon, Andy Avel, and Frank Princiotta.

*—Edward Todd Urbansky*
West Harrison, Indiana

# Contents

# Introduction

On March 2, 1998, the United States Environmental Protection Agency's Office of Water formally added perchlorate ($ClO_4^-$) to the drinking water contaminant candidate list (CCL).[1] As with any species on the CCL, it is unknown whether a National Primary Drinking Water Regulation (NPDWR) will eventually be promulgated. However, the Office of Water did conclude that there was sufficient chance for occurrence throughout the country to merit monitoring perchlorate under the Unregulated Contaminants Monitoring Rule (UCMR).[2]

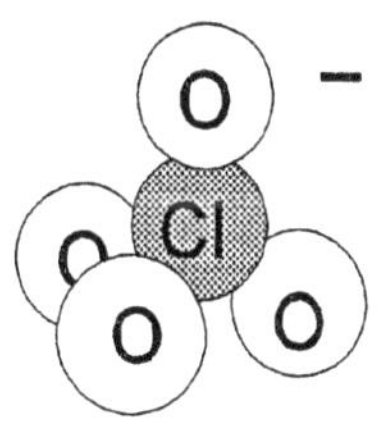


In 1997, perchlorate-contaminated water supplies were found in Nevada, Utah, and California. Contamination extends to the Colorado River at Lake Mead, thereby affecting not only metropolitan Las Vegas, but the entire Southern California region dependent on the trans-desert aqueduct. Based on available information about the risk to human health, a provisional action level of 18 ng $mL^{-1}$ ( = 18 μg $L^{-1}$ = 180 nmol $L^{-1}$) has been adopted by the California Department of Health Services. This has resulted in closing more than 30 wells in California. The concentration in Lake Mead (at the water intake for Las Vegas) has remained under 14 ng $mL^{-1}$, and the Nevada Division of Environmental Protection (part of the Department of Conservation and Natural Resources) has not adopted an action level. Perchlorate continues to be found around the nation, and the sources are not always identifiable. Certainly, waste streams from industrial or military operations that use perchlorate have undergone scrutiny, and this has reduced or eliminated discharge at some sites. Nearly every state has sites where groundwater or surface water contamination is a *possibility*. With the collection of drinking water data from publicly operated treatment works (POTWs) under the UCMR, the implications for potable water supplies will emerge.

The perchlorate ion is an interesting character: full of oxidizing chemical potential but generally without the energy to jump the hurdle of activation and pursue its thermodynamic destiny. In Chapter 1, Espenson explores "The Problem and Perversity of Perchlorate." Perchlorate's chemical and physical properties have been described in some detail elsewhere, and perchlorate can be difficult to quantitate but especially to reduce.[3,4] What makes perchlorate particularly unique is its combination of low (thermodynamic) stability and low (kinetic) lability. Although the driving force for perchlorate to act as an oxidant is very high, perchlorate reduction is slow even with strong, labile reductants, e.g., $Ti^{3+}$, as Earley discusses in Chapter 9.

At present, toxicology studies are incomplete and a safe concentration that ensures no human health effects has not been established. In mammalian physiology, perchlorate disrupts the production of thyroid hormones needed for metabolism. The thyroid gland takes up iodide ion on the basis of ionic size. Unfortunately, perchlorate is very similar in size and can be taken into the thyroid cells. However, unlike iodide which is then incorporated into hormones, the perchlorate simply takes up space. The result is lower intrathyroid iodide and thus lower hormone output. The EPA's National Center for Environmental Assessment set a provisional concentration of 18 ng $mL^{-1}$ ( = 18 μg $L^{-1}$ = 180 nmol $L^{-1}$) based on a minimum of studies in 1992. At that time, no need for additional work was apparent. Efforts to update the reference dose (RfD) and corresponding no observable adverse effects level (NOAEL) are underway. A draft assessment was released in 1999 and underwent external review; the reviewers recommended several refinements. Completion depends on USAF funding, which has paid for most of the health effects studies. Clark elaborates on perchlorate toxicology in Chapter 3.

Most of the perchlorate contamination in the American West appears to be the result of previously legal dumping of wastewater from military installations and defense contractors. For this reason, the U.S. Air Force has taken a lead role in clean-up and research. The original source was probably the ammonium salt, which serves as both a solid oxidant (often reacted with powdered aluminum as a reducing agent) and an energy booster. Below 300 °C, the decomposition of ammonium perchlorate is described by eq 1:[5]

$$4\ NH_4ClO_4 \rightarrow 2\ Cl_2 + 2N_2O + O_2 + 8\ H_2O \qquad (1)$$

Although the ammonium cation itself serves as a reductant in this intramolecular redox reaction, the first three products in eq 1 are oxidants. Even though the ammonium is completely consumed, the desire for electrons remains unsatiated.

Most of the chapters describe processes and strategies for use in water treatment or site remediation. At present, anion exchange and microbiological reduction are the dominant areas of research. What the future holds for nanofiltration or electrochemical reduction remains to unfold. Obviously, the best solution to pollution is to stop contribution. That is, it is far better not to discharge a pollutant in the first place. While dilution may mitigate pollution, it is harder to clean up 1 million gallons (3.78 ML) of water that contains 82 ng $ClO_4^-$ $mL^{-1}$ than to destroy the 310 g of perchlorate (e.g., 382 g of $NaClO_4$) it contains. Likewise, treating a waste effluent that contains 82 μg $ClO_4^-$ $mL^{-1}$ is far easier than trying to treat a downstream drinking water source at 1000-fold dilution, i.e., 82 ng $ClO_4^-$ $mL^{-1}$. This is nothing new.

After a few chapters of background coverage, there are several chapters on the analytical chemistry of perchlorate. Ion chromatography and electrospray ionization mass spectrometry dominate this area. The remainder and majority of the chapters relate to treatment and remediation—dealing with specific technologies or strategies or dealing with what factors influence these processes.

One of the issues to gain prominence recently is the presence of naturally occurring perchlorate in Chile saltpeter (sodium nitrate, $NaNO_3$, also known as soda niter or nitratine). Analyses of Chile saltpeter by EPA's National Risk Management Research Laboratory and DoE's Oak Ridge National Laboratory have found concentrations of ~ 1 g $ClO_4^-$ $(kg\ NaNO_3)^{-1}$. Although this material constitutes a small fraction of total U.S. fertilizer consumption, application of Chile saltpeter is often highly localized; for example, it is particularly favored by tobacco farmers. Application of 1 metric ton corresponds to a dose of 1 kg of $ClO_4^-$ (1.2 kg of $NaClO_4$). To be diluted to 18 ng $ClO_4^-$ $mL^{-1}$ requires 15,000 gal (56,000 L) of water. What does this mean for agricultural use of this product? The answer is unclear.

Natural attenuation of perchlorate by biological processes is still unstudied. However, uncatalyzed chemical reduction is too slow under the dilute and near-neutral conditions encountered environmentally or botanically. Limited information is available on the rate and favorability of uptake by plants; this includes both crop and non-crop plants. If the perchlorate is taken up, does it stay in the edible portion? Suppose tobacco, for instance, were shown to accumulate perchlorate. Would it matter? The perchlorate would literally go up in smoke. Dilution by rain and irrigation is a factor as is the destiny of the run-off. Does it flow to a large, swift river? A deep aquifer? What is the dilution factor? What lives there? Will anyone or any animal drink the water? Environmental problems pose many technical, scientific, societal, and political questions. We will concentrate on the technical and scientific issues, but we must remember that science and technology do not exist in a vacuum, and the fate of perchlorate as an environmental contaminant shall be determined by more than the state of the science.

During the first half of 2000, a survey of fertilizers is planned to ascertain what level of perchlorate exposure might be expected. It is too soon to speculate on exactly how that will take place or what the final outcome might be. I hope you will all keep your eyes on the literature as we all continue to explore and learn more about perchlorate in the environment.

## REFERENCES


1. Perciasepe, R. "Part III. Environmental Protection Agency. Announcement of the drinking water contaminant candidate list; notice." *Federal Register* **1998**, *63* (40), 10273–10287.

2. Browner, C.M. "Part II. Environmental Protection Agency. 40 CFR Parts 9, 141 and 142. Revisions to the unregulated contaminant monitoring regulation for public water systems; final rule." *Federal Register* **1999**, *64* (180), 50555–50620.

3. Urbansky, E.T. "Perchlorate chemistry: implications for analysis and remediation." *Bioremediation Journal* **1998**, *2*, 81–95.

4. Urbansky, E.T.; Schock, M.R. "Issues in managing the risks associated with perchlorate in drinking water." *Journal of Environmental Management* **1999**, *56*, 79–95.

5. Schilt, A.A. *Perchloric Acid and Perchlorates*. GFS Chemicals, Inc.: Columbus, 1979; p. 35, and references therein.

A testament to the power of ammonium perchlorate as an oxidant in the aerospace era—the first launch of the space shuttle *Columbia*, April 12, 1981. Courtesy of NASA, image # 581-30461.

# Chapter 1
# THE PROBLEM AND PERVERSITY OF PERCHLORATE

James H. Espenson

Ames Laboratory and Department of Chemistry, Iowa State University of Science and Technology, Ames, Iowa 50011. E-mail: Espenson@AmesLab.gov

## INTRODUCTION

The title might just have well have contained additional words, such as "peculiarity" and "persistence," for each in its way is further characteristic of this species. Perchlorate is indeed peculiar, in that its reactions in practice are usually not those predicted from reliable thermodynamic calculations; persistent, in that spontaneous reactions do not occur, leaving perchlorate in place, and perverse in that factors other than thermodynamics, kinetics in particular, govern its actual behavior. These issues may for some areas of chemistry create a problem, if the accumulation of perchlorate and its resistance poses a difficulty; for others, the lack of reactivity of perchlorate creates an opportunity, in that perchlorate salts can be used in many situations requiring an inert electrolyte.

The essence of the matter is this: reliable thermodynamic data predict that perchlorate salts, including ammonium perchlorate, would be vigorously reactive substances. They would be unstable in themselves, and in solution perchlorate ions would be a vigorous and reactive oxidizing agent. Practice indicates otherwise, however, another instance of the well-established dictum that thermodynamics establishes which reactions *can* occur, but kinetics governs those that *will*. In the field of perchlorate chemistry, kinetic barriers are a major consideration.

The formula of the perchlorate ion is $ClO_4^-$, which is the negatively-charged ion present in salts such as the common compounds sodium perchlorate, $NaClO_4$, and ammonium perchlorate, $NH_4ClO_4$. The solid salts contain perchlorate ion. Another common commercial form is perchloric acid, $HClO_4$. These salts and perchloric acid are readily soluble in water. Such solutions contain $ClO_4^-$. This ion consists of the unique element, Cl, in the center of a tetrahedral grouping of the four oxygens; the ion is isostructural with the common compound methane. The negative charge is dispersed evenly over the four oxygen atoms, and not concentrated anywhere locally. That dispersion of charge provides one reason for the fact that perchlorate ion is notorious for its feeble ability to bind to positively-charged metallic centers. Perchlorate complexes thus are rare; in dilute aqueous solution, the medium of interest to us here, such complexes are not found.

When dissolved in water, perchlorate salts are not reactive through coordination, and in fact, hardly react at all in any manner. In fact, the situation is just the opposite. Perchlorates are so notorious for their lack of chemical reactivity that they often find use as inert salts ("supporting electrolytes") in aqueous solutions. Under conditions for other reactions, not involving perchlorate, no reaction is found. In this context, the addition of

*Perchlorate in the Environment*, edited by Urbansky.
Kluwer Academic/Plenum Publishers, New York, 2000.

perchlorate is used to provide a constant ionic environment for the reactions of other ionic materials, being used because perchlorate ions are entirely inert.

The purpose here is to review what is known about the chemical reactivity of perchlorate ions, particularly in an aqueous environment.

This contrast between prediction and actuality in perchlorate chemistry provides a cogent reminder that thermodynamics defines the possible; the actuality may fall short. When that is so, it is not because the thermodynamic data or predictions of chemical reactions made on that basis are in any way unreliable. Rather, the situation is that rates of one or more chemical reactions are quite low. The reactions of perchlorate ions have proved to be so slow, in fact, that a long time, perhaps prohibitively long, would be required to complete any of the chemical reactions in question. One says that perchlorate reactions demonstrate high kinetic barriers. An exploration of why that is so, and what one might do about it, is a part of this presentation.

## PERCHLORATE REACTIONS

### Specific Spontaneous Reactions of Perchlorate

Certain thermodynamic data concerning perchlorate ions are summarized in Table 1. Refer to them will be made in due course. One of the reactions that perchlorate *can* [notice that the term is not *will*] undergo is the spontaneous release of oxygen. Depending on what one takes as the chlorine-containing product, different chemical equations can be written; in practice, chloride ions are ultimately formed in such reactions.

To illustrate the point derived from thermodynamic data, take the reaction of perchlorate that evolves molecular oxygen and yields chloride ions. The chemical equations are these:

$$ClO_4^- \rightleftharpoons Cl^- + 2O_2(g) \quad (1)$$

Thermodynamics speaks to the occurrence of this oxygen-releasing reaction under ambient conditions. The usual reference state is 298 K, or 25 °C, which will be used here. A measure of the extent to which this reaction can occur is given by the value of $\Delta G°$. Alternatively, one perfectly exact means of quantitatively expressing the tendency of perchlorate to release oxygen is to state the pressure of oxygen needed to maintain the given reaction at 50% completion.

**Table 1.** Various methods of representing the oxidizing strength of perchlorate ions

| | | |
|---|---|---|
| $ClO_4^- + 2H^+ + 2e^- \rightleftharpoons ClO_3^- + H_2O$ | $\Delta E° = +1.20$ V | (2) |
| $ClO_4^- + H_2 \rightleftharpoons ClO_3^- + H_2O$ | $\Delta G° = -232$ kJ | (3) |
| $ClO_4^- + 8H^+ + 8e^- \rightleftharpoons Cl^- + 4H_2O$ | $\Delta E° = +1.38$ V | (4) |
| $ClO_4^- + 8H_2 \rightleftharpoons Cl^- + 4H_2O$ | $\Delta G° = -1066$ kJ | (5) |

The thermodynamic data themselves are quite reliable, and they give this result:

$$P_{O_2} = 4 \times 10^{27} \text{ atm} \quad (6)$$

as the equilibrium pressure of oxygen at 50% conversion, when perchlorate and chloride ions are present at equal concentrations. This calculation forces one to conclude that an impossibly high pressure of oxygen would be needed to keep perchlorate as such were this reaction a reality under ordinary circumstances.

The practical fact of the matter is otherwise. Perchlorate salts, or solutions of perchlorate ions, do not evolve oxygen at room temperature, or under any other "ordinary" conditions. There is some other factor at work here. That factor has already been noted: the kinetic barrier toward this reaction is so high that perchlorate salts in practice are highly inert and unreactive materials.

A second reaction expected of perchlorate ions is that it should readily oxidize metal salts. To illustrate this point, several metal ions in oxidation state (II) will be used; in the following equation M is, for example, Fe, Cr, Mn, Ni, etc.:

$$ClO_4^- + 8M^{2+} + 4H_2O \rightleftharpoons Cl^- + 8M^{3+} + 8OH^- \quad (7)$$

There are other ways of writing these equations, such as:

$$ClO_4^- + 8M^{2+} + 8H^+ \rightleftharpoons Cl^- + 8M^{3+} + 4H_2O \quad (8)$$

Regardless of the method of writing the chemical equation, the facts of the matter are clear: (1) associated with these reactions are large thermodynamic driving forces for them to proceed as written and (2) the reactions do not occur under ordinary circumstances (short, perhaps, of excessively high temperatures), indicating that substantial kinetic barriers are imposed.

An abbreviated list of certain permissible reactions of perchlorate ions is given in Table 2. The list of perchlorate reactions that are of no practical importance could be made quite extensive, but there is no purpose so served. The point is made: reactions in which perchlorate serves as an oxidizing agent—these being the only ones known that could convert it to a different chemical form—are forbiddingly slow.

## Explosive Reactions of Perchlorate

Let us turn now to an examination of those reactions that stand as exceptions to the generalization just given. A certain category of perchlorate reactions occur with explosive force, often dangerously so. Chemists are warned against systems that contain hot, concentrated perchloric acid, against mixing certain organic compounds (notably, perchlorates with alcohols and dimethylsulfoxide) with perchloric acid, and in isolating certain perchlorate salts of transition metal complexes, particularly those containing organic ligands. A prominent journal in the field cautions against isolating such materials, advising small quantities if there is no suitable perchlorate substitute. This is what is said:[1]

> Perchlorate salts of metal complexes with organic ligands are potentially explosive. In general, when noncoordinating anions are required, every attempt should be made to substitute species such as the fluoro sulfonates for the perchlorates. If a perchlorate must be used, only small amounts of material should be prepared and should be handled with great caution. [Cf. *J. Chem. Educ.* **1973**, *50*, A335; *Chem. Eng. News* **1983**, *61*, 4; **1963**, *41*, 47.]

**Table 2.** Some Spontaneous Reactions of Perchlorate That Do *Not* Occur

From thermodynamics, perchlorate should, but does not:

(1) Oxidize water:

$$2ClO_4^- + 2H_2O \rightleftharpoons Cl_2 + 3O_2 + 4OH^- \quad (9)$$

(2) Oxidize most metal salts:

$$ClO_4^- + 8M^{2+} + 8H^+ \rightleftharpoons Cl^- + 8M^{3+} + 4H_2O \quad (10)$$

where M = Fe, Cr, Mn, Ni, etc.

(3) Oxidize bromide and iodide ions:

$$ClO_4^- + 8Br^- + 4H_2O \rightleftharpoons Cl^- + 4Br_2 + 8OH^- \quad (11)$$

Inevitably, perchlorates are explosive only in combination with oxidizable components, such as organic compounds or ammonium ions present. Here a few examples will be cited:

1. Explosive esters of perchloric acid
   The ethyl ester was made by distilling barium ethylsulfate and barium perchlorate; it is said to be incomparably more explosive than any known substance. [Hare and Boyle, 1841]

2. Perchlorates and certain organic solvents
   Attempts to add $HClO_4$ to $Me_2SO$ [dimethyl sulfoxide] led to violent explosions. [Cram and Lein, 1985]

3. Perchlorate salts of certain metals in certain organic solvents
   Lead perchlorate, $Pb(ClO_4)_2$, is readily soluble in organic solvents, but the alcoholic solution exploded so violently that it was not further investigated. [Willard and Kastner, 1931]

The explosive force of perchlorate oxidations, when they do occur, comes as no surprise, given the thermodynamic factors cited in the previous section coupled with the production of small gas molecules that expand thermally. The harmful nature of these reactions occurs when high concentrations and concentrated solutions are involved. The force of the explosion arises from the quantity of hot gas (molecular oxygen) released. Reactions such as these are "temperamental," and it is often difficult to anticipate circumstances under which a rapid reaction will occur.

**Smooth Reactions of Perchlorate**

There are, however, perchlorate reactions that do take place in a controlled and predictable fashion, even in dilute solutions in water. The known reactions of certain transition metal complexes will be cited, along with a few that do not take place. Table 3 presents a list of several of these reactions and the time required for each reaction under specified conditions of concentration and temperature.

One must ask why certain metal complexes appear in Table 3 as having the practical ability to reduce perchlorate ions. Most metals are absent from this tabulation because, like the two final entries in that table, they essentially fail to react. One factor that is evidently *not* a deciding factor is the thermodynamic driving force. The two listed species have the most negative reduction potential of any on the list: $Cr^{2+}$, –0.42 V and $Eu^{2+}$, –0.43 V. Nonetheless, perchlorate ions are not reduced by them.

**Table 3**. Certain metal ion reagents that do reduce perchlorate ions, and two that do not [a]

| Metal Complex | Reaction Time [b] | Reference |
|---|---|---|
| $[Ru(H_2O)_6]^{2+}$ | 87 h | Ref. 3 |
| $[Ti(H_2O)_6]^{3+}$ | 1 470 h | Ref. 4 |
| $[Ru(NH_3)_6]^{2+}$ | 7 300 h | Ref. 5 |
| $[U(H_2O)_n]^{3+}$ | est. ~11 000 h | Ref. 6 |
| $[V(H_2O)_6]^{2+}$ | 466 000 h | Ref. 7 |
| $[V(H_2O)_6]^{3+}$ | 555 000 h (63 y) | Ref. 8 |
| $[Cr(H_2O)_6]^{2+}$ | > $10^6$ h (no reaction) | a |
| $[Eu(H_2O)_n]^{2+}$ | est. ~ $10^7$ h | Ref. 7 |

[a] See Ref. 9 and the presentations in Refs. 10 and 11
[b] This is the time needed to complete one-half of the reaction at $10^{-3}$ M concentration, 25 °C

One factor common to the five successful reagents is each can exist as a metal-oxo complex (i.e., with a terminal metal-oxygen group) in an oxidation state of the metal two or sometimes one unit higher. Thus these are known and reasonably stable forms: $L_5Ru^{IV}{=}O^{2+}$, $(H_2O)_5Ti^{IV}{=}O^{2+}$, $(H_2O)_nU(O)_2^{+}$, $(H_2O)_5V^{IV}{=}O^{2+}$, and $(H_2O)_4V^{V}(O)_2^{+}$. No oxo complex has been reported for europium, and that for chromium, while known as the tetravalent $(H_2O)_5Cr^{IV}{=}O^{2+}$, has a lower thermodynamic stability.[2]

From that point of view, it is reasonable to suggest, as advocated by Taube,[10] that the mechanism involves the direct transfer of an oxygen atom from perchlorate to the metal. In general terms and using a +2 charge on the metal ion, one can suggest this transformation is involved:

$$ClO_4^- + M(H_2O)_n^{2+} \rightleftharpoons ClO_3^- + M(H_2O)_{n-1}O^{2+} + H_2O \qquad (12)$$

That reaction would be the most difficult and slowest part of the sequence of steps that needs to occur. The product is $ClO_3^-$, the chlorate ion, and most reducing agents convert it is a stepwise manner to $Cl^-$, the stable chloride ion product. Therefore, once the first and difficult step has been realized, the reduction of perchlorate would likely proceed rapidly to completion.

**A New Method for the Catalytic Reduction of Perchlorate**

For reasons unrelated to issues of perchlorate reduction, experiments were carried out with the unusual organometallic oxide $CH_3ReO_3$, methyltrioxorhenium(VII), often abbreviated as MTO. In organic solvents this reagent reacts with certain reagents that can make strong bonds to oxygen (stronger than that in MTO itself). A suitable example is triphenyl phosphine, which gives this reaction in benzene and toluene and presumably other organic solvents:[12,13]

$$CH_3ReO_3 + PPh_3 \rightleftharpoons CH_3ReO_2 + Ph_3P{=}O \qquad (13)$$

The methyldioxorhenium product (sometimes abbreviated as MDO) is a derivative of Re(V). In actuality it is coordinated by two additional phosphine molecules; in the solid state this adduct was obtained and structurally characterized: $MTO \cdot MDO \cdot (PPh_3)_2$.

The aqueous systems, more of interest here, can be carried out by using hypophosphorous acid, $H_2P(O)OH$, as the reagent that extracts an oxygen atom from MTO:

$$CH_3ReO_3 + H_2P(O)OH \rightleftharpoons CH_3ReO_2 + HP(O)(OH)_2 \qquad (14)$$

The exact form of MDO in aqueous solution has not be determined, since it is not too stable against dimerization and oligomerization, but it is perhaps a species in which two water molecules are coordinated:

$CH_3Re(O)_2(OH_2)_2$

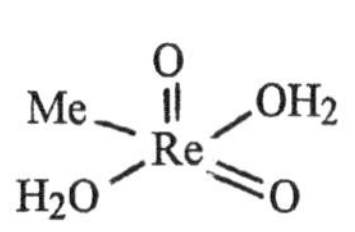


This recently prepared compound is being presented because it holds the record to date as far as perchlorate reduction. The first step of the reaction is, as usual, the slowest. The chemical equation for the first step of its reaction with perchlorate ions is:

$$CH_3ReO_2 + ClO_4^- \rightleftharpoons CH_3ReO_3 + ClO_3^- \quad (15)$$

The rate of this reaction is given by

$$-d[ClO_4^-]/dt = k[ClO_4^-][CH_3ReO_2] \quad (16)$$

where $k = 7.2$ L mol$^{-1}$ s$^{-1}$ at 25 °C, the concentrations being expressed in the units mol L$^{-1}$ and the rate as mol L$^{-1}$ s$^{-1}$. In other terms, one can express the rate reaction by the time needed to complete one-half of the reaction at $10^{-3}$ M concentration. In those terms, the reaction time is 0.039 h = 2.3 min. It is not all that fast as rapid reactions go, but certainly a record-setter for perchlorate, >1000 times faster than any reaction listed in Table 3.

Because chloride ions were detected in quantitative yield, it was presumed that the stripping of the next three oxygen atoms occurred in reactions more rapid than that of perchlorate itself. Experiments were done to assess directly the reaction between $CH_3ReO_2$ and $ClO_3^-$. That study was characterized by a rate constant $3.8 \times 10^4$ L mol$^{-1}$ s$^{-1}$, $10^{3.7}$ times greater than the value for $ClO_4^-$. This supports the assumption that subsequent steps convert perchlorate to chloride.

Molybdenum reagents are also useful.[14] Note should be also be taken that molybdate and tungstate are active catalysts for the reduction of perchlorate ions by stannous chloride, an older finding in need of re-examination.[15,16]

## REMEDIATION

The prospects for sufficiently-facile chemical reduction of perchlorate ions in dilute aqueous solutions do not appear bright. Even the best reactions are sluggish, and for economic and other reasons the heavy-metals (such as rhenium) would need to be replaced by others, or the reaction be made catalytic.

It does seem useful, however, to mention here one distinguishing property of perchlorate ions, lest it be overlooked. Most anion exchange resins have a high affinity for $ClO_4^-$, so much so that this anion is bound more tightly than nearly any other. The actual binding strength varies from one resin composition to another, but the general observation remains valid. This might provide the basis for a useful and practical method, particularly if the used resin, saturated in perchlorate, were to be regenerated or were so inexpensive that it could be discarded.

## REFERENCES

1. "Instructions to authors." *Inorganic Chemistry* **1999**, *38*, 11A.

2. Scott, S.L.; Bakac, A.; Espenson, J.H. "Oxidation of alcohols, aldehydes, and carboxylates by the aquachromium(IV) ion." *Journal of the American Chemical Society* **1992**, *114*, 4205–4213.

3. Kallen, T.W.; Earley, J.E. "Reduction of the perchlorate ion by aquoruthenium(II)." *Inorganic Chemistry* **1971**, *10*, 1152–1155.

4. Cope, V.W.; Miller, R.G.; Fraser, R.T.M., "Titanium(III) as a reductant in electron-transfer mechanisms." *Journal of the Chemical Society (A)* **1967**, 301–306.

5. Endicott, J.F.; Taube, H. "Studies on oxidation-reduction reactions of ruthenium ammines." *Inorganic Chemistry* **1965**, *4*, 437-445.

6. Peretrukhin, V.F.; Krot, N.N.; Gel'man, A.D., "Effect of the oxidation of trivalent uranium in aqueous solutions on the completeness of the electrolytic reduction of uranium(IV)." *Soviet Radiochemistry (Engl. transl.)* **1970**, *12*, 85–88.

7. Adin, A.; Sykes, A.G., "The kinetics of oxidation of europium(II) with vanadium(III) and chromium(III) in aqueous perchloric acid solutions." *Journal of the Chemical Society (A)* **1966**, 1230–1236.

8. King, W.R.; Garner, C.S. "The kinetics of the oxidation of vanadium(II) and vanadium(III) ions by perchlorate ion." *Journal of Physical Chemistry* **1954**, *58*, 29–33.

9. Abu-Omar, M.M.; Espenson, J.H., "Facile abstraction of successive oxygen atoms from perchlorate Ions by methylrhenium dioxide." *Inorganic Chemistry* **1995**, *34*, 6239–6240.

10. Taube, H. "Observations on atom-transfer reactions." *ACS Symposium Series* **1981**, *198*, 151–171.

11. Lappin, A.G. *Redox Mechanisms in Inorganic Chemistry.* Ellis-Horwood: New York, 1994; pp. 263–265.

12. Zhu, Z.; Espenson, J.H. "Methylrhenium trioxide as a catalyst for oxidations with molecular oxygen and oxygen transfer." *Journal of Molecular Catalysis* **1995**, *103*, 87–94.

13. Herrmann, W.A.; Roesky, P.W.; Wang, M.; Scherer, W. "Multiple bonding between main-group elements and transition metals. 135. Oxorhenium(V) catalysts for the olefination of aldehydes." *Organometallics* **1994**, *13*, 4531–4535.

14. Hahn, M.; Wieghardt, K. "Kinetics and mechanism of the oxidation of $[L_2Mo_2(H_2O_2(\mu\text{-}OH)_2]^{4+}$ (L = TACN) with perchlorate." *Inorganic Chemistry* **1984**, *23*, 3977–3982.

15. Haight, G.P. Jr.; Sager, W.F. "Evidence for preferential one-step divalent changes in the molybdate-catalyzed reduction of perchlorate by stannous ion in sulfuric acid solution." *Journal of the American Chemical Society* **1952**, *74*, 6056–6059.

16. Haight, G.P. Jr., "Mechanism of the tungstate catalyzed reduction of perchlorate by stannous chloride." *Journal of the American Chemical Society* **1954**, *76*, 4718–4721.

# Chapter 2

# THE CHEMISTRY OF PERCHLORIC ACID AND PERCHLORATE SALTS: REALIZING THE BENEFITS®

John R. Long*

GFS Chemicals, Inc., Columbus, Ohio 43222

## INTRODUCTION

The association of perchlorates with pyrotechnics, rocket propellant compositions and other unusually energetic chemical processes has fueled the perception that the use and handling of this class of chemicals lies beyond the capabilities of most chemists and technicians. In reality, the properties of perchloric acid and perchlorate salts are extremely well-documented,[1] and situations that have the potential to cause safety concerns have been well researched and publicized in a variety of forums. One thing which has been lacking to this point is a sustained effort to coordinate such information and make it available not only in appropriate scientific publications, but occasionally in other media as well.

Two significant developments have occurred in the last few years to bring perchlorate chemistry into a more critical light. One change has been the enhancement of methodologies that now allow the detection of perchlorate ion at low part per billion levels. While perchorate ion characteristics have been studied in connection with the diagnosis and treatment of Graves' disease patients, the effects of perchlorate at particularly low concentrations has not been a topic of intense research. The redefinition of the analytical threshold has changed that forever.

The second recent point of attention has been the discovery of apparent ground water contamination in several states, most notably California and other western states, which appear to have been in the proximity of major perchlorate manufacturing facilities. In especially environmentally sensitive climates, this situation has drawn understandable attention to the possible effects of trace perchlorate levels on human health.

®This work was presented at the 218th national meeting of the American Chemical Society as part of the Environmental Division symposium Perchlorate in the Environment, held August 22–24, 1999, in New Orleans, Louisiana.

*Phone: 614-224-5345. Fax: 614-225-1175. Electronic mail: jrlong@gfschemicals.com.

Considering the variety of factors and uncertainties that underlie the current situation, there would appear to be no time like the present to bring together the insights available from industrial resources, analytical investigations, academic research, and the experience of the manufacturers of perchloric acid and perchlorate salts to provide a balanced perspective to all the health, safety and environmental issues pertinent to this area of chemistry.

## DISCUSSION OF PROPERTIES AND DATA

### Physical and Chemical Characteristics of Perchloric Acid

Despite widespread caveats that have been extended to the application of perchlorates, the scientific literature of any given year contains thousands of references in which a member of this chemical family plays a significant role. The range of chemistries easily crosses many disciplines; major R&D work of an engineer, clinical technician, environmental analyst, instrumentalist, or other lab specialist may fall upon an individual who has little understanding of the nuances of perchlorate chemistry.

Perhaps the most important property to be considered in any evaluation of perchlorate chemistry is the inherent stability of the perchlorate species. Under most conditions, the perchlorate ion is extremely stable, and very content to exist in its ionic form, $ClO_4^-$. Routine attempts to reduce dilute concentrations of perchlorate under mild conditions are usually met with frustration. This accounts for the intense interest in optimizing conditions under which the biodegradation of perchlorate might be accomplished;[2] chemical reduction has proven to be impractical.

The properties of perchloric acid need to be considered in their entirety, because it frequently is the combination of selected properties of reactant and substrate that provides both a desirably enhanced reactivity (when used properly) and a hazard potential (when used improperly). Many inappropriate assumptions about perchloric acid can easily be corrected by applying any number of several basic principles.

For example, when cold (i.e., modest temperatures) perchloric acid solutions are non-oxidizing at any concentration below 73%; in addition, hot, *dilute* perchloric acid solutions do not exhibit significant oxidizing character. The concentration of perchloric acid needs to be approximately 50% before a hot solution becomes markedly oxidizing. The data represented in Table 1 relate perchloric acid concentration to observed boiling points; a graphical representation of the oxidizing range of perchloric acid is provided in Figure 1.

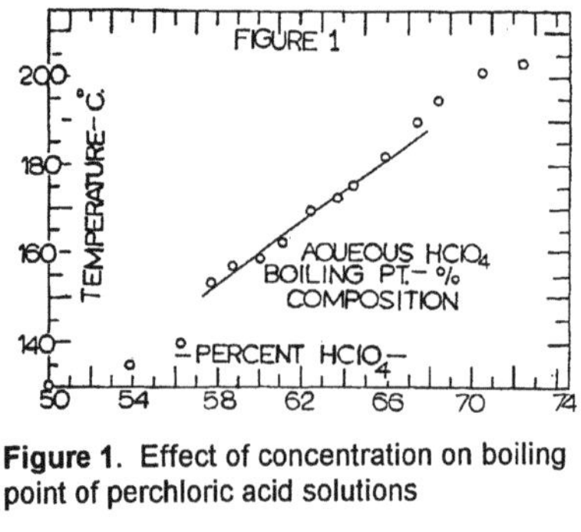


**Figure 1.** Effect of concentration on boiling point of perchloric acid solutions

**Table 1.** Boiling Points of Aqueous Perchloric Acid Solutions

| Acid Conc, % | B.P., °C | Acid Conc, % | B.P. °C |
|---|---|---|---|
| 56.3 | 140 | 62.6 | 170 |
| 57.7 | 154 | 64.5 | 175 |
| 58.8 | 157 | 66.0 | 182 |
| 59.2 | 157 | 67.5 | 190 |
| 60.1 | 160 | 68.4 | 195 |

The commercialization of perchloric acid was accompanied by the generation of a considerable amount of data governing its properties and character. One test of its oxidizing power has typically involved reaction with cellulose under controlled conditions. Table 2 shows the effect of increasing acid concentration (and thereby the boiling temperature) upon cellulose.

**Table 2.** Graded Potential Oxidation of Cellulose

| Sample (0.8 g) | $HClO_4$ Concn. (%) | Time for Complete Oxidation (min) | Remarks |
|---|---|---|---|
| Cellulose | 50.0 | 375 | Reaction still incomplete |
| Cellulose | 57.7 | 60 | Excess foaming |
| Cellulose | 60.1 | 52 | No excess foaming |
| Cellulose | 62.6 | 45 | same |
| Cellulose | 64.0 | 17 | same |
| Cellulose | 66.0 | 9 | same |

Use of catalytic amounts of ammonium vanadate were insufficient to initiate significant oxidation using 50% perchloric acid. Repeating the treatment with 60% acid boiling at 160C resulted in degradation of the cellulose to a chocolate brown residue; removal of the catalyst under these conditions reduced the oxidation to a prohibitively slow rate. At acid concentrations of 64% or higher, the oxidation is complete within minutes. Use of 72% perchloric acid would result in uncontrolled oxidation of the cellulose, in part because the boiling temperature for this concentration exceeds 200 C.

Many other insights into the behavior of perchloric acid are available: the oxidation potential of the hot, concentrated acid is about 2.0, similar to that of ozone; perchloric acid has the highest protonic activity of all the mineral acids; hot, concentrated perchloric acid is a powerful dehydrating agent and solvent; glacial acetic acid can be mixed with perchloric acid in any proportion; and properly stored reagent grade solutions (70% concentration) are stable indefinitely.

**Applications Guidelines**

Perhaps the single most important principle to understand in the application of perchloric acid involves its use in combination with other oxidizing acids such as nitric acid and sulfuric acid. By careful adjustment of both temperature and acid concentrations of an appropriate acid mixture, it is possible to achieve a step-wise, gradual increase in oxidation potential from moderate to very high to extreme levels, according to the effect desired in the given oxidation. At lower temperatures, the less vigorous oxidizing power of the complementary acid is already degrading the organic material; as a result, by the time the conditions of temperature and concentration are reached that permit oxidation by perchlorate, sufficient oxidizable material has been consumed that the reaction of the perchlorate with the substrate remnant is achieved in a controlled fashion.

Since every substrate has its own sensitivity to oxidizing environments, every perchloric acid oxidation must follow an established protocol that is based on data gathered over a suitably wide range of conditions of temperature and concentration. Large scale processes which effectively control the temperature and concentration of perchloric acid solutions have been carried out for decades in major industrial plant operations dealing in chemistries involving catalytic conversions, electrolysis, and general chemical manufacturing.

Many other factors govern the application of perchloric acid and its salts in research and industry. Their excellent solubility and conductivity in aqueous and non-aqueous environments frequently apply to polymeric and electrolyte blends; this includes use as dopants to impart anti-static and conductive properties that improve various electrochemical devices. Perchlorate versatility is apparent in reports of ongoing commercial interest in such materials in battery electrolyte research; their relative inertness in electrochemical reactions is a particular advantage. The vigorous oxidizing power of perchlorates in intimate contact with combustible material can be exploited under controlled conditions requiring initiation by application of sufficiently high temperature.

## Safety and Handling Information

A great deal of anecdotal insight can be offered as testimony to the importance of establishing and maintaining well-defined protocols for the use of perchloric acid and perchlorate salts. Just as their behavior conforms to well defined expectations, experience has shown that the observation of several basic safety precautions will go a long way toward minimizing or eliminating most of the risks associated with the use of these materials. Again, the prime directive must be the observation of temperature and concentration guidelines in all phases of operations with these materials.

Beyond that, common sense will serve as a good base line of precaution. Perchloric acid work, particularly digestions involving fuming perchloric acid, must be carried out in a properly constructed and regularly maintained fume hood, preferably one that features a wash-down system. Any of several designs of a portable vapor-trap apparatus may also be useful in certain circumstances, (Figure 2).

Unit includes fume eradicator, suction flask, rubber stopper, and 500ml Erlenmeyer flask

Unit includes fume eradicator, reflux head, suction flask, rubber stopper and 500ml Erlenmeyer flask

Unit includes fume eradicator with insealed dropping funnel, suction flask, rubber stopper and 500ml Erlenmeyer flask

**Figure 2.** Perchloric acid fume eradicators configured for varying degrees of reaction control.

Hood users should post signs restricting a hood to perchloric acid use, and specifically indicate that organics are prohibited in or under the hood. If the hood must be converted for use for other chemistries, the hood must be totally decontaminated; dismantling a perchloric acid hood must be done according to established safety protocols.

Do not mix perchloric acid waste with any other waste, and store perchloric acid away from organic and flammable materials. Handle perchloric acid using full body protection, including goggles or face shield, gloves, and apron or coat. Dilute perchloric acid by adding acid to water, never the reverse.

**Environmental and Health Guidelines**

Based on available health studies, the state of California imposed in 1997 an emergency perchlorate standard of 18 parts per billion, a value which factored in various uncertainties relating to limits in study duration, protection of sensitive individuals, and certain perceived data deficiencies. Since that standard was imposed, a number of significant literature reports have been issued that argue against a statistically significant health risk attributable to perchlorate.

For example, a cross-sectional occupational health study[3] of thyroid function in perchlorate workers reported that comprehensive measurement of thyroid function across roughly three orders of magnitude of exposure and of dose revealed no anomalies; human thyroid function was "not affected." No clinical evidence of thyroid abnormalities was found in any exposure group, and blood cell counts were normal. The negative results from occupational airborne perchlorate exposure at a mean absorption rate of 34 milligrams per day were indicative of a no-observed-adverse-effect-level (NOAEL) that should be factored into the evaluation of human health risks from environmental perchlorate contamination.

A separate study[4] addressed the potential effect of perchlorate on the rate of congenital hypothyroidism. Data from neonatal screening programs involving nearly 700,000 newborns were reported to indicate no increase in the incidence of congenital hypothyroidism with perchlorate levels within the 4–16 $\mu g\ L^{-1}$ range.

## CONCLUSION

The amalgamation of scientific data gathering and public policy determination is an elusive exercise under the best of circumstances. Fortunately, issues relating to the presence of perchlorate in the environment can be considered in the light of an increasing body of scientific data. An effective environmental evaluation, and conclusions therefrom, must reconcile analytical data with the effects demonstrated by careful epidemiological and occupational studies.

The use of perchloric acid and perchlorate salts in commercial and industrial processes supports a business segment dealing in many tons per year of these valuable materials. The breadth of information available on their character and properties is considerable. In actuality, this class of chemicals remains underused and underappreciated in key areas of application. The prospect of fully realizing the benefits of perchlorates and perchloric acid hinges upon the way that accurate, comprehensive data can be collected and applied.

Greater efforts are now being made by many organizations and individuals to centralize and distribute product information as part of a program of consumer education for both scientists and the general public. Programs such as this would serve two significant purposes: to enhance the understanding of safety and health factors that impact material use; and to foster the improvement of a broad spectrum of processes and protocols involving perchlorate in chemical analysis, research and manufacturing.

## REFERENCES

1. Schilt, A.A. *Perchloric Acid and Perchlorates.* G. Frederick Smith: Columbus, OH, 1979; passim. The 2nd edition is scheduled for publication in 2000.

2. Urbansky, E. T.; Schock, M. R. "Issues in managing the risks associated with perchlorate in drinking water." *Journal of Environmental Management* **1999**, *56*, 79–95.

3. Lamm, S. H.; Braverman, L. E.; Feng, X. L.; Richman, K.; Pino, S.; Howearth, G. "Thyroid health status of ammonium perchlorate workers: a cross-sectional occupational health study." *Journal of Epidemiological and Occupational Medicine* **1999**, *41*, 248–260.

4. Lamm, S. H.; Doemland, M. "Has perchlorate in drinking water increased the rate of congenital hypothyroidism?" *Journal of Epidemiological and Occupational Medicine* **1999**, *41*, 409–411.

---

**Editor's Note:** Dr. Long is correct in asserting the wide applicability of perchlorates and perchloric acid in research. ACS's Chemical Abstracts Service shows thousands of entries a year that mention these species. Some noteworthy examples suggested to me by Dr. Long include the following:

■$LiClO_4{\cdot}Et_2O$ is used in organic synthesis, especially Diels-Alder reactions.

Greico, P. "Organic chemistry in lithium perchlorate-diethyl ether." In *Organic Chemistry: Its Language and Its State of the Art.* M.V. Kisakürek, Ed. New York: VCH, 1993.

Greico, P.; Handy, S.T.; Beck, J.P. "Acid catalyzed intramolecular Diels-Alder reactions in lithium perchlorate–diethyl ether: enhanced reaction rates and diastereoselectivity." *Tetrahedron Lett.* **1994**, *35*, 2663.

Greico, P.; Collins, J.L.; Handy, S.T. "Acid catalyzed ionic Diels-Alder reactions in concentrated solutions of lithium perchlorate in diethyl ether."*Synlett* **1995**, 1155.

Greico, P.; Walker, J.K. "Intramolecular cationic [5+2] cycloaddition reactions promoted by trimethylsilyl triflate in 3.0 M lithium perchlorate–ethyl acetate: application to a formal total synthesis of (±)-isocomene." *Tetrahedron* **1997**, *53*, 8975.

■Perchloric acid finds use in petrochemical and geochemical analysis.

McGowan, C.W.; Diehl, H. "The oxidation of green river oil shale with perchloric acid. Part I. The reaction of green river oil shale with perchloric acid of varying concentration and boiling point." *Fuel Process. Technol.* **1985**, *10*, 169.

McGowan, C.W.; Diehl, H. "The oxidation of green river oil shale with perchloric acid. Part II. The analysis of oxidation products." *Fuel Process. Technol.* **1985**, *10*, 181.

McGowan, C.W.; Pearce, R.C.; Diehl, H. "A comparison of the dissolution of model compounds and the kerogen of green river oil shale by oxidation with perchloric acid—a model for the kerogen of green river oil shale." *Fuel Process. Technol.* **1985**, *10*, 195.

McGowan, C.W.; Markuszewski, R. "Fate of sulfur compounds in coal during oxidative dissolution in perchloric acid." *Fuel Process. Technol.* **1987**, *17*, 29.

McGowan, C.W.; Markuszewski, R. "The direct determination of sulphate, sulphide, pyritic, and organic sulphur in coal by selective, step-wise oxidation with perchloric acid." *Fuel* **1988**, *67*, 1091.

Stanton, B.J.; Morris, R.M.; McGowan, C.W. "The characterization of the kerogen of Chattanooga oil shale. Part I. The oxidation of Chattanooga shale with boiling perchloric acid." *Fuel Process. Technol.* **1991**, *29*, 85.

Ailey-Trent, K.S.; McGowan, C.W.; Lachowicz, E.; Markuszewski, R. "The effect of time, catalysts and standard additions on sulfate produced during the oxidation of coal with perchloric acid." *Fuel* **1993**, *72*, 1197.

# Chapter 3
## TOXICOLOGY OF PERCHLORATE

James J.J. Clark*

Komex•H2O Science•Inc., 11040 Santa Monica Boulevard, Suite 335, Los Angeles, California 90025

### INTRODUCTION

Perchlorate ($ClO_4^-$) is an anion commercially available as a salt with many cations. The most common forms of perchlorate include ammonium perchlorate (used as a solid rocket oxidant and ignitable source in munitions and fireworks) and potassium perchlorate (used in road flares and in air bag inflation systems as well as to treat Graves' Disease [hypothyroidism] since the 1950s). Other forms of perchlorate include ammonium perchlorate, potassium perchlorate, sodium perchlorate, aluminum perchlorate, hydrazinium perchlorate, hydrogen perchlorate, hydroxylammonium perchlorate, lithium perchlorate, magnesium perchlorate, nitronium perchlorate, and perchloric acid. Perchlorate is also formed in laboratory waste as a byproduct of perchloric acid.

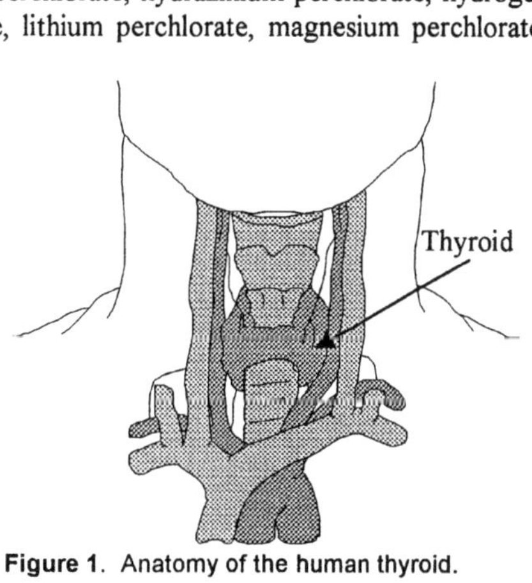

Figure 1. Anatomy of the human thyroid.

The thyroid gland (Figure 1), along with the thymus, suprarenal capsules and spleen, are classified as the ductless glands (i.e., glands that do not possess an excretory duct).[1] The thyroid is located at the front and sides of the neck, and consists of two lateral lobes connected across the middle line by a narrow transverse portion called the *isthmus*. The thyroid gland is the largest of the endocrine organs that function exclusively as an endocrine gland.[2] The thyroid is unique since it is the only gland in

*Phone: 310-914-5901. Fax: 310-914-5959. Electronic mail: jclark@komexh2O.com

*Perchlorate in the Environment*, edited by Urbansky.
Kluwer Academic/Plenum Publishers, New York, 2000.

the body to accumulate iodine in large quantities for the production of hormones. The thyroid produces two hormones: $T_3$ (L-triiodothyronine or monoiodotyrosine) and $T_4$ (L-thyroxine or triiodotyrosine) that regulate growth, cell differentiation, and the metabolisms of lipids, proteins, and carbohydrates.[2,3]

The hypothalamus gland, located in the deep in the brain, controls the complex system of hormone secretion from the thyroid. The hypothalamus secretes a specific hormone, TRH (thyrotropin-releasing hormone), that signals the anterior pituitary gland to produce TSH (thyrotropin stimulating hormone).

TSH signals the thyroid to start taking in iodine from the circulating blood in order to bind it to a protein. Once this is accomplished, the thyroid gland releases the protein-iodine complex into the blood stream. The thyroid gland concentrates iodine from the blood, iodinates tyrosine (an amino acid), and subsequently generates the thyroid hormones ($T_4$ and $T_3$) within the large glycoprotein thyroglobulin. A large store of preformed hormone is present in the thyroid. This protein bound iodine is what was earlier referred to as $T_3$ and $T_4$. The four hormones TRH, TSH, $T_3$, and $T_4$ work in an inverse way. Normal thyroid secretion depends on TSH. High concentrations of TSH stimulate the thyroid, and as a result, $T_3$ and $T_4$ levels increase. If $T_3$ and $T_4$ concentrations increase then TRH and TSH concentrations decrease. The balanced negative feed back regulation of these four hormones keeps the body in a normal metabolic state. There is also a direct relationship between TRH and TSH. If the hypothalamus produces an excess of TRH, then the anterior pituitary produces more TSH. TSH cannot inhibit the hypothalamus. However, high concentrations of $T_3$ and $T_4$ do.

The cellular structure of the thyroid is unique for endocrine glands, consisting of follicles (ranging in size from 20 to 250 μm) that contain colloid produced by the follicular cells. The follicular cells are cuboidal to columnar and secrections are directed forward the lumen of the follicles. A network of inter- and intrafollicular capillaries provides the follicles with an abundant blood supply. Follicluar cells have a long profile of rough endoplasmic reticulum and a large Golgi apparatus to synthesis and package protein that is transported into the follicular lumen. Microvillar projections between the luminal side of the follicular cells and the colloid form an interface. Final assembly of thyroid hormones occurs extracellularly from the thyroid

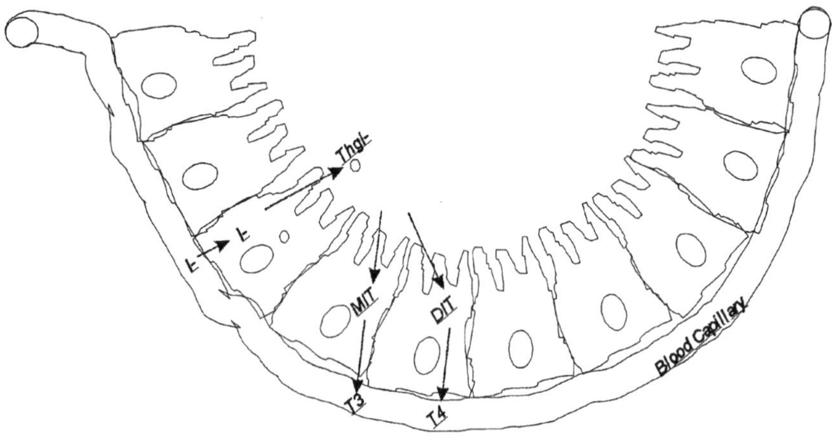


**Figure 2.** Thyroid follicular cells and metabolism of thyroid hormones.

within the follicular lumen. Raw materials such as iodide are trapped in the follicular cells from plasma, transported into the cell against the concentration gradient to the lumen and oxidized by a peroxidase in microvillar membranes to iodine, $I_2$. (See Figure 2).

Assembly of thyroid hormones within the follicular lumen is possible because of thyroglobulin, a high molecular weight glycoprotein synthesized within the endoplasmic reticulum of the follicular cells.[2] The necessary amino acids (tyrosine and others) and carbohydrates such as mannose, fructose, and galactose, come from the circulatory system. Having incorporated tyrosine within the molecular structure of thyroglobulin, iodine is bound to the tyrosyl residues in thyroglobulin at the surface of the follicular cells to form monoiodotyrosine (MIT) and diiodotyrosine (DIT). MIT and DIT combine to form thyroxine-$T_4$ and triiodothyronine-$T_3$, which are secreted by the thyroid gland.

$T_4$ and $T_3$ (see structure of $T_3$ in Figure 3) function in similar fashions, although much of the biological action is the result of the monodeiodination of $T_4$ to 3,5,3′-triiodothyronine before interacting with target cells. Under certain conditions such as protein starvation, neonatal animals, liver and kidney disease, etc., $T_4$ will be preferentially monodeiodinated to 3,3′,5′-triiodothyronine, the biologically inactive form of $T_3$. $T_4$ stimulates oxygen utilization and heat production in every cell of the body. Carbohydrate utilization, protein catabolism, and greater oxidation of fats occur. Direct administration of thyroxine will increase heart rate. Normal function of the central nervous system is dependent on the normal output of thyroxine. When $T_4$ levels are suppressed the CNS does not function normally resulting in lethargic, dull, and mentally deficient animals. Myelin in the fiber tracts is decreased, cortical neurons are smaller and fewer, and the vascularity of the CNS is reduced. The young of animals with thyroxine deficiency have permanent neuron damage. Adult animals with thyroxine deficiency show reversible neuron damage.

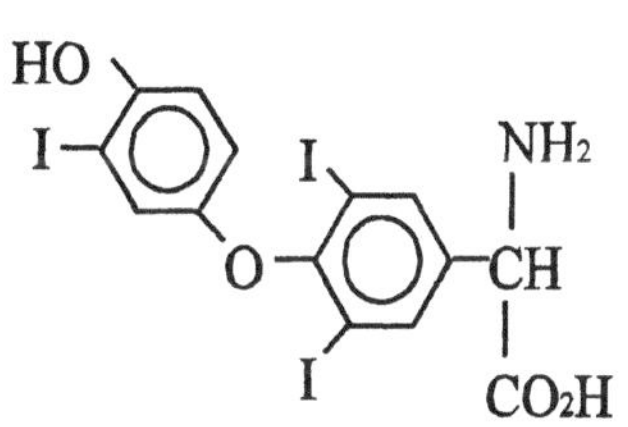

**Figure 3.** Structure of $T_3$.

The subcellular mechanism of thyroid hormone action may involve the mitochondria or nuclear receptors. Free $T_3$ enters the cells and binds to the inner membrane of the mitochondria, activating the mitochondrial energy metabolism or the free $T_3$ may bind to the nuclear receptor increasing transcription of the genetic message to increase new protein synthesis.[4] The overall effects of $T_4$ and $T_3$ are to (1) increase the basal metabolic rate, (2) make more glucose available to meet the elevated metabolic demands by increasing glycolosis, glucogenesis, and glucose absorption from the intestine, (3) stimulate new protein synthesis, (4) increase lipid metabolism and conversion of cholesterol into bile acids and other substances, activation of lipoprotein lipase, and increase sensitivity of adipose tissue to lipolysis by other hormones, (5) stimulate the heart rate, cardiac output, and blood flow; and (6) increase neural transmission, cerebration, and neuronal development in young animals.

According to Wolff[5] the selectivity for iodide or perchlorate in the thyroid gland might be based on ion exchange mechanism using a large cation such as a quaternary amine, e.g., arginine. Perchlorate may be a competitive inhibitor of the sodium/iodide symporter of the thyroid follicular cell, which actively transports iodine from the blood into the thyroid.

Perchlorate is primarily eliminated from the body in urine. Peak blood levels after oral intake are reported to be 3 h. Despite the high redox potential for perchlorate, it appears to be excreted virtually unchanged both in rat[6] and man,[7] as reported by Wolff.[6]

Wyngaarden et al.[8,9] showed that the perchlorate ion inhibited iodide accumulation in propylthiouracil-blocked glands. Ingbar and Freinkel[10] later showed the same effect in unblocked thyroid. The competitive nature of the inhibition was established using thyroid slices and analysis by double reciprocal or Dixon plots.[11,12] Perchlorate promotes a very rapid and nearly complete discharge of iodide from propylthiouracil-treated glands. This discharge may be accomplished by stimulating counter transport of iodide.[13]

Wolff[6] states that perchlorate leads to all the reversible effects expected from the blockade of thyroid hormone synthesis in experimental animals: goiter with histological signs of thyroid stimulation that fails to occur in hypophysectomized animals; thyroidectomy cells in the anterior pituitary; the corresponding changes in thyroid hormone and TSH levels; decrease in $QO_2$, etc.[6,14] All these effects are reversed upon cessation of perchlorate administration or with $T_3$ or $T_4$. According to Eichler[15] perchlorate is excreted unmetabolized, with approximately 95% recovery in the urine over 72 h. Eichler reported that within 6–8 h, the urine contained 50% of a 1 or 2 g oral dose given to an adult male. Durand[16] also reported that within 5–9 h, the urine contained 50% of a 0.8-g oral dose given to an adult male.

## RECOMMENDED DAILY ALLOWANCE OF IODINE

According to Andreoli et al.[17] and the National Research Council[18] (NRC) the recommended daily iodine intake is 150 μg $day^{-1}$ for the average adult. For pregnant women and nursing mothers, the RDA is increased to 200 μg $day^{-1}$ to allow for the needs of the fetus. The average diet in the United States contains 250 to 750 μg $day^{-1}$ of iodine due to the enrichment of foods with iodine.[18] Iodine is converted to iodide in the stomach; after rapid absorption from the gastrointestinal tract, iodide is distributed in the extracellular fluids. The thyroid follicular cells actively transport iodide from the blood stream across the follicular membrane. Iodide is rapidly and almost completely absorbed and transported to the thyroid gland for synthesis into the thyroid hormones, to salivary and gastric glands, and to the kidneys for excretion into the gastrointestinal tract and urine.[19] Since iodide is secreted into the gastrointestinal tract, the primary route of excretion for inorganic iodine is through urine.

According to the NRC, there were an estimated 400 million iodine-deficient persons in the under developed regions of the world[19] and an estimated 112 million in the more developed areas.[20] After the introduction of iodized salt in 1924 in the United States, the incidence of endemic goiter decreased. Residual cases of goiter remain in certain areas of the United States (California, Texas, Kentucky, Louisiana, and South Carolina) and Canada (the prairie regions) —mainly among women and children. Important environmental sources include specific foodstuffs (i.e., seafood, milk, bread), water, fertilizers, feeding practices, and food processing. A minimum intake level of 50–75 μg $day^{-1}$ is needed to maintain the higher level of iodine excretion in a population.[21] According to the NRC, iodine intakes up to 2 mg $day^{-1}$ have caused no adverse reactions in healthy adults, and 1 mg $day^{-1}$ produced no adverse reactions in children.

Although thyroid function is dependent upon an adequate dietary intake of iodine as a substrate for hormone synthesis, the thyroid readily compensates for a modest decrease in iodine intake by enlarging and actively transporting much of the circulating iodine.[22]

## MEDICAL USE OF PERCHLORATE FOR AMIODARONE-INDUCED THYROTOXICOSIS

Amiodarone is an iodinated benzofuran derivative containing nearly 40% iodine by mass (see structure in Figure 4). The drug is stored in many tissues, especially adipose tissue, with an estimated biological half-life of 8 months.[23–25] Amiodarone is a potent drug used mainly to treat cardiac tachyarrythmias. Known side effects of amiodarone include: (1) changes in the peripheral metabolism of thyroxine, (2) direct toxic effect on the thyroid parenchyma, and (3) excess iodide production by deiodination of amiodarone.

Amiodarone interferes with the deiodination of $T_4$ to $T_3$ by the Type I, propyl-thiouracil-sensitive, 5′-deiodinase,[26–30] resulting in the bulk of peripheral $T_3$ production, resulting in an increase in total and free $T_4$, an increase in $rT_3$, and a corresponding decrease in total and free $T_3$ in the plasma and most tissues. Although the mechanism of thyroid toxicity is not understood, cellular damage incurred suggests a destructive thyroiditis.[31,32] The lesions produced resemble the damage that is sometimes seen with excess iodide exposure.[33,34]

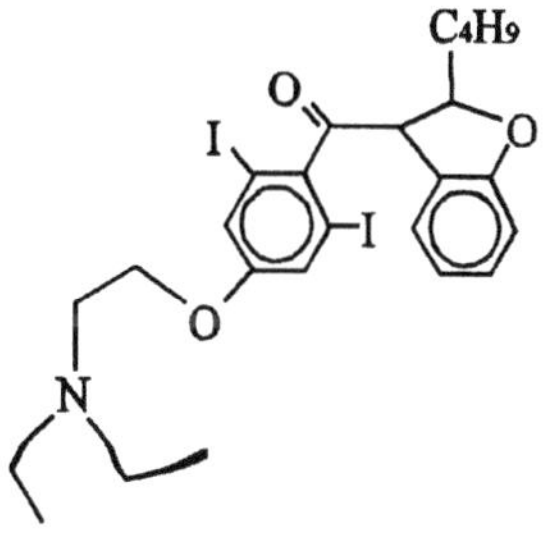

**Figure 4.** Structure of amiodarone.

Perchlorate is now used to treat patients who have iodine-induced thyrotoxicosis (e.g., amiodarone-associated thyrotoxicosis, AAT). Thyrotoxicosis is the clinical syndrome that results when tissues exposed to high levels of circulating thyroid hormones. AAT patients treated for 1 month or longer with perchlorate at doses up to 1 g perchlorate per day had no evidence of agranulocytosis or aplastic anemia.[35–38] According to Martino et al.[39] amiodarone represents one of the most common sources of iodine-induced thyrotoxicosis. They found that the prevalence of AAT was higher in males than females. AAT was found to occur in patients with underlying thyroid disorders and in patients with apparently healthy thyroids.

According to De Visscher,[40] a limitation in the use of perchlorate is the incidence of gastric irritation and toxic reactions. Perchlorate can be used as a diagnostic tool to assess the completion of peroxidase blockage by thionamide. Patients with the thyroid disease called "congenital iodine organification defect" have goiters but are not always hypothyroid. This defect is detected by a positive perchlorate discharge test: After radioactive iodide is given to a patient, patients partially or totally discharge the radioactive labeled iodide after ingestion of perchlorate. Patients without congenital iodine organification defect will not discharge radioactive iodide after ingestion of perchlorate. For normal patients, the radioactive iodide has been rapidly incorporated into thyroglobulin and is not available for discharge from the thyroid gland.

## APLASTIC ANEMIA?

Evidence of aplastic anemia with use of perchlorate is equivocable at best. The first reported case of aplastic anemia associated with the use of potassium perchlorate for the treatment of Graves' Disease is in 1961 in the *Medical Memoranda* by Hobson. Graves' Disease is prevalent in the United States at a rate of approximately 1–2% in females, and the female:male ratio of prevalence is 7:1. It may occur at any age, but the peak incidence is between the third and fourth decades of life. The appearance of seven cases of fatal aplastic anemia from 1961 to 1966—four in Britain,[41] one in the United States,[42] one in Norway,[43] and one in Israel[44]—virtually stopped the use of perchlorate except for single doses used in the perchlorate discharge test.[6]

The doses used in all of the cases reported were lower than 1 g $day^{-1}$ in patients who were treated for periods of 2–6 months before aplastic anemia was detected. It is believed that the duration of potassium perchlorate use may have been contributing factor. Four of the cases occurred in a cluster, raising the question of whether the potassium perchlorate was contaminated. No serious side effects have been reported with the use of perchlorate[45] and one case with a 22-year maintenance dosage of 200 mg $day^{-1}$ had no complications.[46]

Results from recent genotoxicity studies reported by the EPA's National Center for Environmental Assessment (NCEA) show that the perchlorate ion is not genotoxic in any of the assays run for the genotoxicity battery (See EPA NCEA Draft Report below). Previous studies, which did show a positive relationship between the use of perchlorate and thyroid tumors in rats, were performed as long-term bioassays at high doses. As reported by NCEA the rat appears to be more sensitive to thyroid cancer caused by thyroid-pituitary disruption than does man. When these data are considered along with the data from AAT patients treated high doses of perchlorate (1 g $day^{-1}$),[45-47] confidence in the carcinogenic potential of the perchlorate ion in man and the reported cases of aplastic anemia associated with use of potassium perchlorate is very low.

## EPIDEMIOLOGICAL STUDIES

A few studies have been conducted on the effect of perchlorate on healthy human subjects.[48,49] A 1974 study showed that in five healthy volunteers 600 mg $day^{-1}$ for 1 week increased non-thyroxine iodine release by 65% and suggested that it might completely block iodide uptake by the thyroid. In a 1992 study, Brabant et al.[49] were unable to induce a state of iodine depletion in five healthy male volunteers by oral administration of 900 mg $KClO_4$ per day for 4 weeks. Brabant et al. later reported that mild goiters were observed without an increase in thyroid-stimulating hormone (TSH) levels in a 5-week-long repeat of that study. These are the only studies that indicate effect levels from perchlorate exposure in healthy humans.

Results of a recent epidemiological study for counties in California and Nevada that contain a drinking water source impacted by perchlorate do not have higher rates of congenital hypothyroidism, the most likely outcome for fetuses exposed to perchlorate *in utero*.[22] Congenital hypothyroidism occurs when the maternal thyroid and fetal thyroid are unable to supply adequate thyroid hormone to the fetus. Congenital hypothyroidism is known to occur only in presence of severe iodine deficiency, a very rare circumstance in the United States.

Using the data from neonatal screening programs from the States of California and Nevada from the years 1996 and 1997, the authors analyzed the data for any increased incidence of congenital hypothyroidism in the seven counties with elevated perchlorate concentrations in drinking water. Concentrations of perchlorate in drinking water ranged from 4–16 $\mu g\ L^{-1}$ (ppb). Assuming a water-intake level of 2 L per day per person for an adult, this corresponds to a daily dose of up to 32 $\mu g\ day^{-1}$. The minimum therapeutic levels of perchlorate needed to suppress the thyroid in the treatment of hyperthyroidism are approximately 200,000 $\mu g\ day^{-1}$ (200 mg $day^{-1}$). Of the nearly 700,000 newborns screened, only 249 cases of congenital hypothyroidism were identified, where 243 were expected. This corresponds to an overall risk ratio of approximately 1.0 (95% confidence interval, 0.9 to 1.2). The risk ratios for each of the individual counties ranged from 0.6 to 1.1.

Lamm et al.[50] also reported a cross-sectional occupational study of ammonium perchlorate workers, where a *no observable adverse effects level* (NOAEL) of 34 mg $day^{-1}$ was calculated. Two groups of workers at the same perchlorate production plant, 37 workers assigned to the production of perchlorate and 21 assigned to the production of sodium azide ($NaN_3$). Although mean age of the group assigned to the production of perchlorate were five years younger that the group assigned to the production of sodium azide, the two groups did not differ in height and weight, alcohol or tobacco use, in reported medical problems, or in the clinical examination. Workers exposed to airborne concentrations of ammonium perchlorate ranged from 0.004 to 167 mg total particulate perchlorate per day. Changes in urinary output of perchlorate following shift work at the plant by exposed workers demonstrated that inhalation exposure leads to absorption of perchlorate and to its excretion. Estimated internal doses of perchlorate that workers received, based on airborne concentrations, range from 1, 4, 11, or 34 mg perchlorate per day. Thyroid function, as measured by $T_4$, $T_3$, and TSH concentration in blood, showed no difference in the exposed and unexposed groups. The results of the study indicate that workers at the plant were exposed to doses lower than therapeutic levels (200 mg $day^{-1}$) via inhalation, and that thyroid function was unaffected.

## EPA NCEA DRAFT PERCHLORATE REPORT

In May 1999, EPA's NCEA prepared a draft toxicological review document: "Perchlorate Environmental Contamination: Toxicological Review and Risk Characterization Based on Emerging Information." The draft document presented an updated human health risk assessment as well as a screening ecological assessment. In an effort to derive a single oral reference dose (RfD) based on the precursor effects for both non-cancer health effects and thyroid cancer were considered.

NCEA found that perchlorate was readily absorbed from the intestinal tract, and the oral route was considered the major route of exposure. Unsurprisingly, absorption through the skin was not observed. Other than occupational exposure to fumes, mists, dusts, such as in the manufacturing process, inhalation of perchlorate was considered negligible. Potential effects of exposure to perchlorate may include disturbances in the hypothalamic-thyroid axis, included concerns for carcinogenic, neurodevelopmental, developmental, reproductive, and immunotoxic effects. The testing strategy included the following:

- A 90-day drinking water toxicity study in rats with ammonium perchlorate
- A neurobehavioral development study of ammonium perchlorate administered orally in drinking water to rats
- Oral (drinking water) dosage-range developmental toxicity study of ammonium perchlorate in rabbits
- Oral (drinking water) two-generation (one litter per generation) reproduction study of ammonium perchlorate in rats
- Effects of ammonium perchlorate on immunotoxicological, hematological, and thyroid parameters in B6C3F1 mice
- Genotoxicity assays for ammonium perchlorate
- Ecotoxicity assays for ammonium perchlorate

The testing strategy confirmed that the target tissue for perchlorate toxicity is the thyroid gland as indicated by the perturbations of $T_3$, $T_4$, and TSH hormones and by thyroid histopathology in both adult and postnatal rats. The hormone effect was observed at the lower ranges of the exposures tested, from 0.01–1.0 mg $kg^{-1}$ $day^{-1}$, while the histopathology effects occurred at higher doses. Neurobehavioral effects and effects in the brains of offspring occurred at higher concentrations. Preliminary data indicated a potential effect on reproductive parameters and immunotoxicity although no effects were observed in rabbits of the developmental study. Although thyroid tumors had been observed previously in rats exposed in long-term bioassays at high doses, perchlorate was not found to be genotoxic in any assay of the genotoxicity battery.

NCEA's draft report recommended that because of the strong correlation between $T_3$, $T_4$, and TSH, and between changes in $T_3$, $T_4$, and TSH as the precursor lesions to subsequent effects on thyroid hyperplasia that could potentially lead to thyroid tumors or to altered development. The proposed NOAEL was established at the precursor lesion. The rat model was considered relevant yet conservative for human health risk assessment of potential thyroid neoplasia, since the rat appears to be more sensitive to thyroid cancer caused by thyroid-pituitary disruption. Adverse noncancer thyroid effects, such as thyroid enlargement and histopathology, were presumed to pose a human noncancer health hazard.

The proposed revised RfD, assumed to be protective for potential cacinogenicity was derived using effects in the thyroid histopathology observed in rat pups on PND5 in the neurodevelopmental study at 0.1 mg $kg^{-1}$ $day^{-1}$. A composite uncertainty factor of 100 was included to address uncertainties resulting from data gaps and for extrapolation of a minimal *lowest observable adverse effects level* (LOAEL) and intrahuman pharmocodynamic differences and for interspecies differences. The RfD was further adjusted by 0.85 to represent the perchlorate anion, rather than the test chemical ammonium perchlorate. The revised draft RfD for perchlorate was set at 0.9 μg $kg^{-1}$ $day^{-1}$ and confidence in the RfD was designated as medium.

The screening ecological risk assessment pointed out the significant data gaps present. No bioaccumulation data were available to indicate whether perchlorate accumulates in animal tissue. Limited data suggested that perchlorate may be taken up and concentrated in aerial plant parts, especially leaves. A secondary acute value of 5 mg $L^{-1}$ (as perchlorate) was derived to be protective of 95% of aquatic organisms during short-term exposures with 80% confidence. The secondary chronic value of 0.6 (as perchlorate) was derived to be 95% of aquatic organisms during short-term exposures with 80% confidence. For terrestrial plants, a screening benchmark of 4 mg $kg^{-1}$ was derived assuming an uncertainty factor of 10 for interspecies variance from the quartile inhibitory concentration in sand of 41 mg $kg^{-1}$. With limited data on the effects of perchlorate on soil invertebrates, a conservative estimate of a threshold was derived at 1 mg $kg^{-1}$. A factor of 10 for interspecies variance and LOAEL-to-NOAEL extrapolation was applied to the human health risk LOAEL estimate to obtain a screening benchmark of 0.01 mg $kg^{-1}$ for the representative herbivore, the meadow vole (a rodent). Weaknesses in the assessment included the following:

- There is a need for more accurate exposure information that should include transport and transformation processes, notably the fate of perchlorate in irrigated soils.

- There is a need for more accurate information regarding the linkage between biologically effective internal dose in adults and fetus. Confidence in human health risk assessments could be increased by more definitive studies on the amount of disturbance seen in the hypothalamic-pituitary-thyroid axis by perchlorate and neurobehavioral effects.

- Initial assumptions used in the screening ecological assessment still require data to confirm or deny their validity.

The lack of exposure information hampers the comparison of human health and ecological toxicity assessment for accurate characterization of risk. The proposed RfD has been based on precursor effects considered protective for both the thyroid neoplasia and neurodevelopmental effects.

## EXTERNAL REVIEW OF THE EPA NCEA DRAFT REPORT

EPA's Office of Solid Waste and Emergency Response (OSWER) funded a scientific peer review of the Draft Toxicological Review Document on perchlorate. A team of scientists with expertise in general toxicology, thyroid function and toxicology, developmental toxicology, neurotoxicology, immunotoxicology, pharmacology, genetic toxicology, medical endocrinology with an emphasis on thyroid function, biostatistics, assessment of risks due to non-cancer and cancer health effects, and assessment of risks due to ecological effects were assemble and reviewed the completed studies.

A peer review workshop was held February 10–11, 1999, in San Bernadino, California. The peer review panel concluded that the presentation of data in the draft toxicological review document was well-done, but further work was needed before a final RfD could be definitively evaluated. Thyroid hyperplasia (increase in cell number) rather than thyroid hypertrophy (increase in cell size) was recommended by the panel as the as the end point for the RfD. It was concluded that hypertrophy of the thyroid is an adaptive effect and not an adverse effect.

The panel further recommended that a pathology-working group be convened to review the thyroid and brain tissue from all previous and pending studies. In addition the Panel recommended that a common nomenclature for lesions be used and that a consistent pathology review across studies be performed. The external reviewers concluded that the RfD proposed by EPA in the toxicological review document (0.9 $\mu g\ kg^{-1}\ day^{-1}$) is likely to be conservative, based upon the existing toxicological database.

The panel found that the results of the ecological studies strongly supported the screening ecological risk assessment. The major flaw of the screening ecological risk assessment was the limited data on the current levels of perchlorate in the environment and the potential long-term effects. These data limitations resulted in a conservative threshold level and suggested that further studies be performed. The types of fish, wildlife and plants at risk from perchlorate are unknown.

Until further evaluations are performed EPA will recommend that the existing provisional reference dose range of 0.1 to 0.5 $\mu g\ kg^{-1}\ day^{-1}$ continue to be used until a final benchmark is approved.

## WHERE DO WE GO FROM HERE?

After two years of intense scrutiny by the scientific, industrial, military, and regulatory communities into the issue of perchlorate, and the completion of a draft toxicological profile of perchlorate and a proposed revised RfD for perchlorate, the question currently facing all of the parties involved becomes "Where do we go from here?" With the limited resources available to all parties involved and a rapidly growing number of chemicals that must have determinations made regarding their potential to cause harm to a community, regulatory agencies can no longer take the position to "regulate" without accepting some responsibility to help find creative solutions to contamination problems. In fact, EPA's Office of Water is required to set MCLs based on best available technologies—essentially the ability to meet the proposed regulation.

The success of the perchlorate initiative over the last two years has come directly from the partnership created by the scientific, industrial, military, and regulatory communities and what each partner brings to the process via funding, protocol review, and planning. The charter of the IPSC was to facilitate and coordinate activities related to technological issues (occurrence, health effects, treatability and waste stream handling, analytical detection, and ecological impacts) and to create information transfer links for interagency and intergovernmental activities regarding these areas of concern. With the current and pending information that has been gathered a robust health risk estimate may be made regarding the potential for perchlorate to impact various communities.

*Where do we go from here?* The obvious answer is that we allow the regulatory process to continue with the new information generated. Perchlorate has been added to the EPA's Drinking Water Contaminant Candidate List (DWCCL) under the Safe Drinking Water Act. EPA's Office of Water must make a determination as to whether perchlorate must be regulated or whether more information must be gathered before a determination can be made.

It appears at this time that perchlorate is likely to be listed as needing additional research in the areas of analytical methods, more complete occurrence data, health effects, and treatment technologies, before EPA can make determination as to whether its regulation in drinking water is needed. If a NPDWR for perchlorate is to be promulgated under SDWA, USEPA must prepare a health risk reduction and cost analysis to evaluate health risk, benefits, and costs of alternative maximum contaminant levels (MCLs), adding several years to the process.

The EPA is required to decide whether or not to regulate at least 5 contaminants from the DWCCL no later than August 6, 2001. The fact that additional research is needed regarding perchlorate makes it unlikely that EPA would select it to make a determination to regulate by the 2001 deadline. Nevertheless, EPA could choose to regulate perchlorate at that time or earlier, if data indicate that regulation is necessary. An interim NPDWR could also be issued by EPA if new data indicates that perchlorate presents an urgent threat to public health. At this time however, the decision to set action levels rests with state and local authorities.

## EXPOSURE ROUTES

Once released to the environment, perchlorate is a very inert ion that may be taken up by humans and other animals through multiple pathways including airborne dust, water, plant materials, and fertilizers. Regardless of pathway, the target organ remains the same: the thyroid gland. As seen in a recent epidemiological study of workers in an ammonium perchlorate manufacturing plant,[50] inhalation of ammonium perchlorate dust can lead to a measurable internal dose of perchlorate. The vapor pressure of perchlorate salts and acids is very low at normal temperatures; therefore, gas-phase exposure is negligible.

Ingestion of toxic doses of perchlorate can lead to severe gastrointestinal pain, vomiting and diarrhea. Respiratory distress may also develop due to the conversion of hemoglobin to methemoglobin. The lethal dose for an adult is ~15 g or 0.214 g $kg^{-1}$.[51] Late toxic nephritis due to toxic doses of perchlorate has been reported. Ingestion of water containing perchlorate, in the concentrations reported in various locations of the United States, which are significantly lower than the lethal dose of 15 g, are unlikely to lead to measurable health effects.

Recent debate has focused on the potential for perchlorate to be translocated from the environment to plants. Reports by Nzengung[52] and others[53,54] provide evidence for the uptake of perchlorate by various plants. In addition, phytoremediation of perchlorate using woody plants may involve rhizotransformation in soil and short-term uptake into plant.[52] Phytoremediation with wetland plants and herbs may also involve external degradation of perchlorate in the soil.[52] If consumed, plants grown in areas with perchlorate contamination could be a viable route of exposure to humans.

## WHAT MCLG?

The Safe Drinking Water Act authorizes the EPA to set a maximum contaminant level goal (MCLG) and a National Primary Drinking Water Regulation (NPDWR) for contaminants that may have an adverse effect on the health of persons, that are known to occur or are likely to occur in public water systems at levels of public health concern, and where a meaningful opportunity for health risk reduction exists.

Currently, there is no NPDWR for perchlorate. Maximum contaminant level goals (MCLG) and maximum contaminant levels (MCL) are likely to be higher than the current California Preliminary Action Level (PAL) of 18 $\mu g\ L^{-1}$ (18 ppb) based on the draft revised RfD. Initial estimates are that the MCLG will be as high as 32 ppb.

Assuming an oral RfD of 0.9 $\mu g\ kg^{-1}\ day^{-1}$, drinking water consumption of 2 L $day^{-1}$, 70 kg adult, and 100% exposure from drinking water, and MCLG of about 32 ppb would be expected. At this same RfD, assuming a 10 kg child consuming 1 L $day^{-1}$, an MCLG of 9 ppb would result. Additional research is needed to assess the feasibility and costs for removing perchlorate to these low levels before an MCL could be established.

## ACKNOWLEDGMENT

The author wishes to thank the editor for his input, guidance, and assistance in the preparation of this manuscript.

## REFERENCES

1. Gray, H. *Gray's Anatomy*. Running: Philadelphia, PA, 1974, passim.

2. Capen, C.C.; Martin, S.L. "The Effects of xenobiotics on the structure and function of thyroid follicular and c-cells." *Toxicologic Pathology* **1989**, *17*, 266–293.

3. Lehninger, A.L.; Nelson, D.L.; Cox, M.M. *Principles of Biochemistry*. New York: Worth, 1993; p. 752.

4. Hammarstrom S., K. Sterling, K.; Milch, P.O. "Thyroid hormone action: the mitochondrial pathway." *Science* **1977**, *197*, 966–999.

5. Wolff, J. "Perchlorate and the thyroid gland." *Pharmacological Reviews* **1998**, *50*, 89–105.

6. Eichler, O.; Hackenthal, E. "Über ausscheidung und stoffwechsel von perchlorate gemessen mit $^{36}ClO_4^-$." *Naunyn-Schmiedebergs Archiv für Experimentelle Pathologie und Pharmakologie* **1962**, *243*, 554–565.

7. Anbar, M.; Guttmann, S.; Lewitus, Z. "The mode of action of perchlorate ions on the iodine uptake of the thyroid gland." *International Journal of Applied Radiation Isotopes* **1959**, *7*, 87–96.

8. Wyngaarden, J.B.; Wright, B.M.; Ways, P. "The effects of certain anions upon the accumulation and retention of iodide by the thyroid gland." *Endocrinology* **1952**, *50*, 537–549.

9. Wyngaarden, J.B.; Stanbury, J.B.; Rapp, B. "The effects of iodide, perchlorate, thiocyanate, and nitrate administration upon the iodide concentrating mechanism of the thyroid." *Endocrinology* **1953**, *52*, 568–574.

10. Ingbar, S.H.; Freinkel, N. "Concentration gradients for radioiodide in unblocked thyroid glands of rats: effects of perchlorate." *Endocrinology* **1956**, *58*, 95–103.

11. Wolff, J.; Maurey, J.R. "Thyroidal iodide transport: comparison of iodide with anions of periodic group VIIA." *Biochimica et Biophysica Acta* **1962**, *57*, 422–426.

12. Wolff, J.; Maurey, J.R. "Thyroidal iodide transport: the role of ion size." *Biochimica et Biophysica Acta* **1963**, *69*, 58–67.

13. Rocmans, P.A.; Penel, J.C.; Cantrain, F.R.; Dumont, J.E. "Kinetic analysis of iodide transport in dog thyroid slices: perchlorate-induced discharges." *American Journal of Physiology* **1977**, *232*, E343–E352.

14. Krüskemper, H.L.; Kleinsorg, H. "Antithyroideal wirkung von perchlorate." *Naunyn-Schmiedebergs Archiv für Experimentelle Pathologie und Pharmakologie* **1954**, *223*, 460–480.

15. Eichler, O. "On the pharmacology of perchlorate" (in German). *Naunyn-Schmiedebergs Archiv für Experimentelle Pathologie und Pharmakologie* **1929**, *144*, 251–260.

16. Durand, M.J. "Reserches sur l'elimination des perchlorates, sur leur repartition dans les organs et sur leur toxicite." *Bulletin de la Societe de Chimique Biologique* **1938**, *20*, 428–435.

17. Andreoli, T.E.; Carpenter, C.C.J.; Bennett, J.C.; Plum, F., Eds. *Cecil Essentials of Medicine*. Saunders: New York, 1997, passim.

18. National Research Council. *Recommended Daily Allowances*, 10th ed. National Academy Press: Washington, D.C., 1989.

19. Hetzel, B.S.; Maberly, G.F. "Iodine." In W. Mertz, Ed., *Trace Elements in Human and Animal Nutrition,* 5th ed. Academic, New York, 1986, passim.

20. Matovinovic, J. "Endemic goiter and cretinism at the dawn of the third millennium." *Annual Review of Nutrition* 1983, 3, 341–412.

21. National Research Council. *Iodine Nutriture in the United States.* Summary of a Conference, October 31, 1970. Report of the Committee on Food Protection, Food and Nutrition Board, National Academy of Sciences: Washington D.C., 1970.

22. Lamm, S.H.; Doemland, M. "Has perchlorate in drinking water increased the rate of congenital hypothyroidism?" *Journal of Occupational and Environmental Medicine* **1999**, *41*, 409–411.

23. Holt, D.W.; Tucker, G.T.; Jackson, P.R. Storey, G.C.A. "Amiodarone pharmacokinetics." *American Heart Journal* **1983**, *106*, 840–847.

24. Latini, R.; Tognoni, G.; Kates, R.E. "Clinical pharmacokinetics of amiodarone." *Clinical Pharmacokinetics* **1984**, *9*, 136–156.

25. Léger, F.A.; Fragu, P.; Rougier, P.; Laurent, M.F.; Vincens, M.; Auriol, M.; Helal, O.B.; Chouette, G.; Savoie, J.C. "Iodine-induced thyrotoxicosis: analysis of 85 consecutive cases." *European Journal of Clinical Investigation* **1984**, *14*, 449–455.

26. Burger, A., Dinichert, D.; Nicod, P.; Jenny, M.; Lemarchand-Béraud, T.; Vallotton, M.D. "Effect of amiodarone on serum $T_3$, $rT_3$, $T_4$, and TSH." *Journal of Clinical Investigation* **1976**, *58*, 255–259.

27. Franklyn, J.A.; Davis, J.R.; Gamage, M.D.; Little, W.A.; Ramsden, D.B.; Sheppard, M.C. "Amiodarone and thyroid hormone action." *Clinical Endocrinolology* **1985**, *22*, 257–264.

28. Melmed, S.; Nademanee, K.; Reed, A.W.; Hendrickson, J.; Singh, B.N.; Hershman, J.M. "Hyperthyroxinemia woth bradycardia and normal TSH secretion following chronic amiodarone administration." *Journal of Clinical Endocrinology and Metabolism* **1981**, *83*, 997–1001.

29. Nademanee, K.; Piwonka, R.W.; Singh, B.N.; Hershman, J.M. "Amiodarone and thyroid function." *Progress in Cardiovascular Diseases* **1989**, *31*, 427–437.

30. Singh, B.N.; Nademanee, K. "Amiodarone and thyroid function: clinical implications during antiarrythmic therapy." *American Heart Journal* **1983**, *106*, 857–869.

31. Capiello, F.; Boldorini, R.; Tosoni, A.; Praneo, S.; Bernasconi, R.; Raggi, U. "Ultrastructural evidence of thyroid damage in amiodarone-induced thyrotoxicosis." *Journal of Endocrinology Investigation* **1995**, *18*, 862–868.

32. Brennan M.D.; Erickson, D.Z.; Carney, J.A.; Bahn, R.S. "Nongoitrous amiodarone-associated thyrotoxicosis: evidence of follicular disruption in vitro and in vivo." *Thyroid* **1995**, *5*, 177–183.

33. Belshaw, B.E.; Becker, D.V. " Necrosis of follicular cells and discharge of iodine induced by administering iodide to iodine-deficient dogs." *Journal of Clinical Endocrinology and Metabolism* **1973**, *36*, 466–474.

34. Many, M.C.; Mestdagh, C.; Van Den Hove, M.F.; Denef, J.F. "In vitro study of acute toxic effects of high iodine doses in human thyroid follicles." *Endocrinology* **1992**, *131*, 621–630.

35. Reichert, L.J.; DeRooy, H.A.M. Treatment of amiodarone induced hypothyroidism with potassium perchlorate and methimazideideole during amiodarone treatment. *British Medical Journal* **1989**, *298*, 1547–1548.

36. Martino, E.; Mariotti, S.; Aghini-Lombardi, F. Mariotti, S.; Lenziardi, M.; Baschieri, L.; Braverman, L.E.; Pinchera, A. "Short term administration of potassium perchlorate

restores euthyroidism in amiodarone iodine-induced hypothyroidism." *Journal of Endocrinology Investigation* **1986**, *63*, 1233–1236.

37. Martino, E.; Aghini-Lombardi, F.; Mariotti, S. "Treatment of amiodarone associated thyrotoxicosis by simultaneous administration of potassium perchlorate and methimasideideole." *Journal of Endocrinology Investigation* **1986**, *9*, 201–207.

38. Trip, M.D.; Duren, D.R.; Wiersinga, W.M. "Two cases of amiodarone-induced thyrotoxicosis successfully treated with a short course of antithyroid drugs while amiodarone was continued." *British Heart Journal* **1994**, 72, 266–268.

39. Martino, E.; Aghini-Lombardi, F.; Mariotti, S.; Bartalena, L.; Braverman, L.; Pinchera, A. "Amiodarone: a common source of iodine-induced thyrotoxicosis." *Hormone Research* **1987**, 26, 158–171.

40. De Visscher, M., Ed. *Comprehensive Endocinology: The Thyroid Gland.* Raven: New York, 1980; passim.

41. Trotter, W.R. "The relative toxicity of antithyroid drugs." *Journal of New Drugs* **1962**, *2*, 333–343.

42. Krevans, J.R.; Asper, S.P.; Reinhoff, W.F. "Fatal aplastic anemia following use of potassium perchlorate in thyrotoxicosis." *JAMA: Journal of the American Medical Association* **1962**, *181*, 162–164.

43. Gjemdal, N. "Fatal aplastic anemia following use of potassium perchlorate in thyrotoxicosis." *Acta Medica Scandinavica* **1963**, *174*, 129–131.

44. Barzalai, D.A.; Sheinfeld, M. "Fatal complications following the use of potassium perchlorate in thyrotoxicosis." *Israel Journal of Medical Sciences* **1966**, *2*, 453–456.

45. Bartalena, L.; Brogioni, S.; Grasso, L.; Bogazzi, F.; Burelli, A.A.; Martino, E. "Treatment of amiodrane-induced thyrotoxicosis, a difficult challenge: results of a prospective study." *Journal of Clinical Endocrinology and Metabolism* **1996**, *81*, 2930–2933.

46. Connel, J.M.C. "Long-term use of perchlorate." *Postgraduate Medicine* **1981**, *57*, 516–517.

47. Reichert, L.J.; DeRooy, H.A.M.. "Treatment of amiodarone induced hypothyroidism with potassium perchlorate and methimazideideole during amiodarone treatment." *British Medical Journal* **1989**, *298*, 1547–1548.

48. Burgi, H.; Benguerel, M.; Knopp, J.; Kohler, H.; Studer, H. "Influence of perchlorate on the secretion on non-thyroxine iodine by the normal human thyroid gland." *European Journal of Clinical Investigations* *4*, 65–69.

49. Brabant, G.; Bergmann, P.; Kirsch, C.M.; Kohrle, J.; Hesch, R.D.; Von Zur Muhlem, A. "Early adaptation of thyrotropin and thyroglobulin secretion to experimentally decreased iodine supply in man." *Metabolism* **1992**, *4*, 1093–1096.

50. Lamm, S.H.; Braverman, L.E.; Li, F.X.; Richman, K.; Pino, S.; Howearth, G. "Thyroid health status of ammonium perchlorate workers: a cross-sectional occupational health study." *Journal of Occupational and Environmental Medicine* **1999**, *41*, 248–260.

51. Deichmann, W.; Gerade, H., Eds. *Toxicology of Drugs and Chemicals*. Academic: New York, 1969; passim.

52. Nzengung, V.; Wang, C. "Influences on phytoremediation for perchlorate contaminated water." *Preprints of Extended Abstracts, Division of Environmental Chemistry.* 218th ACS National Meeting. ACS: **1999**, *39* (2), New Orleans, LA, August 22–26, 1999, 95–98.

53. Bacchus, S.; Susarla, S.; Wolfe, L.; Harvey, G.; McCutcheon, S. "Predicting field performance of herbaceous species for phytoremediation of perchlorate." *Preprints of Extended Abstracts, Division of Environmental Chemistry.* 218th ACS National Meeting. ACS: **1999**, *39* (2), New Orleans, LA, August 22–26, 1999, 98–100.

54. Susarla, S.; Wolfe, N.; McCutcheon, S. "Perchlorate uptake in lettuce seedlings." *Preprints of Extended Abstracts, Division of Environmental Chemistry.* 218th ACS National Meeting. ACS: **1999**, *39* (2), New Orleans, LA, August 22–26, 1999, 66–68.

---

**Editor's Note:** In this chapter, Dr. Clark refers to the EPA NCEA draft assessment report, which predicts a potable water NOAEL of 32 ng $mL^{-1}$. It is important to realize that this number may change as the results of additional studies become known. As of December 1999, the level recognized by the EPA as safe for consumption is 4–18 ng $mL^{-1}$; however, there are no regulations based on this concentration. EPA's Assistant Administrator for the Office of Research and Development, Norine Noonan, stresses that EPA does not use incomplete assessments to develop policy. NCEA's draft assessment was marked "do not quote or cite," but the document has already been widely cited as a draft. Until a new assessment is approved by external review and EPA management, the previous level stands. A binding toxicological assessment is expected in late 2000 at the earliest.

It is worth pointing out that the draft report gives a reference dose of 0.9 μg $kg^{-1}$ $day^{-1}$ (one significant digit), but the authors compute an NOAEL of 32 ng $mL^{-1}$. Given the uncertainty intrinsic in the quantitation of perchlorate ion in potable water, it seems preferable to round the number to 30 ng $mL^{-1}$ or 3 μg $dL^{-1}$ (emphasizing one significant digit). At its best, ion chromatography will have an uncertainty of ± 1–2 ng $mL^{-1}$.

# Chapter 4

## REGULATING PERCHLORATE IN DRINKING WATER[⊗]

Frederick W. Pontius,[*①] Paul Damian,[②] and Andrew D. Eaton[③]

①Regulatory Affairs Consultant, Lakewood, Colorado 80226
②Earth Tech, Englewood, Colorado 80111
③Montgomery Watson Laboratories, Pasadena, California 91101

### INTRODUCTION

Perchlorate ($ClO_4^-$) has emerged in recent years to become a significant new threat to drinking water supplies and the environment. It is an oxidizing anion that originates as contaminant in the environment from the dissolution of ammonium, potassium, magnesium, or sodium salts. Perchlorate is very mobile in aqueous systems and can persist for many decades under typical ground and surface water conditions.

The Safe Drinking Water Act (SDWA) authorizes the Environmental Protection Agency (EPA) to set a maximum contaminant level goal (MCLG) and a National Primary Drinking Water Regulation (NPDWR) for contaminants that may have an adverse effect on the health of persons, that are known to occur or is likely to occur in public water systems at levels of public health concern, and where a meaningful opportunity for health risk reduction exists. The Administrative Procedure Act defines the minimum legal procedural requirements that the agency must meet when setting a drinking water regulation. In addition, the EPA has authority under the Negotiated Rulemaking Act of 1990 to involve stakeholders in the formulation of a proposed rule in a process known as regulatory negotiation.

[⊗]This work was presented at the 218th national meeting of the American Chemical Society as part of the Environmental Division symposium Perchlorate in the Environment, held August 22–24, 1999, in New Orleans, Louisiana.

[*]Author to whom correspondence should be directed. Phone: 303-986-9923. Electronic mail: Pontius@polnow.net

*Perchlorate in the Environment*, edited by Urbansky.
Kluwer Academic/Plenum Publishers, New York, 2000.

## SDWA REQUIREMENTS

Perchlorate is included on the EPA's Drinking Water Contaminant Candidate List (CCL) prepared under the SDWA.[1] The CCL includes 50 chemical and 10 microbiological contaminants that may require regulation under the SDWA. It has been divided into two categories: (1) contaminants for which sufficient information exists to begin to make regulatory determinations in 2001 as required by the SDWA, and (2) contaminants for which additional research and occurrence information is necessary before regulatory determinations can be made. Perchlorate is listed as needing additional research in the areas of analytical methods, more complete occurrence data, health effects, and treatment technologies, before EPA can make a determination as to whether its regulation in drinking water is needed.

If regulated, the SDWA specifies how an NPDWR for perchlorate is to be set. EPA is required to prepare a health risk reduction and cost analysis to evaluate the health risks, benefits, and costs of alternative maximum contaminant levels (MCLs). An MCLG, a nonenforceable health goal, is set. The MCL is set as close to the MCLG as is feasible using best technology or other means available, as listed by EPA. EPA is required to make a determination as to whether the benefits of the MCL justify or do not justify the costs. If the agency determines that the benefits of an MCL do not justify the costs, then the agency has flexibility to set the MCL at a level where the health benefits do justify the cost of regulation. In addition, the agency may consider risk-risk trade-offs in setting the MCL. The SDWA provides EPA the authority to issue an interim NPDWR for a contaminant in situations representing an urgent threat to public health.

## ANALYTICAL METHODS

Currently, the most common analytical technique for perchlorate in waters is suppressed ion chromatography (IC) using an AS11 column with conductivity detection. This sensitive method was first developed by the California Dept. of Health Services in early 1997 (using a slightly different IC column, the AS5) to investigate perchlorate in California ground water systems. Dionex developed the AS11 column, which has a much greater capacity than the AS5, improving the ability to analyze samples with higher solids content and allowing the use of a simpler eluent (sodium hydroxide) compared to the AS5 column (sodium hydroxide plus *p*-cyanophenol). An interlaboratory study conducted in 1998 by the Interagency Perchlorate Steering Committee found both methods to be relatively rugged for relatively low total dissolved solids samples.[2] A practical quantitation level of approximately 6 $\mu g\ L^{-1}$ was achievable on an interlaboratory basis. Improvements in analytical methods for perchlorate are continuing.

## OCCURRENCE

Ammonium perchlorate is manufactured for use as the oxidizer component and primary ingredient in solid propellant for rockets, missiles, and fireworks. In addition to other minor uses, perchlorate salts are also used on a large scale as a component of air bag inflators.

Large-scale production of perchlorate-containing chemicals in the U.S. began in the mid-1940s. Because of its shelf life, perchlorate in the U.S. missile and rocket inventory must be removed and replaced with a fresh supply. Hence, large volumes have been disposed of in Nevada, California, Utah, and various states since the 1950s.

Perchlorate began to be discovered at various manufacturing sites and in well water and drinking water supplies in 1997, shortly after development of a low-level analytical method. Forty-four states have confirmed perchlorate manufacturers or users. Releases into ground or surface water have been confirmed in 14 states. The majority of locations where perchlorate has been detected in groundwater are 14 facilities in California associated with manufacture or testing of solid rocket fuels for the military or the National Aeronautics and Space Administration (NASA). Seven National Priority List sites in California are affected by these releases.

Two facilities that manufactured ammonium perchlorate in Nevada were found to have released perchlorate to groundwater that is the source for low levels (4 to 16 $\mu g\ L^{-1}$) in Lake Mead and the Colorado River. This water is used for drinking water, irrigation, and recreation by millions of people in Nevada, California, Arizona, and Native American tribes.

Perchlorate concentrations reported in wells and surface water vary widely. At one facility near Henderson, Nevada, concentrations in groundwater monitoring wells was measured as high as 0.37% ($3.7 \times 10^6\ \mu g\ L^{-1}$). Water suppliers in both northern and southern California have detected perchlorate in 144 public water supply wells, with 38 of these above California's provisional action level of 18 $\mu g\ L^{-1}$. The highest level of perchlorate reported in any public water supply well was 280 $\mu g\ L^{-1}$, with few others testing greater than 100 $\mu g\ L^{-1}$.

The American Water Works Service Company conducted sampling and analysis of 425 drinking water supply wells in 16 states. Of these, perchlorate was found above 4 $\mu g\ L^{-1}$ in 7 wells (1.6%), with the highest level at 6.4 $\mu g\ L^{-1}$. Wells testing positive for perchlorate were located in CA, IN, IA, and PA. Drinking water wells in the following states had no detections of perchlorate: AZ, CT, IL, MD, MA, MI, MO, NJ, NM, NY, OH, and WV.

A systematic national survey of perchlorate occurrence has not been conducted. The AWWA Research Foundation is coordinating a study to characterize possible perchlorate contamination of drinking water sources in areas of high risk. The extent of actual or even potential perchlorate contamination is unclear for many parts of the country, and for some areas of the country perchlorate contamination may not be an issue.

EPA recently promulgated the Unregulated Contaminant Monitoring Rule (UCMR) that will require public water systems to monitor for perchlorate.[3] Both Community Water Systems (CWSs) and Nontransient NonCommunity Water Systems (NTNCWSs) serving greater than 10,000 persons (about 2774 systems) will be required to monitor for perchlorate and report the results to the state primacy agency and EPA. Monitoring will be conducted within the first 3 years of a 5-year monitoring cycle (2001 to 2003). Small systems designated in state monitoring plans will be required to forward a sample to EPA for analysis. The UMCR will provide additional occurrence data that will be used by EPA to help determine whether a drinking water regulation for perchlorate is needed.

## HEALTH EFFECTS

Perchlorate is readily absorbed from the gastrointestinal tract upon ingestion. Absorption through the skin is expected to be minimal due to the high polarity of the chemical. Inhalation uptake of pure perchlorate or its salts is also expected to be limited due to the low volatility of these compounds, however, significant inhalation exposure to particulate perchlorate may occur where perchlorate dusts have become suspended in air. Once absorbed into the body, perchlorate is excreted unchanged rapidly and virtually completely in the urine. There is no evidence indicating that perchlorate is metabolized to any significant extent.

The primary mechanism of toxicity of perchlorate is inhibition of normal iodide uptake by the thyroid.[4] In addition, perchlorate has also been shown to stimulate excessive release of iodide stored in the thyroid. Thus, the thyroid can be considered the target organ of toxicity. Perchlorate's inhibition of iodide-uptake by the thyroid is the basis for its historical use in the treatment of Graves' disease, a disease characterized by excessive production of the iodide-containing thyroid hormones: thyroxine ($T_3$) and triiodothyronine ($T_4$). By blocking iodide uptake by the thyroid, perchlorate inhibits the excess production of these hormones. In healthy individuals, exposure to sufficient levels of perchlorate would be expected to produce symptoms of hypothyroidism including such effects as dry and itchy skin, dry and brittle hair, sluggishness, muscle and joint pain, and headaches.

Potentially more significant, however, is the possibility of adverse effects on normal growth and development in children or the fetus. In the case of the child or developing fetus, temporary disturbances of thyroid function such as might result from excess perchlorate exposure, may result in permanent effects. The thyroid hormone is critical to normal brain development, which begins in the uterus and extends to 3 years of age. Perchlorate can cross the placenta, potentially disrupting development of the fetus. Perchlorate exposure would be expected to have drastically different effects on fetuses and infants than on humans. These effects could include mental retardation, speech impairment, impaired fine motor skills, deaf-mutism, and spasticity.[4] Questions remain as to the perchlorate exposure necessary to cause adverse effects in fetuses, infants, and adults, and whether these effects are actually occurring as a result of perchlorate contamination.[5,6,7]

Other than thyroid effects, very little published information exists regarding the health effects of perchlorate. Indeed, this paucity of health effects data has been one of the most problematic issues associated with perchlorate contamination. EPA has recommended an oral Reference Dose (RfD) of 1 to 5 $\mu g^{-1}\ kg^{-1}\ day^{-1}$ for risk assessment. The RfD is the exposure level thought to be without significant risk to humans, including sensitive subgroups, when the contaminant is ingested daily over a lifetime. EPA released a draft toxicological review and risk characterization for external review,[4] with a tentative revised RfD value of 0.9 $\mu g^{-1}\ kg^{-1}\ day^{-1}$. The review panel recommended additional studies be completed to answer key questions and a final recommendation for an oral RfD is not expected until additional health effects studies are completed.

## TREATMENT TECHNOLOGIES

Perchlorate is nonvolatile and highly soluble in water, so it cannot be removed from water by conventional filtration, sedimentation, or air stripping.[8] Urbansky and Schock[9] have recently reviewed treatment technologies for removing perchlorate from drinking water. They must be cost-effective, acceptable to regulatory agencies and the public, cause no other water quality problems, and not be limited by waste disposal problems. Another option for lowering perchlorate concentrations in contaminated water supplies is by blending with uncontaminated water supplies.

Ion exchange and membrane processes can remove perchlorate from water. Ion exchange is being evaluated in California's San Gabriel Valley for removing approximately 30 to 200 $\mu g\ L^{-1}$ perchlorate from groundwater. Ion exchange resins that can selectively remove perchlorate rather than competing ions (e.g., chloride, sulfate, bicarbonate) that may be present at higher concentrations are needed.

Nanofiltration and reverse osmosis will also remove perchlorate, but these technologies are costly. Ion exchange and membrane processes also generate concentrated perchlorate-containing waste brines that may be difficult to dispose. Treatment of the brine may be needed to lower its volume or toxicity before disposal.

Ozone-peroxide treatment has minimal effect on perchlorate in water. Ozone-peroxide followed by granular activated carbon has been found promising at one site in the San Gabriel Valley, but additional studies are needed to evaluate the long-term effectiveness, reliability, and cost.

Most attention to date has been focused on developing an anaerobic biochemical reduction process, whereby microbes are used to convert perchlorate to a less toxic or innocuous form.[10] The Air Force Research Laboratory, Materials and Manufacturing Directorate began developing biochemical reactor systems for treating high level perchlorate contaminated wastewater (i.e., 1000 to 10,000 $\mu g\ L^{-1}$) in the early 1990s. A continuous-stirred-tank-reactor system began treating wastewater from rocket motor production operations in Utah in 1997. Additional pilot tests have been conducted to evaluate removal of low level perchlorate contamination and although results are promising, additional studies are needed to evaluate the cost, reliability and public acceptance of this technology.

## REGULATING PERCHLORATE

EPA must address several key issues when deciding whether or not to regulate perchlorate. Analytical methods appear capable of detecting perchlorate at concentrations of concern. Does perchlorate occur with such frequency nationally, or would it be expected to occur at such frequency, to justify EPA issuing a NPDWR? Additional occurrence data are needed to answer this question.

Assuming an oral RfD of 0.9 $\mu g\ kg^{-1}\ day^{-1}$, drinking water consumption of 2 L $day^{-1}$, 70 kg person, and 100% exposure from drinking water, an MCLG of about 32 $\mu g\ L^{-1}$ would be expected. At this same RfD, assuming a 10 kg child consuming 1 L $day^{-1}$, an MCLG of 9 $\mu g\ L^{-1}$ would result. Additional research is needed to determine a final oral RfD, and to assess the feasibility and costs for removing perchlorate to these low levels before an MCL could be established.

EPA is required to decide whether or not to regulate at least 5 contaminants from the CCL no later than August 6, 2001. If results from additional research are available regarding perchlorate, then EPA could select it for a determination to regulate by the 2001 deadline. The agency must revise the CCL no later than February 2003, and perchlorate could be deferred to the revised list. A determination whether or not to regulate at least 5 contaminants from this second CCL is required no later than August 2006. If perchlor-ate is regulated from this second CCL, then a proposed NPDWR must be issued no later than 2 years later (August 2008). The final rule is required 18 months later (February 2010), with a 9 month extension possible if needed by EPA. Under the SDWA, water systems have 3 years to comply with an NPDWR, with 2 additional years possible if capital improvements are needed. Following this regulatory timeline, compliance with an MCL for perchlorate would not be required until 2013.

Public concern regarding the potential effects of short-term exposure to perchlorate on fetuses and infants could dictate that EPA act to address perchlorate in regulation prior to the timeline described above. EPA could choose to regulate perchlorate at any time, if data indicates that regulation is necessary. An interim NPDWR could be issued by EPA if health effects data indicate that perchlorate presents an urgent threat to public health. At this time, however, the authority to address perchlorate contamination, determine action levels, and set drinking water limits rests with state regulatory agencies.

## REFERENCES

1. Perciasepe, R. "Part III. Environmental Protection Agency. Announcement of the drinking water contaminant candidate list; notice." *Federal Register* **1998**, *63*, 10274–10287.

2. Jackson, P.E.; Gokhale, S.; Rohrer, J.S. "Recent developments in the analysis of perchlorate using ion chromatography." In *Perchlorate in the Environment*, E.T. Urbansky, Ed. Plenum: New York, 2000; Ch. 5.

3. Browner, C. "Part II. Environmental Protection Agency. 40 CFR Parts 9, 141 and 142. Revisions to the unregulated contaminant monitoring regulation; final rule. *Federal Register*, **1999**, *64*, 50556–50620.

4. *Perchlorate Environmental Contamination: Toxicological Review and Risk Characterization Based on Emerging Information*. External Review Draft. Environmental Protection Agency, Office of Research and Development, National Center for Environmental Assessment: Washington, D.C., Dec. 1998; EPA Doc. No. NCEA-1-0503. URL: http://www.epa.gov/ncea/perch.htm.

5. Lamm, S.H.; Braverman, L.E.; Li, F.X.; Pino, S.; Howearth, G. "Thyroid health status of ammonium perchlorate workers: a cross-sectional occupational health study." *Journal of Occupational and Environmental Medicine* **1999**, *41*, 248–260.

6. Lamm, S.H.; Doemland, M. "Has perchlorate in drinking water increased the rate of congenital hypothyroidism?" *Journal of Occupational and Environmental Medicine* **1999**, *41*, 409–411.

7. Gibbs, J.P.; Ahmad, R.; Crump, K.S.; Houck, D.P.; Leveille, T.S.; Findley, J.E.; Francis, M. "Evaluation of a population with occupational exposure to airborne ammonium perchlorate for possible acute or chronic effects on thyroid function." *Journal of Occupational and Environmental Medicine* **1998**, *40*, 1072–1082.

8. Urbansky, E.T. "Perchlorate chemistry: implications for analysis and remediation." *Bioremediation Journal* **1998**, *2*, 81–95.

9. Urbansky, E.T.; Schock, M.R. "Issues in managing the risks associated with perchlorate in drinking water." *Journal of Environmental Management* **1999**, *56*, 79–95.

10. Logan, B.E. "A review of chlorate- and perchlorate-respiring microorganisms." *Bioremediation Journal* **1998**, *2*, 69–79.

# Chapter 5

# RECENT DEVELOPMENTS IN THE ANALYSIS OF PERCHLORATE USING ION CHROMATOGRAPHY[⊗]

Peter E. Jackson*, Swati Gokhale, and Jeff S. Rohrer

Dionex Corporation, 1228 Titan Way, Sunnyvale, California 94088-3606

## INTRODUCTION

Ammonium perchlorate, a key ingredient in solid rocket propellants, has recently been found in ground waters in regions of the U.S. where aerospace materials, munitions, and fireworks were developed, tested, or manufactured. Perchlorate has been found in ground and surface waters in California, Nevada, Utah, Texas, New York, Maryland, Arkansas, and West Virginia, although the total extent of the contamination problem is not known.[1,2] Perchlorate poses a human health concern as it interferes with ability of the thyroid gland to utilize iodine to produce thyroid hormones. Current data from the EPA indicate that exposure to a concentration less than 4–18 $\mu g\ L^{-1}$ (ppb) perchlorate provides adequate health protection.[2] Recently, the EPA National Center for Environmental Assessment has announced a new provisional oral reference dose (RfD) for perchlorate of 9 $\mu g\ kg^{-1}\ day^{-1}$, which would correspond to an action level of 32 $\mu g\ L^{-1}$ in drinking water.[3] A final action (e.g., health advisory) level is expected sometime in 2000.

Perchlorate contamination of public drinking water wells has become a serious problem in California. While perchlorate is listed on the EPA Contaminant Candidate List as a research priority, it is not currently regulated under the Federal Safe Drinking Water Act.[4] However, the California Department of Health Services (CDHS) has adopted an action level for perchlorate in drinking water of 18 $\mu g\ L^{-1}$. The CDHS recommends that utilities remove drinking water sources that contain perchlorate concentrations above the18 $\mu g\ L^{-1}$ action level from service. To date, perchlorate has been detected in over 100 public drinking water wells in California, with more than 20 wells now closed due to contamination.[2]

[⊗]This work was presented at the 218th national meeting of the American Chemical Society as part of the Environmental Division symposium Perchlorate in the Environment, held August 22–24, 1999, in New Orleans, Louisiana.

*Author to whom correspondence should be directed. Phone: 408-481-4262. Fax: 408-737-2470. Electronic mail: peter_jackson@dionex.com.

*Perchlorate in the Environment*, edited by Urbansky.
Kluwer Academic/Plenum Publishers, New York, 2000.

The California Department of Health Services developed an ion chromatographic (IC) method for the analysis of trace perchlorate in 1997 to support the CDHS action level of 18 $\mu g\ L^{-1}$ in drinking water.[5] The CDHS method uses a large loop injection with a Dionex IonPac AS5 column and a hydroxide eluent containing *p*-cyanophenol. Detection is achieved by suppressed conductivity using a chemically regenerated AMMS suppressor. An updated IC method employing an IonPac AS11 column, hydroxide eluent, and suppressed conductivity detection with a self regenerating ASRS suppressor, was developed in 1998.[6] Also, the EPA Office of Solid Waste has recently announced a new perchlorate method (Method 9058), which includes conditions for using either the IonPac AS5 or AS11 columns.[7] This method will be published in the EPA OSWER Methods Manual—Draft Update IVB.

This paper will report on a recent developments for the determination of trace level perchlorate using ion chromatography. The performance of the AS5 and AS11 methods will be discussed and their application to the analysis of perchlorate in a variety of environmental samples, including drinking water, and ground and surface waters will be demonstrated. The application of a new polarizable anion analysis column, the IonPac AS16, for the determination of perchlorate in high ionic strength samples will be also discussed.

## EXPERIMENTAL

### Instrumentation

Either 4500 or DX-500 ion chromatographs (Dionex Corporation, Sunnyvale, CA) were used for this work. Separations were carried using Dionex IonPac® AS5, AS11 and AS16 (250 × 4.0 mm) analytical columns and IonPac AG5, AG11 and AG16 (50 × 4.0 mm) guard columns. Anions were detected by suppressed conductivity detection using either an Anion Micro-Membrane Suppressor, AMMS® with a regenerant of 35 mN (18 mM) sulfuric acid at 10 mL $min^{-1}$; or an Anion Self-Regenerating Suppressor, ASRS®-ULTRA operated at 300 mA in the external water mode.

### Reagents and Procedures

All water used was deionized water, Type I reagent grade, 18 MΩ cm resistivity or better. Sodium hydroxide, 50% w/w aqueous solution was obtained from Fisher Scientific (Pittsburgh, PA). Sodium perchlorate, 99% ACS reagent grade was obtained from Aldrich (Milwaukee, WI), as was 95% *p*-cyanophenol. ACS reagent grade chemicals were used for the preparation of the standards for the interference and recovery studies, with the exception of humic acid and selenate standards, which were prepared from technical grade reagents.

## RESULTS AND DISCUSSION

In order to quantify perchlorate at low $\mu g\ L^{-1}$ levels, it is essential to optimize chromatographic conditions in terms of retention time, peak shape and baseline noise. The perchlorate ion is a "polarizable" anion, consequently it should be chromatographed on a hydrophilic anion exchanger to minimize peak tailing. In addition, perchlorate is highly retained on anion exchange resins and requires a strong mobile phase to elute it within a reasonable timeframe, which is desirable for lower detection limits. Initial investigations on an IonPac AS5 column showed that an eluent of 120 mM hydroxide containing an organic modifier, such as *p*-cyanophenol, was required to elute perchlorate from the AS5 column. The effect of *p*-cyanophenol over the range of 0–3 mM on perchlorate retention was investigated, with an eluent of 120 mM NaOH containing 2.0 mM *p*-cyanophenol providing optimal peak shape and a retention time for perchlorate of approximately 7 minutes.[5] The perchlorate anion is well resolved from common inorganic anions, which essentially elute at the column void volume under these conditions.

A large loop injection (740 μL) is required for this application in order to achieve sub-ppb detection limits for perchlorate. The method detection limit (MDL) using the IonPac AS5 column was determined by spiking perchlorate at concentrations of 1.0, 2.5, and 4.0 $\mu g\ L^{-1}$ into reagent water, as shown below in Table 1.

**Table 1.** Method detection limit in reagent water with the IonPac AS5 column[5]

| Perchlorate Spike Conc. (µg $L^{-1}$) | No. of Spiked Replicates | Mean Recovery (µg $L^{-1}$) | Standard Deviation (µg $L^{-1}$) | Calculated MDL (µg $L^{-1}$) |
|---|---|---|---|---|
| 1.0 | 14 | 0.87 | 0.11 | 0.6 |
| 2.5 | 16 | 2.3 | 0.12 | 0.8 |
| 4.0 | 16 | 3.9 | 0.11 | 0.7 |
| | | **Pooled MDL (*dof* = 43)** | | **0.7 µg $L^{-1}$** |
| | | **MRL (5 × MDL)** | | **4 µg $L^{-1}$** |

A linearity study was performed to ensure accurate quantification of perchlorate in the low µg $L^{-1}$ range. A correlation coefficient of 0.9998 was obtained for a plot of peak area versus concentration in the 2–100 µg $L^{-1}$ range, demonstrating that calibration is linear at the levels required for the quantification of perchlorate in drinking and ground waters.

In addition to the AS5 column, it has also been shown that perchlorate can be successfully chromatographed on an IonPac AS11 column.[6] The major advantage of using this more hydrophilic column is that *p*-cyanophenol is not required in the eluent in order to achieve reasonable peak shape for perchlorate. This approach enables the use of electrolytic self regererating suppressors (e.g., ASRS-ULTRA), which add considerable convenience to the operation of the ion chromatograph, as such devices are not recommended for use with eluents containing electroactive modifiers, such as *p*-cyanophenol. The use of an IonPac AS11 column with an eluent of 100 mM sodium hydroxide permits the elution of perchlorate in less than 10 minutes. Figure 1 shows a typical chromatogram of a 20 µg $L^{-1}$ perchlorate standard obtained using the AS11 column.

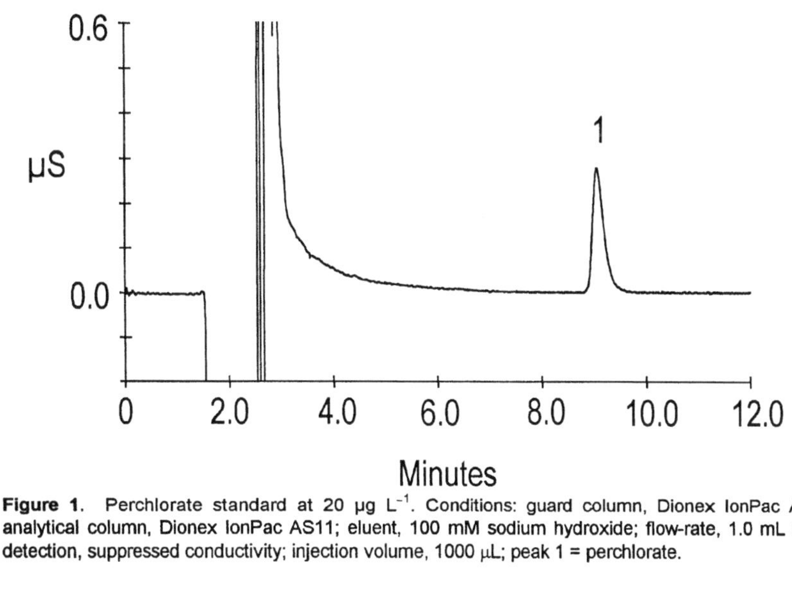


**Figure 1.** Perchlorate standard at 20 µg $L^{-1}$. Conditions: guard column, Dionex IonPac AG11; analytical column, Dionex IonPac AS11; eluent, 100 mM sodium hydroxide; flow-rate, 1.0 mL $min^{-1}$; detection, suppressed conductivity; injection volume, 1000 µL; peak 1 = perchlorate.

The method detection limit was determined for the AS11 column using seven replicates of 2.5 μg $L^{-1}$ perchlorate spiked into reagent water according to the procedure outlined in EPA Method 300.0.[8] The single-operator MDL was was calculated to be 0.3 μg $L^{-1}$ using the conditions shown in Figure 1.

Both the IonPac AS5 and AS11 columns were tested for interferences by injecting low μg $L^{-1}$ levels of perchlorate in the presence of 100 μg $L^{-1}$ solutions of 22 common anions. Of the anions investigated, only cyanide, iodide, and thiocyanate display any significant retention on either column using the elution conditions described above. Perchlorate is resolved by at least 2 minutes from the nearest eluting anion, thiocyanate, which is not typically found at high levels in drinking or ground waters.[1,2]

Despite the use of a large loop injection, no evidence of column overloading was observed when injecting drinking water samples.[1] Figure 2 shows a chromatogram of Sunnyvale, CA, tap water spiked with 6.0 μg $L^{-1}$ perchlorate obtained using the IonPac AS11 column.

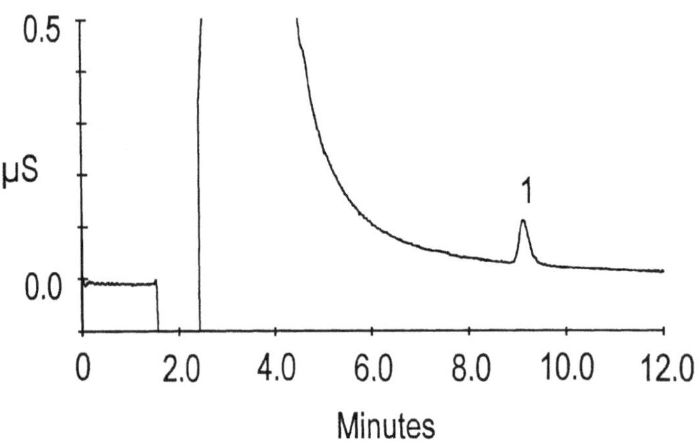


**Figure 2.** Sunnyvale, CA, tap water spiked with 6.0 μg $L^{-1}$ perchlorate. Conditions: as for Figure 1, except; peak 1 – perchlorate (5.9 μg $L^{-1}$).

Ground water samples may contain high concentrations of common anions, particularly chloride, carbonate, and sulfate. The effect of mg $L^{-1}$ levels of common anions on perchlorate recovery was investigated by injecting solutions of low μg $L^{-1}$ levels of perchlorate in the presence of 50, 200, 600 and 1000 mg $L^{-1}$ chloride, carbonate and sulfate, respectively. Quantitative recoveries were obtained for perchlorate in all cases using either the IonPac AS5 or AS11 columns, demonstrating that mg $L^{-1}$ levels of common anions have no significant effect on the recovery of low μg $L^{-1}$ levels of perchlorate. Table 2 shows the results obtained using the IonPac AS11 column.

**Table 2.** Effect of mg $L^{-1}$ levels of common anions on perchlorate recovery (20 µg $L^{-1}$) on the IonPac AS11 column[1]

| Anion | Concentration (mg $L^{-1}$) | Perchlorate Recovery (%) |
|---|---|---|
| Carbonate | 50 | 96.6 |
| Carbonate | 200 | 98.8 |
| Carbonate | 600 | 92.1 |
| Carbonate | 1000 | 94.2 |
| Chloride | 50 | 92.2 |
| Chloride | 200 | 99.2 |
| Chloride | 600 | 98.7 |
| Chloride | 1000 | 97.4 |
| Sulfate | 50 | 94.4 |
| Sulfate | 200 | 100.0 |
| Sulfate | 600 | 93.4 |
| Sulfate | 1000 | 97.4 |

Essentially, both the AS5 or AS11 columns give similar performance, in terms of linearity, MDLs, freedom from interferences, and spiked recoveries, as was demonstrated in the recent IPSC collaborative study.[9] The IPSC study, which involved 19 laboratories, was organized to quantitatively evaluate the performance of IC methods for the measurement of perchlorate in drinking and ground water. The study samples consisted of wellwater at three total dissolved solids levels of 72, 144, and 288 mg $L^{-1}$, which were spiked with perchlorate at concentrations of 6, 18 ppb and 36 µg $L^{-1}$. Both the AS5 and AS11 columns were found to be satisfactory for perchlorate analysis in typical ground and surface water samples, as shown below in Table 3.

**Table 3.** Comparison of interlaboratory recovery and bias obtained using the IonPac AS5 and AS11 columns[9]

| | RECOVERY (%) | | BIAS (%) | | CV (%) | CV (%) |
|---|---|---|---|---|---|---|
| **Sample** | **AS-11** | **AS-5** | **AS-11** | **AS-5** | **AS-11** | **AS-5** |
| C2T1 | 95.9 | 99.3 | -4.1 | -0.7 | 11.5 | 10.7 |
| C2T2 | 95.8 | 99.3 | -4.2 | -0.7 | 10.3 | 8.6 |
| C2T3 | 95.2 | 99.4 | -4.8 | -0.6 | 12.1 | 5.4 |
| C3T1 | 99.2 | 101.0 | -0.8 | 1.0 | 6.7 | 5.5 |
| C3T2 | 97.7 | 101.3 | -2.3 | 1.3 | 8.0 | 4.4 |
| C3T3 | 99.7 | 98.4 | -0.3 | -1.6 | 9.5 | 3.4 |
| C4T1 | 96.2 | 99.3 | -3.8 | -0.7 | 7.6 | 4.1 |
| C4T2 | 97.3 | 101.5 | -2.7 | 1.5 | 6.0 | 6.6 |
| C4T3 | 97.5 | 98.7 | -2.5 | -1.3 | 7.4 | 5.9 |
| STO | 101.4 | 101.3 | 1.4 | 1.3 | 3.3 | 3.1 |
| **AVERAGE** | **97.6** | **99.9** | **-2.4** | **-0.1** | **8.2** | **5.8** |

STO; 50 µg $L^{-1}$standard, T1-3; TDS range of 72–288 mg $L^{-1}$, C2-4; spiked (6–36 µg $L^{-1}$) wellwater samples.

Dionex has recently developed a new column for the analysis of polarizable anions, such as perchlorate, iodide, and thiocyanate. The IonPac AS16 column is more hydrophilic and has a significantly higher ion exchange capacity than either the AS5 or AS11 columns. This column allows the injection of higher ionic strength samples and is also compatible with the new EG40 automated KOH eluent generator.[1] Figure 3 shows a chromatogram of a 20 $\mu g\ L^{-1}$ perchlorate standard obtained using IC with an AS16 column, the EG40 Eluent Generator and suppressed conductivity detection.

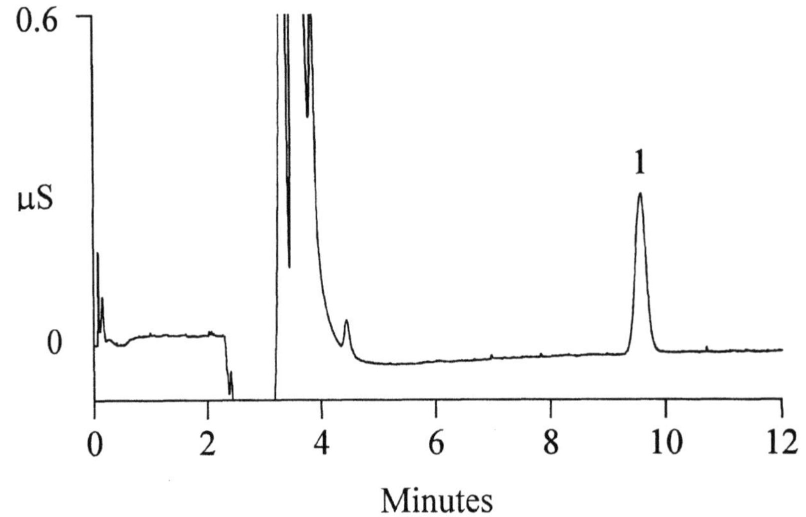


**Figure 3.** Perchlorate standard at 20 $\mu g\ L^{-1}$. Conditions: guard column, Dionex IonPac AG16; analytical column, Dionex IonPac AS16; eluent source, EG40; eluent, 65 mM KOH; flow-rate, 1.2 mL $min^{-1}$; detection, suppressed conductivity; injection volume, 1000 µL; peak 1 - perchlorate.

The IonPac AS16 column was also tested for interferences by injecting low $\mu g\ L^{-1}$ levels of perchlorate in the presence of 100 $\mu g\ L^{-1}$ solutions of 22 common anions. In this case, of the anions investigated, only iodide, and thiocyanate display any significant retention on the AS16 column. Perchlorate was again resolved by approximately 2 minutes from the nearest eluting anion, thiocyanate, which is not typically found at high levels in drinking or ground waters. The effect of mg $L^{-1}$ levels of common anions on perchlorate recovery using the AS16 column was also investigated by injecting solutions of low $\mu g\ L^{-1}$ levels of perchlorate in the presence of 50, 200, 600 and 1000 mg $L^{-1}$ chloride, carbonate and sulfate, respectively. Quantitative recoveries were obtained for perchlorate in all cases with the IonPac AS16 column, demonstrating that mg $L^{-1}$ levels of common anions have no significant effect on the recovery of low $\mu g\ L^{-1}$ levels of perchlorate. Table 4 below shows the results obtained using the IonPac AS16 column.

**Table 4.** Effect of mg $L^{-1}$ levels of common anions on perchlorate recovery (20 µg $L^{-1}$) on the IonPac AS16 column

| Anion | Concentration (mg $L^{-1}$) | Perchlorate Recovery (%) |
|---|---|---|
| Carbonate | 50 | 97.6 |
| | 200 | 94.4 |
| | 600 | 95.4 |
| | 1000 | 93.5 |
| Chloride | 50 | 96.1 |
| | 200 | 96.7 |
| | 600 | 109.6 |
| | 1000 | 97.4 |
| Sulfate | 50 | 94.4 |
| | 200 | 96.3 |
| | 600 | 94.7 |
| | 1000 | 95.5 |

## CONCLUSIONS

The use of ion chromatography with the IonPac AS5 or AS11 columns, large loop injection and suppressed conductivity detection provides a simple, interference free method for the determination of perchlorate at low µg $L^{-1}$ levels in drinking and ground waters. The method is linear over the range of 2–100 µg $L^{-1}$ and quantitative recoveries were obtained for perchlorate in spiked drinking and ground water samples. The MDLs permit quantification of perchlorate below the levels which ensure adequate health protection (4 – 18 µg $L^{-1}$), as recommended by the EPA. The new IonPac AS16 column provides similar performance to the AS5 and AS11 columns for drinking water samples, although its higher capacity makes it more suitable for the analysis of trace perchlorate in high ionic strength matrices. Current work on perchlorate analysis by IC involves extending the range of applications to more complex samples, such as wastewaters, soil extracts, fertilizers, and other high ionic strength samples (> 2000 µS $cm^{-1}$) after dilution or additional pretreatment.

## REFERENCES


1. Jackson, P.E.; Laikhtman, M.; Rohrer, J. "Determination of trace level perchlorate in drinking water and ground water by ion chromatography." *Journal of Chromatography A* **1999**, *850*, 131–135.

2. California Department of Health Services, Perchlorate in California Drinking Water, Update July, 1999.

3. Jarabek, A. Paper presented at the Perchlorate Conference, Ontario, CA, March, 1999.

4. *Federal Register* **1997**, 62 FR 52193.

5. California Department of Health Services. Determination of Perchlorate by Ion Chromatography, 1997.

6. Wirt, K.; Laikhtman, M.; Rohrer, J.; Jackson, P.E. "Low-level perchlorate analysis in drinking water and groundwater by ion chromatography." *American Environmental Laboratory* **1998**, *10*, 1, 5.

7. "Method 9058. SW-846 Test Methods for Evaluating Solid Waste Physical/Chemical Methods." Draft Update IVB, Environmental Protection Agency: Washington, DC, 1999.

8. "Method 300.0. Determination of Inorganic Anions in Water by Ion Chromatography." Environmental Protection Agency: Cincinnati, OH, 1993.

9. Chaudhuri, S.; Okamoto, H.; Pia, S.; Tsui, D. Interagency Perchlorate Steering Committee Analytical Subcommittee Report, 1999.

---

**Editor's Note:** The EPA's National Exposure Research Laboratory, in conjunction with the EPA's Office of Water Technical Support Center (both in Cincinnati, Ohio), plans to establish an ion chromatography method for perchlorate sometime in early 2000. The new method is expected to be released as Method 314, but the number designation is not finalized.

# Chapter 6

# ANALYSIS OF TRACE LEVEL PERCHLORATE IN DRINKING WATER AND GROUND WATER BY ELECTROSPRAY MASS SPECTROMETRY⊗

Rebecca A. Clewell,*[①] Sanwat Chaudhuri,[②] Steve Dickson,[②] Rachael S. Cassady,[③] William M. Wallner,[③] J. Eric Eldridge,[①] and David T. Tsui[①]

[①] U.S. Air Force Research Laboratory, Human Effectiveness Directorate, Toxicology Branch, Building 79, 2856 G Street, Wright-Patterson AFB, Ohio 45433-7400

[②] Utah Department of Health Laboratory, 46 North Medical Drive, Salt Lake City, Utah 84113

[③] Utah Department of Environmental Quality, 288 North 1460 West, Salt Lake City, Utah 84114-2102

## INTRODUCTION

The recent discovery of perchlorate contamination in the ground water of several western states has resulted in widespread concern over the quality of drinking water supplies. Ammonium perchlorate has been found in commercial fertilizers and is used as the oxidizer and main ingredient in solid rocket propellants, fireworks, and munitions. The current acceptable level for the presence of perchlorate in drinking water is 18 parts per billion (ppb). However, it has been found at concentrations as high as 0.37% in ground water near munitions manufacturing and testing facilities.[1–11] Perchlorate contamination has also been found in areas such as Texas, where fertilizers are used to maintain land for cattle farming. Two of the main ingredients in these commercial fertilizers, potash and Chilean nitrate, have been shown by Air Force Research Laboratory to contain up to 0.57% by weight perchlorate. Although these deposits contain the necessary ingredients for fertilizers, they are also a very rich source of perchlorate.[9–11]

*Authors to whom correspondence should be directed. Phone: 937-255-5150. Fax: 937-255-1474. E-mail: Rebecca.Clewell@wpafb.af.mil.

⊗This work was presented at the 218th national meeting of the American Chemical Society as part of the Environmental Division symposium Perchlorate in the Environment, held August 22–24, 1999, in New Orleans, Louisiana.

Perchlorate is known to interfere with the uptake of iodide by the thyroid resulting in a decreased ability to produce necessary hormones. Many toxicological studies are currently in progress leading to the development of a reference dose for perchlorate in humans.[8] The growing interest in the presence of perchlorate in ground water and drinking water has brought about a need for viable methods of detection that possess both a high sensitivity and selectivity for perchlorate.

The perchlorate anion has a mass-to-charge ratio of 99.1 u. As a result of the relatively large diameter and the small, delocalized charge of the perchlorate anion, it is only weakly basic toward most Lewis acids. Consequently, most perchlorate salts are highly soluble in water. However, the perchlorate anion is polarizable. The small hydrated radius and low hydration energy allows the anion to form strong complexes with large delocalized organic cations.

Early techniques for perchlorate analysis, such as gravimetric analysis,[12–16] liquid-liquid extraction, and spectrophotometry, [17–22] relied on the ability of perchlorate to form complexes with large organic dyes, such as brilliant green and methylene blue. However, these methods are not selective for the determination of perchlorate. Other anions commonly found in drinking water, such as phosphate, nitrate, and chlorate, can also complex with these dyes. Ion pair high performance liquid chromatography (HPLC),[23–24] capillary electrophoresis (CE),[25–31] and ion selective electrode (ISE)[31–32] have also been used in perchlorate analysis. However, these methods do not have the necessary sensitivity at trace levels.

The current method of choice involves the use of ion chromatography coupled with a conductivity detector.[33–37] This method has an accepted sensitivity of 4 parts per billion (ppb). However, there are some substances found in drinking water which may interfere with the chromatographic separation of perchlorate, and the conductivity detector is not selective for the determination of perchlorate. Ion chromatography relies on retention time as a unique identifier for perchlorate.[38–40] It would therefore be advantageous to develop a method for perchlorate analysis that would verify the presence of perchlorate in water samples analyzed by IC. Additionally, questions have been raised as to the sensitivity and robustness of this technique. For example, it has been demonstrated by Air Force Research Laboratory, Operational Toxicology Branch (AFRL/HEST), that high levels of total dissolved solid (TDS) can completely block the signal of perchlorate in conductivity measurements. Therefore, it is necessary to develop a method for perchlorate analysis that would be selective for perchlorate and would have sensitivity equal to or greater than that of the ion chromatography methods.

Electrospray mass spectrometry is an ideal option for perchlorate analysis, due to the inherent selectivity, sensitivity, and efficiency of the instrument. Through electrospray mass spectrometry, it is possible to selectively monitor the ion of choice. When the spectrometer is tuned to detect a specific mass to charge ratio, the other ions are filtered out as they pass through the quadrupoles. Because the interference anions never reach the detector, the background signal is reduced and the sensitivity of the measurement is enhanced. The selected ion monitoring also allows the species of interest to be analyzed without requiring chromatographic separation. Thus, analysis time for prepared samples is reduced from 20 minutes to about 30 seconds. Additionally, electrospray ionization is useful for perchlorate analysis, because it is able to achieve ionization and vaporization of non-volatile perchlorate as a result of the process in which this technique transfers solvated ions into the gas phase.

Electrospray ionization is an atmospheric pressure ionization (API) technique. Within the API chamber a voltage of 3–8 kV is applied to the electrospray probe. Consequently, the droplets are electrically charged as liquid sample is ejected from the probe. The positive ions are driven to the surface of the droplet, which disrupts the surface tension and disperses the drop into a fine spray of droplets. These droplets are then driven toward the heated capillary by the voltage gradient set between the capillary and the electrospray probe, and by the inert sheath gas, which surrounds the probe and pushes the droplets forward. As the droplets approach the heated capillary, the increased temperature causes the volatile solvent to evaporate. Eventually, the electrical charge within the droplets reaches the Rayleigh Stability Limit. This is the point where the intermolecular forces overcome the surface tension, and the droplet explodes again. This process is repeated either until the analyte is ejected into the gas phase by electrostatic forces, or until the solvent is completely stripped from the anion. The ions subsequently pass through the heated capillary and tube lens into an intermediate vacuum region, where they are focused by the ion optics.[42-43] The perchlorate anions eventually pass into the quadrupole mass filters.

The quadrupole mass filter consists of four parallel rods that serve as electrodes. Two of the rods are connected to the positive end of a DC terminal, and two of the rods are connected to the negative end. Additionally, a transverse AC potential is also superimposed upon each pair of rods. When the spectrometer is in negative ion mode, the negative poles filter the lighter ions. In the negative poles, the AC current pulls the lighter anions into the poles, where they will be neutralized, while the DC potential helps to stabilize the trajectory of the heavier anions. In the positive poles, the first half of the AC cycle directs the lighter ions to the center of the channel. This offsets the movement of the negative anions toward the oppositely charged poles. Since the AC current does not as easily affect the heavier ions, they are drawn into the poles and neutralized. Hence, the positive poles filter the lighter anions, and the negative poles filter the heavier anions, and only ions within a very narrow range of mass to charge ratios are actually allowed to reach the detector.[42,44]

The ability of the quadrupoles to selectively monitor the perchlorate anion and to filter out the interference anions commonly contained in water, and the ability of the electrospray ionization to volatilize non-volatile perchlorate for analysis by mass spectrometry allow the sensitive and selective determination of perchlorate. The purpose of this study was to develop an alternative method for the selective analysis of perchlorate, with equal or better sensitivity than ion chromatography methods that would be applicable to typical ground water samples.

## EXPERIMENTAL

### Test Materials and Reagents

Ammonium perchlorate [*7790-98-9*], acetonitrile [*75-05-8*], and glacial acetic acid [*64-19-7*] were purchased from Sigma-Aldrich (St. Louis, MO). The mobile phase consisted of 0.5% acetic acid in acetonitrile. The stock perchlorate solution of 10 mg $mL^{-1}$ perchlorate was prepared in distilled, deionized water gravimetrically from the ammonium salt of perchlorate.

### Utah and Las Vegas Water Samples

In order to compare the electrospray mass spectrometry method existing ion chromatographic techniques and also to determine applicability to groundwater samples, two sets of real-life water samples were collected from the States of Utah and Nevada. Samples were collected and contributed by the State of Utah, Department of Environmental Quality, Division of Solid and Hazardous Waste and Division of Drinking Water, Salt Lake City, Utah and the U.S. Environmental Protection Agency, National Exposure Research Laboratory, Las Vegas, Nevada.

The Utah samples were collected from three different locations. All samples were ground water samples collected in duplicate, and containing TDS levels between the range of 300 and 800 ppm. Samples were run in the laboratory as unknowns at the time of analysis for perchlorate. Samples 4844–4847 and 4836–4839, were collected from the Harkers Canyon alluvium which consists mostly of unconsolidated sand and gravel. The United States

Geological Survey (USGS) has identified this area of Salt Lake Valley as primary recharge. Samples 4834 and 4835 were obtained from the Coon Creek drainage, which is in the fault margin, situated along the toe of the Oquirrh Mountains. Samples 4842 and 4843 were gathered from a blending point in a drinking water system, which blends water from shallow artesian wells and deep pumped wells. The later samples were taken after chlorination.

The water used to prepare the Nevada study samples was collected in February 1998, from a well identified as 4CP-1. The well is located on the Nevada Test Site (NTS). This well was selected because it was known to have been isolated from atmospheric and ground processes that contribute to the migration of surface compounds into the aquifer. The background tritium concentration in the raw water (< 2 pCi $L^{-1}$) is significantly lower than ground water which is recharged from surface sources, rain, and snow melt (> 30 pCi $L^{-1}$). Because of the long isolation of the water from processes likely to introduce perchlorate, it was unlikely that perchlorate would be present. Using the raw water from 4CP-1, the study samples were prepared at three concentrations (C2, C3, C4) and three TDS levels (T1, T2, T3), in addition to sample C1, which was a blank at the three TDS levels, and a spiked distilled water sample, ST0. The concentration of perchlorate was 6, 18, and 36 parts per billion (ppb) for C2, C3, and C4 respectively and 51 ppb for ST0. Sample C1 was a blank. The TDS concentrations as a percent for T1, T2, and T3, were 25, 50, and 100 percent raw waters, respectively. The balance of the volume for T1 and T2 was distilled raw water.[38]

**Extracted Reference Standards and Sample Preparation**

Perchlorate standards at 0, 1, 5, 10, and 25, ng $mL^{-1}$ were prepared in 2.5 mL distilled, deionized water by serial dilution from the stock (10 mg $mL^{-1}$) ammonium perchlorate solution. Samples were stored in a –25 °C freezer and then an –86 °C freezer for one hour each, in order to thoroughly freeze the samples. The frozen samples were then placed in a lyophilizer overnight, to remove the water and volatile contaminants at a low pressure and temperature, in order to avoid the loss of perchlorate. Following lyophilization, the samples were reconstituted in 1 mL 0.5% acetic acid in acetonitrile mobile phase. Prior to analysis by electrospray mass spectrometry, the reference standards were filtered with 0.45 μm Millipore (Bedford, MA) Millex-HV13 syringe filters.

For water samples collected from Salt Lake City and Las Vegas, 2.5 mL of the water was transferred to polypropylene test tubes without dilution. The samples were stored in a –25 °C freezer and then an –86 °C freezer for one hour each. The frozen samples were then lyophilized overnight. Following lyophilization, the samples were reconstituted in 1 mL 0.5% acetic acid in acetonitrile mobile phase and filtered with Millipore Millex-HV13 (0.45 μm) syringe filters. Samples were then analyzed by electrospray mass spectrometry. Unknown concentrations were found by comparing the peak area to the standard curve generated from the lyophilized standards.

**Unextracted Standards and Instrument Sensitivity Procedure**

Unextracted Standards were prepared from the stock solution of 10 mg $mL^{-1}$ perchlorate in distilled, deionized water. Stock solution was diluted serially without lyophilization in the mobile phase (0.5% acetic acid in acetonitrile) for final concentrations of 0.5, 1.0, 5.0, 10, 25, 50 and 100 ppb. Samples were then injected directly into the mass spectrometer. In order to monitor the sensitivity of the mass spectrometer, an unextracted standard curve was run at the start of each day. A signal to noise ratio of 3:1 or higher was required for the 0.5 ppb standard in order to continue with sample analysis.

**Analytical Method**

Electrospray mass spectrometry. Electrospray mass spectrometry was performed on a Finnigan-Mat TSQ 700 (San Jose, CA). A Harvard Apparatus (South Natick, MA) Model-22 syringe pump was used to deliver the mobile phase at a constant flow rate of 75 μL $min^{-1}$ through an 82.5 cm × 1.14 mm i.d. Intramedic polyethylene tubing. The polyethylene tubing was connected to the mass spectrometer by 22.3 cm of 0.10 μm i.d. fused silica capillary tubing. The heated capillary was set at 200 °C, with an applied voltage of −10 V. The electrospray probe had an applied voltage of 5 kV. Nitrogen was used for the sheath gas, and was set at a pressure of 40 psi. The tube lens was set at −103 V.

The samples were injected directly into the electrospray mass spectrometer through a 10 μL sample loop. Samples were analyzed with the Finnigan ULTRIX 4.4 software. Mass Spectra were collected by scanning mass-to-charge ratios from 50 to 350 u, in a solution of 10 ppm perchlorate in 0.5% acetic acid in acetonitrile mobile phase. Perchlorate samples were selectively monitored at a mass to charge ratio of 99.1 u, using the negative ion MS mode. All other parameters were optimized for the detection of perchlorate by tuning the instrument specifically for perchlorate with a solution of 5 ppm perchlorate in 0.5% acetic acid in acetonitrile.

The method detection limit of the electrospray mass spectrometer was determined by analyzing ten perchlorate standards at concentrations of 1 ppb. Recovery was determined by four extracted standards for concentrations at both 5 and 25 ppb. The measured concentrations of the standards were found by comparing the peak area of the lyophilized standards to an unextracted standard curve.

**Ion Chromatography**

The results obtained by electrospray mass spectrometry were compared to that obtained on ion chromatography. All measurements using ion chromatograph studies (performed by Utah Health Lab and AFRL/HEST) were performed using a Dionex (Sunnyvale, CA) DX 500 ion chromatograph configured with a GP 40 gradient pump, CD 20 Conductivity Detector, and a AS40 Automated sampler. Separation was obtained using a Dionex IonPac AS11 analytical column and an AS11 guard column. Anions were detected with suppressed conductivity detection using an ASRS ULTRA suppressor, an Anion Self-Regenerating Suppressor. The eluent used was a 57 mM sodium hydroxide solution. All water used was de-ionized, reagent grade with 18 MΩ cm resistivity. All samples were analyzed in duplicate in order to confirm analysis and assess matrix effect. A reagent water blank, reagent water blank fortified with known concentration of perchlorate, a sample fortified with known concentration of perchlorate, and standards at three different perchlorate concentrations were analyzed with the samples, in order to assure the quality of analysis.

## RESULTS

**Method Development and Validation**

The mass spectrum of perchlorate in 1% w/w acetic acid/acetonitrile (Figure 1) shows chlorine-containing peaks at 99.1, 140.1, and 159.12 u, which correspond to the $ClO_4^-$ anion, and the $CH_3CN$-$ClO_4^-$, and $CH_3COOH$-$ClO_4^-$ adducts. The *m* + 2 peaks at an abundance of 32.5% support the presence of chlorine.[45] Figure 2 shows the background spectrum for the mobile phase (1% acetic acid in acetonitrile). This background was subtracted from the perchlorate scan in order to show only the response to the ions that contain the perchlorate anion.

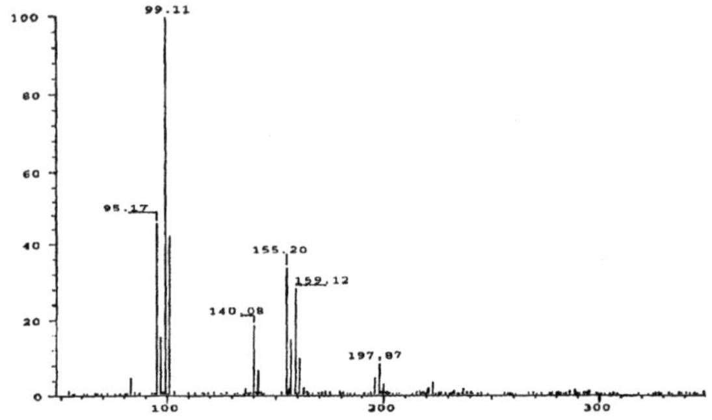


**Figure 1.** Mass spectrum of 10 µg $mL^{-1}$ perchlorate solution in a 1% acetic acid in acetonitrile mobile phase.

The presence of perchlorate in the peaks at 99, 140, and 159 u was verified by the presence of these peaks after subtraction of the background spectrum in Figure 2. The presence of chlorine in the ions was also verified by the presence of $m + 2$ isotope peaks at 32.5% abundance.[45] In order to obtain a calibration curve for the perchlorate samples, a series of standards were injected to the mass spectrometer (Figure 3). Results are obtained as selected ion current profiles (SICP), where each peak indicates an individual injection of a perchlorate standard. Calibration curves (Figure 5) were then obtained for perchlorate by plotting the peak area versus the known perchlorate concentration of the standards for the selected ion monitoring of $m/z = 99.1$

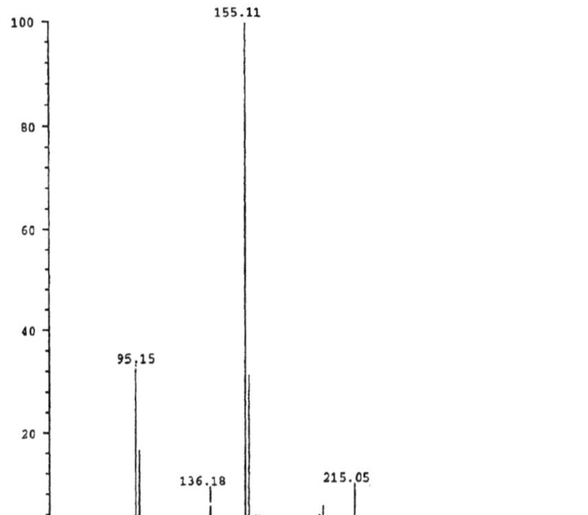


**Figure 2.** Background mass spectrum for 1% w/w acetic acid in acetonitrile mobile phase.

As shown in Figures 3 and 4, a linear correlation was found between the peak area and the perchlorate standards from 0.5 to 100 ppb, with a correlation coefficient of 0.992. Therefore, the electrospray mass spectrometer was programmed to selectively monitor the mass to charge ratio of 99.1. The acetonitrile/perchlorate peaks and the acetic acid-perchlorate peaks did not show the same linear relationship between the detector response and the concentration of perchlorate.

In order to determine the method detection limit (MDL) for lyophilized standards, ten samples at a concentration of 1 ppb were measured in triplicate. The method detection limit was determined according to EPA guidelines,[46] with the following calculation:

$$\mathrm{MDL} = t \cdot s \tag{1}$$

where Student's $t = 2.821$ for a 99% level of confidence ($\nu = 9$) and $s$ is the estimated standard deviation at $[ClO_4^-] = 1$ ppb. The method detection limit for lyophilized samples was calculated to be 0.34 ppb for $S/N = 3$. The minimum reporting limit (MRL) was found to be 1.02 ppb:

$$\mathrm{MRL} = 3 \times \mathrm{MDL} \tag{2}$$

The instrument detection level (IDL) represents the sensitivity of the instrument being used for analysis toward the compound of interest. The use of the instrument detection level provides a basis for the comparison of the sensitivity of different instruments for the same compound. The instrument detection limit was found at 3.8 ppb (eq 3), where $V$ is the volume of the sample injected for analysis.

$$\mathrm{IDL} = \mathrm{MDL} \times V \tag{3}$$

The recovery for the method was determined with samples at concentrations of 5 and 25 ppb. The average percent recovery was calculated to be 82.4% using eq 4:

$$\mathrm{Recovery} = (C_s - C^\circ)/C_{\mathrm{true}} \tag{4}$$

where $C_s$ is the measured concentration of the standard, $C^\circ$ is the measured concentration of the blank, and $C_{\mathrm{true}}$ is the known concentration of the standard. The acceptable range for recovery is 100 ± 20%, i.e., 80–120%.

Results from the collaborative study performed on Las Vegas drinking water are shown in Table 1. Samples are listed in order of increasing concentration of total dissolved solid (TDS). The water samples with low levels of TDS agree with the collaborative ion chromatography results within the acceptable variance of ± 20%. Only one sample (sample number 12) is outside of the acceptable range by –8%. The results for the Utah ground water samples are listed in Table 2, in order of increasing concentration of TDS. Again, samples with low levels of TDS show agreement between electrospray mass spectrometric analysis, and the ion chromatography analysis performed by AFRL. However, as the concentration of total dissolved solid increases, the variance between the two methods also increases.

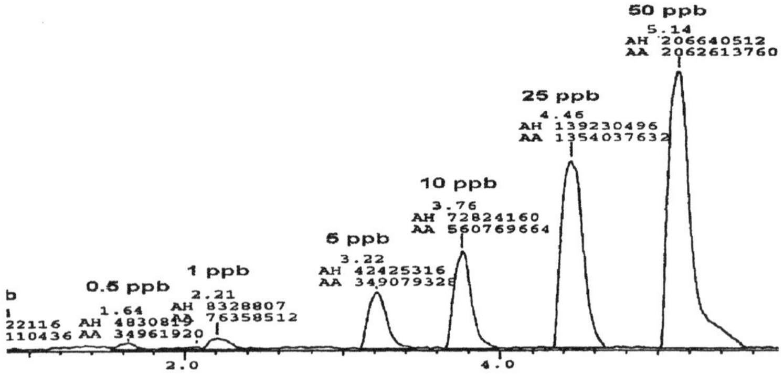


**Figure 3.** ESI-MS response for unextracted standards monitored at $m/z$ = 99.1 u.

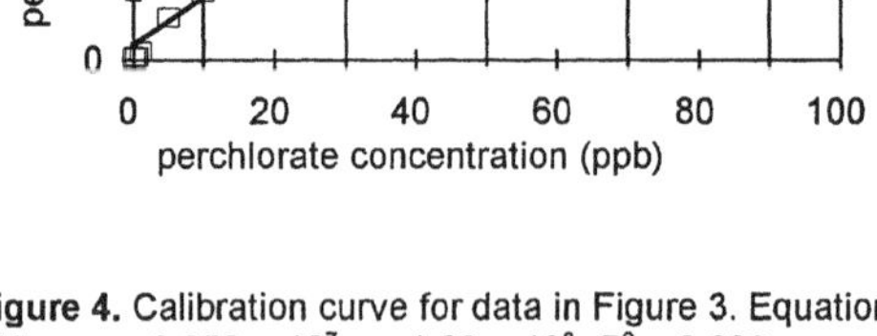


**Figure 4.** Calibration curve for data in Figure 3. Equation of line: $a = 3.856 \times 10^7\, c + 1.23 \times 10^8$; $R^2 = 0.992$.

**Table 1.** Nevada Ground Water Samples

| Sample | TDS (ppm) | TSQ (ppb) | IC (ppb) | % Recovery |
|---|---|---|---|---|
| 1 | 0 | 49.4 | 51 | 97% |
| 2 | 71 | 0 | 0 | 100% |
| 3 | 71 | 0 | 0 | 100% |
| 4 | 71 | 0 | 0 | 100% |
| 5 | 142 | 4.4 | 5 | 88% |
| 6 | 142 | 4.3 | 5 | 86% |
| 7 | 142 | 14.1 | 15 | 94% |
| 8 | 142 | 15.1 | 15 | 101% |
| 9 | 282 | 15.8 | 15 | 105% |
| 10 | 282 | 35.5 | 31 | 115% |
| 11 | 282 | 27.3 | 29 | 94% |
| 12 | 282 | 21.0 | 29 | 72% |

**Table 2.** Utah Ground Water Samples

| Sample | TDS (ppm) | TSQ (ppb) | IC (ppb) | % Recovery |
|---|---|---|---|---|
| 4838 | 302 | 38.2 | 38.3 | 100% |
| 4839 | 303 | 36.3 | 37.5 | 97% |
| 4836 | 318 | 18.7 | 19.3 | 97% |
| 4847 | 319 | 32.8 | 35.5 | 92% |
| 4837 | 320 | 14.5 | 19.4 | 75% |
| 4846 | 321 | 24.9 | 34.7 | 72% |
| 4844 | 369 | 348 | 299 | 116% |
| 4845 | 374 | 392 | 307 | 128% |
| 4834 | 518 | 68.3 | 72.8 | 94% |
| 4835 | 521 | 68.6 | 73.7 | 93% |
| 4842 | 736 | 5.4 | 15.2 | 36% |
| 4843 | 766 | 6.3 | 15.6 | 40% |

## DISCUSSION

Electrospray mass spectrometry is a viable method for the determination of perchlorate in drinking water and ground water. This method has demonstrated equal or greater selectivity and sensitivity for the analysis of perchlorate than existing ion chromatography methods. The ability of electrospray mass spectrometry to selectively monitor the perchlorate anion results in a method that is more selective for perchlorate than the current ion chromatography method, which relies on retention time to identify perchlorate. Furthermore, analysis by electrospray mass spectrometry shows a higher sensitivity than that of the current accepted IC method. Intra-laboratory studies have shown the method detection limit for the electrospray mass spectrometric determination of perchlorate to be 0.38 ppb, while current ion chromatography techniques have an accepted method detection limit of 4 parts per billion. Electrospray mass spectrometry increases the current level of detection by an order of magnitude. The sensitivity of electrospray mass spectrometry toward perchlorate analysis shows an improvement of more than three orders of magnitude over the current ion chromatography method, which requires a 1000 μL injection volume. The instrument detection level for electrospray mass spectrometry was calculated to be 3.8 ppb, which is a significant improvement over the accepted ID of 4000 ppb for ion chromatography.

This method can also be used to successfully quantitatively detect perchlorate in real world drinking water and ground water. Inter-laboratory studies performed on ground water samples obtained from both Utah and Las Vegas water supplies demonstrated the ability of electrospray mass spectrometry to accurately determine the amount of perchlorate present in water samples with typical levels of total dissolved solid. Electrospray mass spectrometric results for typical water samples with TDS concentrations of less than 700 ppm were within ± 20% of corresponding ion chromatography results. However, samples with unusually high levels of total dissolved solid showed a significant variance from values obtained through ion chromatography. Samples with TDS levels greater than 700 ppm showed a difference of 60–70% from the corresponding ion chromatography values.

Especially high levels of total dissolved solid interfere with both ion chromatography and electrospray mass spectrometry. High TDS levels interfere with the signal to the conductivity detector that is used with ion chromatography. In the mass spectrometer, the dissolved solids in particularly dirty water samples can plate out on the heated capillary. This can cause both difficulties in cleaning the instrument and capillary failure. Capillary failure causes an increased background noise and inaccurate quantitation. It is also possible for the large concentration of interference anions to overwhelm the quadrupoles. Consequently, more interference anions reach the detector, and the background noise is significantly increased. The increase in background noise within the detector results in decreased sensitivity and less accurate determination of perchlorate concentrations. It is necessary, then, to develop a sample preparation method which would effectively remove high concentrations of dissolved solids from exceptionally dirty water samples without interfering with the analysis of perchlorate.

AFRL/HEST has developed a sample preparation method for use with both electrospray mass spectrometry and the current ion chromatography methods. This sample preparation method increases the sensitivity for both the IC and electrospray mass spectrometry methods by removing interference anions from the sample, and thereby reducing the background noise during analysis. The sample preparation method involves the use of a silver saturated cation exchange resin to facilitate the removal of the other ions. Silver cations are allowed to exchange onto the cation exchange resin before the sample is introduced. Passing the water sample through the resin allows the less soluble silver salts of the interference anions to precipitate out. However, since perchlorate does not form stable Lewis salts, it remains in solution. The development of this method, which will allow a more sensitive analysis of perchlorate in water with high levels of total dissolved solids, is described in greater detail in a separate report.[47]

## CONCLUSION

Electrospray mass spectrometry is a viable option for the detection and quantitative analysis of trace level perchlorate in drinking water matrices. Analysis can be accomplished without the use of special pairing agents, chemicals, or separation by HPLC. Under typical water conditions, where the TDS concentration is below 700 ppm, the proposed method yields accurate and sensitive measurements of perchlorate contamination. However, when TDS concentrations exceed 700 ppm, it is necessary to utilize a more extensive sample preparation technique before analysis.

## ACKNOWLEDGMENTS

A special thanks is extended to Dr. Charles Brokopp (Utah Department of Health Laboratory, Division of Epidemiology and Laboratory Services), Dr. Dave Mattie (AFRL/HEST), and Lt. Col. Daniel E. Rogers (AFMC LO/JAV) for their support of this project.

## REFERENCES

1. Manning, M. *Las Vegas Sun*, September 23, 1997.

2. Manning, M. *Las Vegas Sun*, September 24–25, 1997.

3. Manning, M. *Las Vegas Sun*, September 20, 1997.

4. Manning, M. *Las Vegas Sun*, September 8, 1997.

5. Manning, M. *Las Vegas Sun*, October 3, 1997.

6. Manning, M. *Las Vegas Sun*, January 27, 1998.

7. "Perchlorate in California Drinking Water." California Department of Health Services, September 1997; http://www/dhs.cahwnet.gov/perevsrv/ddwem/perchl.htm#advice.

8. Mattie, D.R.; Jarabek, A.M. "Perchlorate environmental contamination: testing strategy based on mode of action." *The Toxicologist. Toxicological Sciences* **1999**, *48*, 113.

9. Ericksen, G.E. "Geology and origin of the chilean nitrate deposits." United States Government Printing Office: Washington, DC, 1981.

10. Van Moort, J.C. "Natural enrichment processes of nitrate, sulfate, chloride, iodate, borate, perchlorate, and chromate in the caliches of Northern Chile." IV Congreso Geologico Chileno: Universidad del Norte Chileno, 1985.

11. Eldridge, J.E.; Tsui, D.T. Personal communication, May 1999.

12. Welcher, F. J. *Organic Analytical Reagents.* Van Nostrand: New York, NY, 1947; Vol. 3, pp. 138–146, and references therein.

13. Welcher, F. J. *Organic Analytical Reagents.* Van Nostrand: New York, NY, 1948; Vol. 3., pp. 326–327.

14. Harris, D. C. *Quantitative Chemical Analysis*, 3rd ed. Freeman: New York, NY, 1991; pp. 146, 722–723.

15. Hayes, O. B. "Studies in qualitative inorganic analysis. Part XXXII." *Mikrochimica Acta* **1968**, *3*, 647–648.

16. Chadwick, T. C. "2.4.6-Triphenylpyrylium chloride. A new organic analytical reagent for the determination of certain anions." *Analytical Chemistry* **1973**, *45*, 985–986.

17. Burns, D. T.; Chimpalee, N.; Harriot, M. "Flow-injection extraction-spectrophotometric determination of perchlorate with brilliant green." *Analytica Chimica Acta*, **1989**, *217*, 177–181.

18. Burns, D. T.; Hanprasopwattana, P. "The spectrophotometric and spectro-fluorimetric determination of perchlorate by extraction with amiloride hydrochloride." *Analytica Chimica Acta* **1980**, *118*, 185–189.

19. Weiss, J.A.; Stanbury, J.B. "Spectrophotometric determination of micro amounts of perchlorate in biological fluids." *Analytical Chemistry* **1972**, *44*, 619–620.

20. Kawase, J.; Nakae, A.; Yamanaka, M. "Determination of anionic surfactants by flow injection analysis based on ion-pair extraction." *Analytical Chemistry* **1979**, *51*, 1640.

21. Kawase, J. "Automated determination of cationic surfactants by flow injection analysis based on ion-pair extraction." *Analytical Chemistry* **1980**, *52*, 2124–2127.

22. Yamamoto, Y.; Okamoto, N.; Tao, E. "Spectrophotometric determination of anions by solvent extraction with cuproin or neocuproin-copper(I) chelate cations." *Analytica Chimica Acta* **1969**, *47*, 127–137.

23. Avdalovic, N.; Pohl, C.A.; Rocklin, R.D.; Stillian, J.R. "Determination of cations and anions by capillary electrophoresis combined with suppressed conductivity detection." *Analytical Chemistry* **1993**, *65*, 1470–1475.

24. Nann, A.; Pretsch, E. "Potentiometric detection of anions separated by capillary electrophoresis using an ion-selective microelectrode." *Journal of Chromatography A* **1994**, *676*, 437–442.

25. Gross, L.; Yeung, E.S. "Indirect fluorimetric detection and quantification in capillary zone electrophoresis of inorganic anions and nucleotides." *Journal of Chromatography A* **1989**, *480*, 169–178.

26. Okada, T. "Electrophoretic detection and evaluation of heteroconjugate anion formation between Bronsted acids and the perchlorate ion in acetonitrile." *Chemical Communications* **1996**, *6*, 1779–1780.

27. Holderbeke, M.V.; Vanhoe, H.; Moens, L.; Dams, R. "Determination of ten inorganic anions in drinking and waste waters with capillary zone electrophoresis (CZE) via indirect UV-detection." *Biomedical Chromatography* **1995**, *9*, 281–282.

28. De Backer, B. L.; Nagels, L. J.; Alderweireldt, F. C. "Liquid chromatographic determination of acids and anions using liquid membrane ion-selective electrodes in a potentiometric flow-through detector." *Analytica Chimica Acta* **1993**, *273*, 449–456.

29. Hauser, P.C.; Renner, N.D.; Hong, A.P.C. "Anion detection in capillary electrophoresis with ion-selective microelectrodes." *Analytica Chimica Acta* **1994**, *295*, 181–186.

30. Hauser, P.C.; Hong, A.P.C.; Renner, N.D. "Surface charge reversal for inorganic anion determination in capillary electrophoresis with an ion-selective microelectrode as detector." *Journal of Capillary Electrophoresis* **1995**, *5*, 209–212.

31. Guilbault, G.G.; Rohm, T. "Ion-selective electrodes and enzyme electrodes in environmental and clinical studies." *International Journal of Environmental and Analytical Chemistry* **1975**, *4*, 51–64.

32. Krokhin, O.V.; Elefterov, A.I.; Obrezkov, O.N.; Shpigun, O.A. "Increase in the sensitivity of the ion chromatographic determination of strongly retained anions." *Zhurnal Analiticheskoi Khimii* **1993**, *48*, 111–116.

33. Williams, R.J. "Determination of inorganic anions by ion chromatography with ultraviolet absorbance detection." *Analytical Chemistry* **1983**, *55*, 851–854.

34. California Department of Health Services, Sanitation and Radiation Laboratories Branch. Determination of Perchlorate by Ion Chromatography, Rev. 0. June 3, 1997.

35. Dionex Application Note 121. Dionex Corporation: Sunnyvale, CA, 1998.

36. Wirt, K.; Laikhtman, M.; Rohrer, J.; Jackson, P.E. *American Environmental Laboratory* **1998**, *10*, 1,5.

37. "Standard Operating Procedure for Perchlorate." EPA Office of Water. URL: http://www.epa.gov/OGWDW/ccl/perchlor/perchlo.html.

38. Chaudhuri, S.; Okamoto, H.; Pia, S.; Tsui, D. "Interagency perchlorate steering committee analytical subcommittee report." Environmental Protection Agency Collaborative Study, 1999.

39. Skoog, D. A.; Leary, J.J. *Principles of Instrumental Analysis*, 4th ed. Harcourt Brace: Fort Worth, TX, 1992; 639–640, 654–656.

40. Clewell, R.; Tsui, D.T.; Mattie, D.R. "Feasibility study for the reduction of perchlorate, iodide, and other aqueous anions." In press.

41. TSQ 7000 ESI/APCI Techniques—Course Manual, Revision A. Finnigan-Mat Institute: San Rafael, CA, March 1994.

42. Watson, T.J. *Introduction to Mass Spectrometry*, 3rd ed. Lippincott-Raven: Philadelphia, PA, 1997; 303–313, 337–338, 432–450, and references therein.

43. Desidero, D.M. *Mass Spectrometry: Clinical and Biomedical Applications*. Plenum: New York, NY, 1992; Vol. 1., pp. 1–33, and references therein.

44. Marchand, R.E.; Hughes, R.J. *Quadrupole Storage Mass Spectrometry*. Wiley: New York, NY, 1989, passim.

45. McLafferty, F.W. *Interpretation of Mass Spectra*, 3rd ed. University Science Books: Mill Valley, CA, 1980, passim.

46. Code of Federal Regulations 40, Ch. 1, Pt. 136, Appendix B. Definition and Procedure For the Determination of the Method Detection Limit, Rev. 1.1.

47. Tsui, D.T.; Clewell, R.; Eldridge, J.E.; Mattie, D.R. "Perchlorate Analysis with the AS16 Separation Column." THIS VOLUME: *Perchlorate in the Environment*, E.T. Urbansky, Ed. Kluwer/Plenum: New York, NY, 2000; Ch. 7.

# Chapter 7

# PERCHLORATE ANALYSIS WITH THE AS16 SEPARATION COLUMN[⊗,†]

David T. Tsui, Rebecca A. Clewell, J. Eric Eldridge, and David R. Mattie

U.S. Air Force Research Laboratory, Human Effectiveness Directorate, Toxicology Branch, Building 79, 2856 G Street, Wright-Patterson AFB, Ohio 45433-7400

## INTRODUCTION

Perchlorate is a powerful oxidant used in solid-rocket propellant mixtures, fireworks, and munitions. The presence of trace level perchlorate in drinking water poses a potential health risk, resulting from perchlorate's ability to interfere with the thyroid gland's uptake of iodide to produce thyroid hormones. The current EPA's recommended acceptable level for perchlorate in drinking water is 18 parts per billion (ppb). Since perchlorate is better known for it's commercial and industrial applications, past occurrence studies have focused on water wells near regions where munitions, aerospace components, and fireworks were manufactured, developed and tested. Perchlorate has been reported in drinking water sources in the states of California, Utah, Nevada, West Virginia, and Texas.[1–6] However, recent reports have suggested that naturally occurring perchlorate is also present in fertilizers, nitrate deposits from northern Chile, and minerals from arid environments. These new findings have sparked intense efforts in the study of natural perchlorate occurrence in non-aqueous matrices, and in the development of new methodologies to support these studies.

The determination of perchlorate in drinking water at low (ppb) levels is a difficult analytical task. Ion chromatography (IC) is currently accepted to be the best available

[⊗] This work was not presented at the 218th national meeting of the American Chemical Society held August 22–24, 1999, in New Orleans, Lousianna.



Author to whom correspondence should be directed. Phone: 937-255-5150. Fax: 937-255-1474. Electronic mail: David.Tsui@he.wpafb.af.mil.

technology for perchlorate analysis in drinking water.[7, 8] In 1997, the California Department of Health Services (CDHS) developed an IC method for the determination of trace level perchlorate in drinking water.[9] The method utilized a Dionex IonPac® AS5 column, a mobile phase consisting of 120 mM sodium hydroxide and 2 mM *p*-cyanophenol (*p*-cyanophenoxide), and suppressed conductivity detection with an Anion MicroMembrane Suppressor. The method detection limit, when using a 740 μL injection loop volume, was determined to be 4 ppb. In April 1998, the Dionex Application Laboratory developed an improved IC method for perchlorate analysis.[10,11] This method uses a 1 mL injection loop volume with an IonPac AS11 column, 100 mM hydroxide eluent, and suppressed conductivity detection with an Anion Self-Regenerating Suppressor (ASRS). The AS11 method detection limit was reported to be 1 ppb. The performances of both methods for drinking water matrix were validated in by inter-laboratory collaborative study, sponsored by the Interagency Perchlorate Steering Committee (IPSC). For drinking water, the study found no differences between the two methods in terms of bias and accuracy.

The detection of perchlorate in complex environmental, industrial, and wastewater matrices poses an greater challenge than even that of drinking water, inviting difficulty from matrix effects, the presence of organic solvents, and high levels of total dissolved solid (TDS). In addition to the method validation, the IPSC collaborative study also evaluated the application of ion chromatography to trace level perchlorate analysis in a variety of environmental and occupational matrices. The collaborative study showed that high TDS concentrations limit the application of ion chromatography for the detection of low level perchlorate, at or near the method detection limit of 4 ppb, in water. Neither the AS5 nor the AS11 method was robust enough to be used with confidence in non-drinking water matrices. Both the electrical conductivity and TDS values for a given sample were defined as an essential prescreening measure. Furthermore, it was determined that preparative techniques for the removal of dissolved solids should be further investigated for the cases where TDS may pose a problem for IC analysis.

The purpose of this study is to develop a method for perchlorate analysis that would be applicable in non-drinking water matrices, such as wastewater, fertilizers, leafy vegetables, and mineral deposits, where the current methods fail. Intra-laboratory studies were performed to evaluate the ability of the Dionex AS16 Separation column to account for the specific difficulties encountered when analyzing low level (ppb) perchlorate in various environmental and industrial matrices. The effects of solvents, eluent strengths, and the proposed sample preparation method were measured in order to compare the efficiency and sensitivity of the AS16 method to those of the recommended AS11 and AS5 methods.

Environmental and industrial wastewater samples are likely to contain a variety of organic solvents, which can affect the analysis of perchlorate by altering the retention time, the sensitivity to perchlorate, and the life of the column. In order to compensate for these effects, a method was developed for perchlorate analysis with the Dionex AS16 ion exchange column that would be more sensitive and have a longer column lifetime. As opposed to the AS11 column, the AS16 is specially designed for use with polarizable anions, which would theoretically increase the sensitivity of the method for perchlorate. The AS16 was also designed to have a higher capacity for exchange than previous columns. This higher capacity should prevent the organic solvents, such as methanol, from binding to the backbone of the resin and reducing the lifetime of the column.

For those environmental and industrial samples containing high TDS concentrations, a sample preparation method was developed that would allow the detection of low level perchlorate with ion chromatography. The sample preparation method was based on the unusually high solubility of perchlorate salts. It was determined that the silver salt of perchlorate would remain in solution, while most of the interference anions would precipitate out as the silver salts (Table 1). In order to prevent any silver from remaining in the sample during analysis, a silver saturated resin was used to facilitate cation exchange. As the water sample passed through the resin, the silver would bind with the interference anions to form the insoluble salts, and the dissolved cations would exchange onto the available sites on the resin. The solid salts could then be filtered out of solution. In this way, the background conductivity could be dramatically reduced, and any excess silver would remain on the resin. The decreased background noise should yield increased sensitivity and improved method detection limits for perchlorate analysis.

**Table 1.** Solubility of inorganic salts of silver

| Salts | Solubility (g x 100 mL$^{-1}$ 4 °C water) | Solubility (ppm in 4 °C water) | Salts | Solubility (g x 100 mL$^{-1}$ 4°C water) | Solubility (ppm in 4°C water) |
|---|---|---|---|---|---|
| $AgClO_4$ | 557 | 5570 | $AgNO_2$ | 0.155 | 1.55 |
| $AgNO_3$ | 122 | 1220 | $Ag_3PO_4$ | 6.5 x $10^{-4}$ | 6.5 x $10^{-3}$ |
| $AgClO_3$ | 10 | 100 | AgCl | 8.9 x $10^{-5}$ | 8.9 x $10^{-4}$ |
| $Ag_2SO_4$ | 0.57 | 5.7 | AgBr | 8.4 x $10^{-6}$ | 8.4 x $10^{-5}$ |
| $AgBrO_3$ | 0.196 | 1.96 | AgI | 2.8 x $10^{-7.25}$ | 2.8 x $10^{-6.25}$ |

## EXPERIMENTAL

### Test Materials

Primary source of perchlorate was ammonium perchlorate (lot 03907LF) purchased from Aldrich (St. Louis, MO). Secondary ammonium perchlorate (lot K15G11) check standards were purchased from Alfa Chemical Company (Ward Hill, MA). Test materials were used without further purification.

### Reagents

The following organic reagents were purchased from Sigma-Aldrich-Supelco: 1,1,1-trichloroethane, 1,1-dichloroethane, 1,2-dibromomethane, 1,2-dichloroethane, bromochloromethane, dibromochloromethane, dichlorobormomethane, dichloromethane, trifluoromethane, trichloroethylene, bromoform, chloroform, methanol, ethanol, freon 113, methyl ethyl ketone, methyl isobuthyl ketone, methyl *t*-butyl ketone, carbon tetrachloride, methylene chloride, methylbenzene, ethylbenzene, benzene, *m*-, *p*-, *o*-xylene, styrene, and toluene. Haloacetonitriles (dichloroacetonitrile, dibromoacetonitrile, bromochloroacetonitrile, trichloroacetonitrile, 1,1-dichloropropanone, 1,1,1-trichloropropanone) and haloacetic acids (monobromoacetic acid, dichloroacetic acid, trichloroacetic acid, dibromoacetic acid) were provided by Dr. Sanwat Chaudhuri (Utah Department of Health, Division of Epidemiology and Laboratory Services, Salt Lake City, UT).

Reagent grade sodium hydroxide was purchased from Sigma-Aldrich. Type I reagent water (18.0 to 18.3 MΩ $cm^{-1}$) was collected from a Barnstead Model D4751 Ultra Pure water system. Inorganic reagents and suppliers used in the interference study are shown in Table 2. All chemicals were reagent grade purity, with the exception of sodium humate and sodium selenate, which were technical grade purity.

**Table 2. List of inorganic reagents and suppliers**

| Reagent | Supplier | Reagent | Supplier |
|---|---|---|---|
| Arsenate | J. T. Baker | Potassium Iodide | Sigma |
| Arsenite | Fisher | Ammonium Molybdate | Sigma |
| Sodium Bromate | Aldrich | Ammonium Nitrate | Sigma |
| Sodium Bromide | Sigma | Sodium Nitrite | Aldrich |
| Sodium Carbonate | Sigma | Potassium Phosphate | Sigma |
| Sodium Chlorate | Sigma | Phthalate | Sigma |
| Sodium Chloride | Fisher | Sodium Selenate | Sigma |
| Potassium Chromate | Aldrich | Potassium Sulfate | Sigma |
| Potassium Cyanide | Sigma | Sodium Sulfite | Fisher |
| Sodium Humate | Aldrich | Potassium Thiocyanate | Sigma |
| Potassium Iodate | Aldrich | Sodium Thiosulfate | Sigma |

**Calibration and Control Standards**

Ammonium perchlorate stock standard solutions at 50 mg $mL^{-1}$ were prepared gravimetrically (Mettler model PE-360 analytical balance, ± 0.1 mg) from pure neat standards. Working standard solutions were prepared at 1000 μg $L^{-1}$ from the individual stock standard solutions. From the working standard-solutions, calibration or control standards at 5, 10, 25, 50, 100, 200, and 500 μg $L^{-1}$ were prepared by serial dilution.

Stock solutions of $BrO_3^-$, $Cl^-$, $ClO_3^-$, $I^-$, $PO_4^{3-}$, $NO_2^-$, $NO_3^-$, $CO_3^{2-}$, $SO_3^{2-}$, $S_2O_3^{2-}$ and $SO_4^{2-}$, and were individually prepared at concentrations of 50,000 ppm (50 g $L^{-1}$) from the solid salts in distilled, deionized water. Solutions at 100, 500, 1000, 5000, and 10,000 ppm of mixed anions were then prepared by serial dilution from the 50,000 ppm stock solutions. All water used was deionized, reagent grade with 18 MΩ cm resistivity.

**Instrumentation and Analytical Method**

**Perchlorate analysis.** Ion chromatography was performed on a Dionex DX-500 (Sunnyvale, CA) ion chromatograph system with a GP-40 gradient pump, CD-20 conductivity detector, a LC-20 chromatography enclosure, and an AS40 automated sampler. The injection volume was 1 mL. Anion separation was obtained on a Dionex IonPac AS16 separation column (4.0 × 250 mm) with an AG16 guard column (4.0 × 50 mm) and an ATC1 anion trap column. An EG40 Eluent Generator with a potassium hydroxide cartridge were used to generate the 35 mM KOH mobile phase. The mobile phase flow rate was set at 1.25 mL $min^{-1}$. Background suppression was achieved by using an Anion Self-Regenerating Suppressor (ASRS)-ULTRA suppressor.

The performance of the AS16 was compared to that the AS11 column. Anion separation was obtained on a Dionex IonPac AS11 separation column (4.0 × 250 mm) with an AG11 guard column (4.0 × 50 mm) and an ATC1 anion trap column. The mobile phase for the AS11 column consisted of 100 mM NaOH in water, with a flow rate of 1.0 mL $min^{-1}$.

For both AS16 and AS11 columns, the regenerant flow rate to the suppressor was 10 mL $min^{-1}$. The suppression current was set 300 mA, and analysis was performed at room temperature. All water used was deionized, reagent grade with 18 MΩ cm resistivity or better. Dionex Peaknet software was used to perform the data processing. An Orion 701A digital Ionanalyzer was used for pH measurements.

**Anion analysis.** Analysis of all anions other than perchlorate was performed on a Dionex DX-500 Microbore system, configured with a GP40 gradient pump, an ASRS-ULTRA suppressor, an ED40 conductivity detector and an AS3500 Autosampler. The suppressor was set at 100 mA and was operated in external water mode. Samples were injected manually through a 25 μL loop at a flow rate of 0.25 mL $min^{-1}$.

For the mixed anion study, the separation of $BrO_3^-$, $Cl^-$, $ClO_3^-$, $I^-$, $PO_4^{3-}$, $NO_2^-$, $NO_3^-$, $SO_3^{2-}$ and $SO_4^{2-}$ was attained with a Dionex IonPac AS16 microbore analytical column (2 mm × 250 mm) and an AG16 guard column (2 mm × 50 mm). The mobile phase consisted of 35 mM NaOH in distilled, deionized water.

For the individual anion study, $Br^-$, $ClO_3^-$, and $Cl^-$ were analyzed on a Dionex IonPac AS11 analytical microbore column (2 mm × 250 mm) and an AG11 guard column (2 mm × 50 mm). Separation was achieved with a gradient program: 0.5 mM NaOH for 7 minutes, and then gradually increased to 38.5 mM NaOH over a period of 20 minutes.

### Effects of Sample Preparation Method on Mixed Anions

For the mixed anion study, a stock artificial water matrix was prepared by combining 10 mg of $S_2O_3^{2-}$ and 600 mg of $BrO_3^-$, $Cl^-$, $ClO_3^-$, $I^-$, $PO_4^{3-}$, $NO_2^-$, $NO_3^-$, $CO_3^{2-}$, $SO_3^{2-}$, and $SO_4^{2-}$ in 1 L of distilled, deionized water for a total TDS of 6,610 ppm. Mixed anion samples at TDS levels of 100, 500, and 1,000 ppm were prepared by serial dilution from the stock artificial water matrix.

Prior to sample analysis, a 2 mL mixed anion sample was filtered serially through an Ag OnGuard (Dionex) cartridge to exchange the interference cations onto the resin and form the insoluble silver salts of the interference anions. The samples were then filtered with the hydrogen OnGuard cartridges to remove excess silver cations that may have been released from the silver cartridge. A Millipore 0.45 μM x 13 mm Millex-$HV_{13}$ Filter was used to filter out any resin or precipitated salts that remained in the sample. Samples were subsequently analyzed for perchlorate and anion concentrations using ion chromatography.

## RESULTS

### Method Development and Validation

Optimized chromatographic conditions were described in Section II. Baseline noise was kept minimal. Prior to analysis, the ion chromatography system was equilibrated to produce a background conductance less than 2 μS. To establish the baseline, system blanks (deionized water) were injected onto the system. The system blanks demonstrated that the IC system was free from contamination, showing negative baseline deflection between 2.2 to 3.2 minutes due to the void volume followed immediately by strong positive baseline deflection from the unretained species. The chromatogram of a 100 ppb perchlorate standard obtained on a new standard bore AS16 column, with EG40 generated 35 mM KOH mobile phase flowing at 1.25 mL $min^{-1}$, showed one distinct peak around 12.5 minutes on a new column. As the column ages, perchlorate elution times decrease and eventually stabilize between 10.5–12.5

minutes. When 35 mM NaOH is used, rather than the EG40 generated KOH, perchlorate eludes from the column between 17 and 18 minutes. Perchlorate retention time was established by injecting standards with increasing concentrations of perchlorate and observing the corresponding increase in the peak area counts. Peak identity was also verified by analyzing perchlorate standards from a second source.

The method detection limit data for perchlorate was obtained on a new column with an EG40 eluent generator. To determine the methods development limit, eight replicates of 5 ppb perchlorate standards were analyzed over 72 hours. Per guidelines and procedures set forth in Code of Federal Regulations 40, Chapter 1, Pt. 136, Appendix B, the calculated method detection limit (MDL) is 0.19 μg $L^{-1}$ (ppb). Student's $t$ for eight samples ($n = 8$, $\nu = 7$ degrees of freedom) at the 99% confidence limit is 2.998. At the calculated method detection limit, the perchlorate peak is well resolved and the signal to noise ratio is greater than 3. The calculated oncolumn limit is 0.19 ng (0.1 μg $L^{-1}$ × 1000 μL = 1.94 x $10^{-4}$ μg). The practical quantitation limit at 10 times the MDL is 1.94 μg $L^{-1}$.

**Table 3.** Method detection limit data for perchlorate analysis on an AS16 column. 35 mM KOH mobile phase, flowing at 1.25 mL $min^{-1}$

| Datum no. | Concentration (ppb) | Area | Retention time (min) |
|---|---|---|---|
| 1 | 5 | 9751 | 12.25 |
| 2 | 5 | 9687 | 12.32 |
| 3 | 5 | 9762 | 12.32 |
| 4 | 5 | 9556 | 12.38 |
| 5 | 5 | 9935 | 12.17 |
| 6 | 5 | 9874 | 12.17 |
| 7 | 5 | 9835 | 12.17 |
| 8 | 5 | 9635 | 12.17 |
| **Average** | | 9754 | 12.24 |
| **Standard Deviation** | | 118 | 0.08 |
| **Per** | | 1% | 0.5% |
| **MDL** | | 0.194 | |

Calibration curves were generated by plotting the concentrations of each standard against the peak area count obtained. A typical calibration curve, obtained on an aged AS16 column (>1,200 injections), was linear through the calibration range from the MDL to 10,000× the MDL, over 5 orders of magnitude. The calibration line was typically described by the equation: $y = 1840.4x$. The correlation coefficient values were > 0.9998. The slope of the line is directly related to the sensitivity of the ion chromatography system. The slopes of calibrations generated from sample batches were monitored over two months. During that period, over 1,200 standards, controls, and water samples were analyzed, and the slopes of the calibration lines were within 1% of the original calibration line, indicating the system response had little or no change. During the same two month period, method accuracy and method precision were also monitored. The method accuracy, which is measured as a percentage of the known value, was between 90 and 110%. The method precision, which is measured as relative percent standard deviation, was better than 1%.

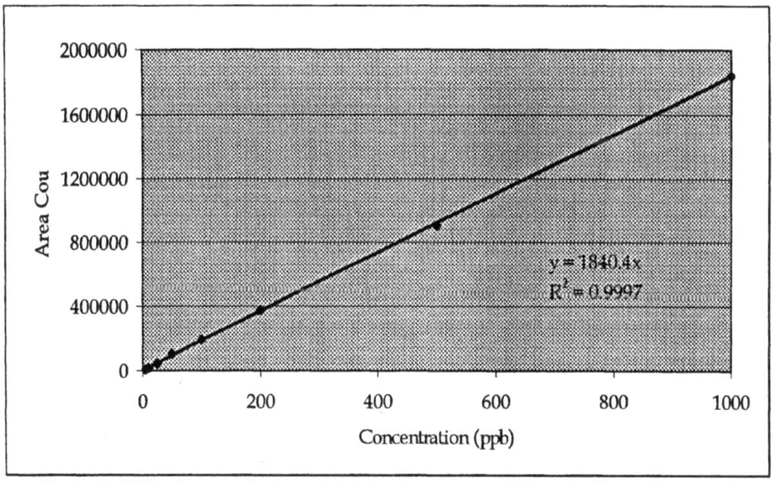


**Figure 1.** Calibration curve generated from standard curve shown in Figure 2.

Stack plots of the ion chromatograms obtained for perchlorate standards with concentrations from 5 to 1000 μg $L^{-1}$ on AS16 and AS11 columns are shown in Figures 2 and 3, respectively. For both the AS16 and AS11 columns, the plots were obtained after the columns had been in use for more than two months. The peak shapes of perchlorate standards from 5 to 1,000 μg $L^{-1}$ on the AS16 column (Figure 2) were highly symmetrical and well resolved, with little or no observed retention time shift in the chromatograms. In comparison, the AS11 column showed increased peak tailing with increasing perchlorate concentrations. On the AS11 column, the perchlorate retention time decreased as much as one minute, with a corresponding increase in perchlorate concentration, from 5 to 1000 μg $L^{-1}$.

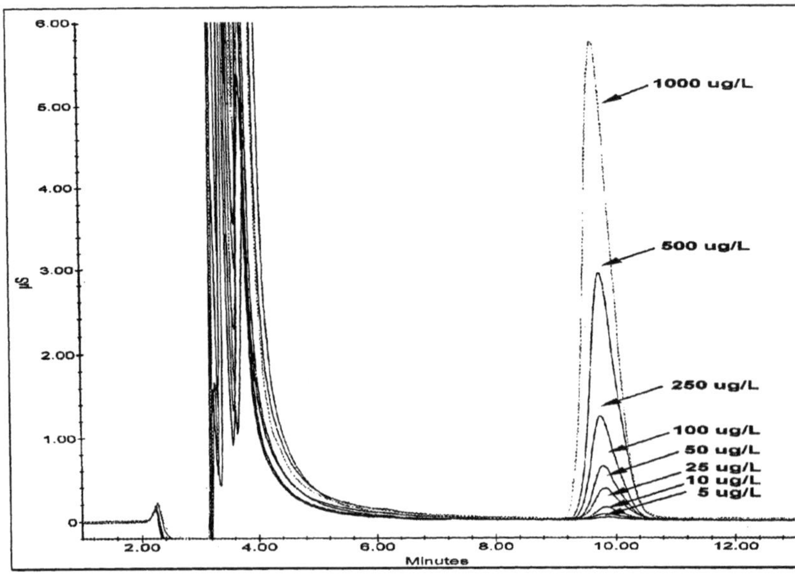


**Figure 2.** A stack plot of 5 to 1000 µg $L^{-1}$ perchlorate calibration standards on an AS16 column. Chromatographic conditions: 35 mM KOH with flow rate at 1.25 mL $min^{-1}$, 1000 µL injection volume.

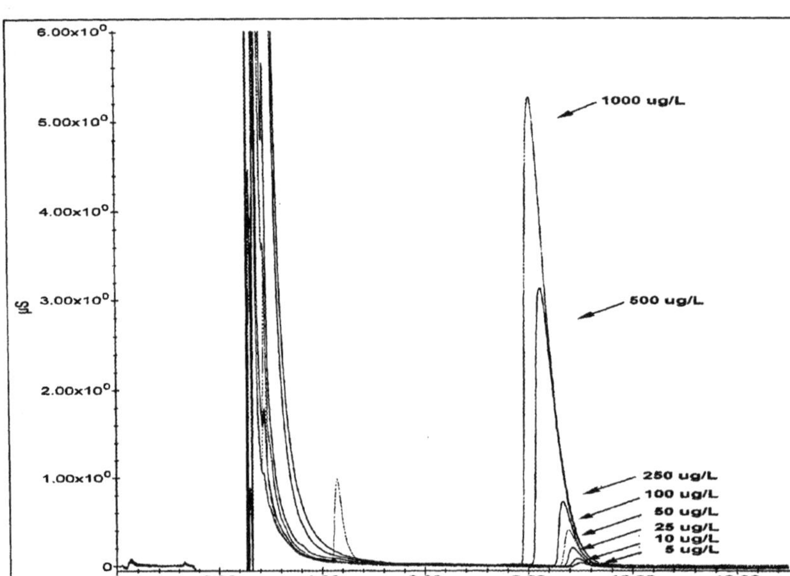


**Figure 3.** A stack plot of 5 to 1000 µg $L^{-1}$ perchlorate calibration standards on an AS11 column. Chromatographic conditions: 100 mM NaOH with flow rate at 1.0 mL $min^{-1}$, 1000 µL injection volume.

**Effects of KOH Mobile Phase on Perchlorate Retention Time and Response**

The effects of the potassium hydroxide mobile phase strength on the chromatography of 100 ppb perchlorate standards were examined on a new AS16 separation column, AG16 guard column, ATC1 trap column, and an ASRS-Ultra suppressor. A stack plot of three 100 ppb perchlorate standards, analyzed with mobile phase concentrations at 25, 35, and 50 mM and flowing at 1.0 mL $min^{-1}$, is shown in Figure 4. With increasing KOH concentrations from 25 to 50 mM, perchlorate retention times were found to decrease from 22.9 to 13.35 minutes (Table 4). Peak area counts remain constant and detector response is unaffected with increasing KOH concentration, as peak height increases from 5,390 to 10,005, while peak width at half height decrease from 0.75 to 0.3 minutes. Figure 4 shows that at a flow rate of 1.0 mL $min^{-1}$, the best chromatographic results are obtained with 35 mM KOH, resulting in good symmetry and no apparent artifact. In comparison, perchlorate peak shape exhibits unacceptable peak broadening at 25 mM and peak fronting at 50 mM. The 17.25 minute perchlorate retention time on the AS16 with 35 mM KOH at 1.0 mL $min^{-1}$ is about 5 minutes longer than the 12 minute retention time obtained with recommended AS11 method. However, the perchlorate retention time on the AS16 was further optimized by increasing the flow rate of the mobile phase.

**Table 4.** Effects of KOH mobile phase strength on 100 ppb perchlorate retention time, area count, and peak height at 1.0 mL $min^{-1}$ flow rate

| | Retention Time | Retention Time | Retention Time | Average | Std Dev | %CV |
|---|---|---|---|---|---|---|
| 25mM | 22.98 | 22.93 | 22.85 | 22.92 | 0.1 | 0.2% |
| 35mM | 17.28 | 17.25 | 17.23 | 17.25 | 0.0 | 0.1% |
| 50mM | 13.37 | 13.35 | 13.33 | 13.35 | 0.0 | 0.1% |
| | **Area Count** | **Area Count** | **Area Count** | **Average** | **Std Dev** | **%CV** |
| 25mM | 204914 | 211280 | 211805 | 209333 | 3132 | 1.5% |
| 35mM | 228414 | 228617 | 229411 | 228814 | 430 | 0.2% |
| 50mM | 212344 | 211152 | 213629 | 212375 | 1011 | 0.5% |
| | **Peak Height** | **Peak Height** | **Peak Height** | **Average** | **Std Dev** | **%CV** |
| 25mM | 5259 | 5439 | 5473 | 5390 | 93.9 | 1.7% |
| 35mM | 8043 | 8053 | 8106 | 8067 | 27.6 | 0.3% |
| 50mM | 10031 | 9865 | 10119 | 10005 | 105.3 | 1.1% |

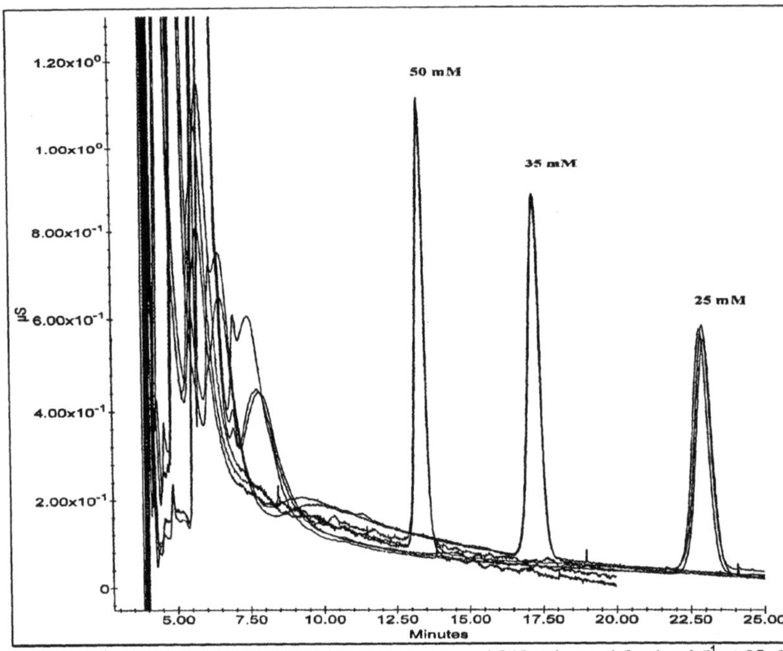


**Figure 4.** Elution profile of 100 ppb perchlorate on an AS16 column, 1.0 mL min$^{-1}$ at 25, 35, and 50 mM KOH mobile phase.

A stack plot of three 100 ppb perchlorate standards analyzed with KOH concentrations of 25, 35, and 50 mM, flowing at 1.0 mL min$^{-1}$ is shown in Figure 5. The raw data is presented in Table 5. When the KOH concetration of the mobile phase was increased from 25 to 50 mM, the retention time of perchlorate decreased significantly from 15.45 to 9.05 minutes. The peak area counts for perchlorate remained constant with changing KOH concentrations, though the shape of the perchlorate peak sharpened dramatically with increased KOH. Although the chromatography showed good peak symmetry at 1.5 mL min$^{-1}$ in both the 35 mM and 50 mM KOH mobile phase, the 50 mM NaOH mobile phase causes the perchlorate peak to elute too quickly from the column. At 50 mM KOH, the perchlorate is not sufficiently separated from the unretained anions. Hence, the optimized AS16 chromatography condition was determined to be between 1.0 and 1.5 mL min$^{-1}$, or 1.25 mL min$^{-1}$, with a 35 mM KOH mobile phase. With the optimized conditions on a new AS16 column, the perchlorate retention time should be 12.5 minutes, which is comparable to that of the AS11 method. With aging, perchlorate retention time decreases to about 10 minutes.

**Table 5.** Effects of KOH mobile phase strength on 100 ppb perchlorate retention time, area count, and peak height at 1.5 mL $min^{-1}$ flow rate

| | Retention Time | Retention Time | Retention Time | Average | Standard Deviation | % CV |
|---|---|---|---|---|---|---|
| 25mM | 15.47 | 15.43 | 15.45 | 15.45 | 0.0 | 0.1% |
| 35mM | 11.77 | 11.78 | 11.78 | 11.78 | 0.0 | 0.0% |
| 50mM | 9.05 | 9.05 | 9.05 | 9.05 | 0.0 | 0.0% |

| | Area Count | Area Count | Area Count | Average | Std Dev | %CV |
|---|---|---|---|---|---|---|
| 25mM | 152131 | 143417 | 148873 | 148140 | 3595 | 2.4% |
| 35mM | 155642 | 155253 | 140904 | 150600 | 6858 | 4.6% |
| 50mM | 152830 | 149243 | 152157 | 151410 | 1557 | 1.0% |

| | Peak Height | Peak Height | Peak Height | Average | Std Dev | %CV |
|---|---|---|---|---|---|---|
| 25mM | 5351 | 5071 | 5323 | 5248 | 125.9 | 2.4% |
| 35mM | 7145 | 7399 | 6728 | 7091 | 276.6 | 3.9% |
| 50mM | 9542 | 9595 | 9671 | 9603 | 52.9 | 0.6% |

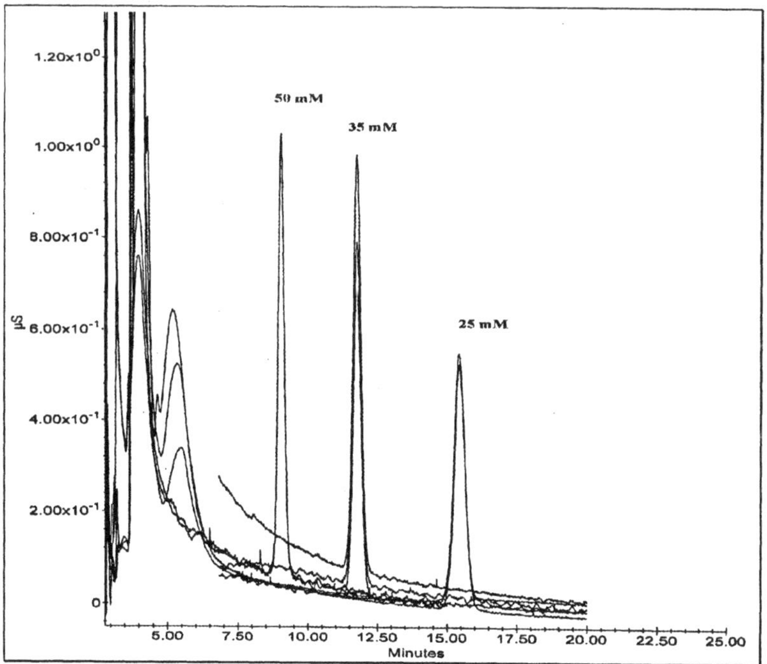


**Figure 5.** Elution profile of 100 ppb perchlorate on an AS16 column, 1.5 mL $min^{-1}$ at 25, 35, and 50 mM KOH mobile phase.

**Effects of Sample pH on Perchlorate Recovery and Retention Time**

The effect of sample pH on perchlorate recovery and retention time on the AS16 column was examined by analyzing 50 ppb perchlorate standards (n = 3) prepared in 1, 5, 10, 20 mM NaOH and HCl. As shown in Table 6 and Figure 6, the retention times of the 50 ppb perchlorate standards decreased from 12.72 to 12.27 minutes, or 3.7% when the sodium hydroxide concentration was increased in the sample from 0 to 20 mM. An increase in sodium hydroxide sample concentration results in lower peak area counts, indicating decreased detector response. The effects of HCl on 50 ppb perchlorate standards are shown in Table 7 and Figure 7. When the hydrochloric acid concentration in the samples was increased from 0 to 20 mM, the retention time of perchlorate increased from 12.72 to 13.20 minutes, or 3.7% and detector response was diminished.

**Table 6.** Effects of NaOH concentrations on 50 ppb perchlorate response and retention time on an AS16 column, 1.25 mL $min^{-1}$, and 35 mM KOH mobile phase

| [NaOH] (mM) | Retention Time (min) | Area Count | Peak Height | Percent Area Count Difference (%) |
|---|---|---|---|---|
| 0 | 12.72 | 114809 | 4285 | 0% |
| 1 | 12.53 | 88928 | 3585 | -23% |
| 5 | 12.45 | 84974 | 3225 | -26% |
| 10 | 12.40 | 89691 | 3048 | -22% |
| 20 | 12.27 | 91218 | 2394 | -21% |

**Table 7.** Effects of HCl concentrations on 50 ppb perchlorate response and retention time on an AS16 column, 1.25 mL $min^{-1}$, and 35 mM KOH mobile phase

| [HCl] (mM) | Retention Time (min) | Area Count | Peak Height | Percent Area Count Difference (%) |
|---|---|---|---|---|
| 0 | 12.72 | 114809 | 4285 | 0% |
| 1 | 12.75 | 89611 | 3474 | -22% |
| 5 | 12.80 | 86485 | 3339 | -25% |
| 10 | 12.96 | 72743 | 3215 | -37% |
| 20 | 13.20 | 71151 | 3364 | -38% |

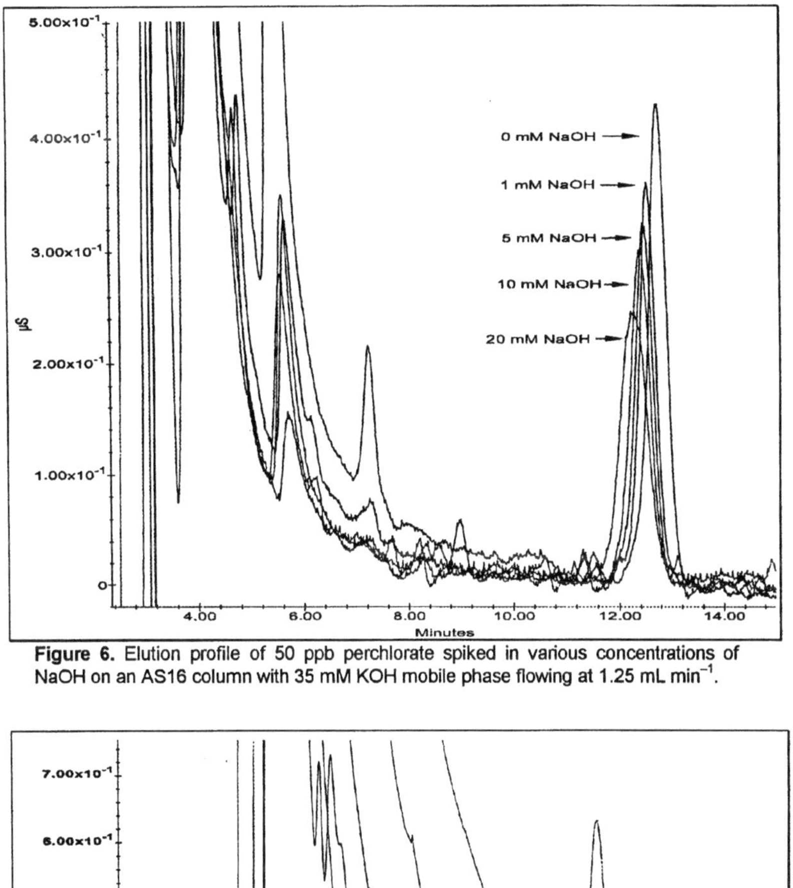


**Figure 6.** Elution profile of 50 ppb perchlorate spiked in various concentrations of NaOH on an AS16 column with 35 mM KOH mobile phase flowing at 1.25 mL min$^{-1}$.

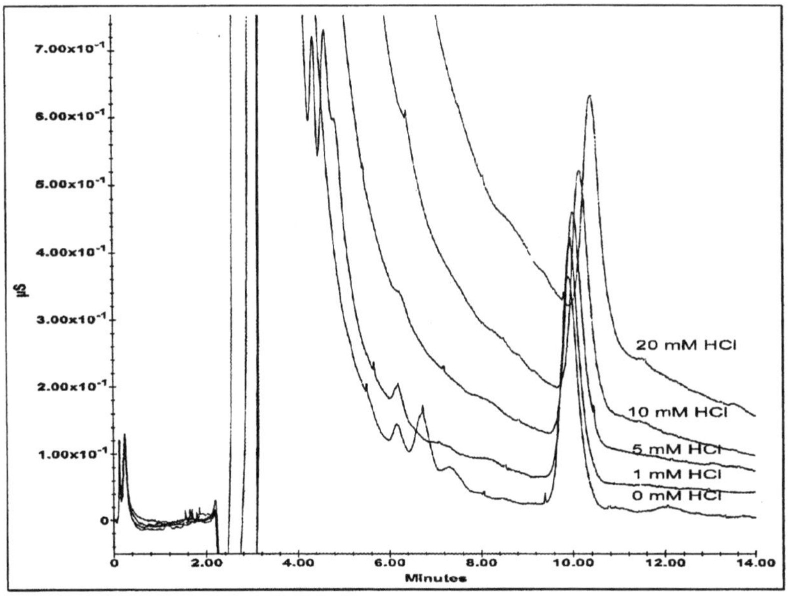


**Figure 7.** Elution profile of 50 ppb perchlorate spiked in various concentrations of NaOH on an AS16 column with 35 mM KOH mobile phase flowing at 1.25 mL min$^{-1}$.

**Interference Study**

Two different interference studies similar to those used by the ISPC when evaluating the AS5 and AS11 columns, were performed with the AS16 method. In the first study, the same set of 22 anions tested by the IPSC were injected at the 100 ppb level (in the presence of 20 ppb perchlorate) on the AS16 column, using the conditions described in the experimental section. Under the optimized conditions for perchlorate analysis on the AS16 column, only cyanide, iodide and thiocyanate showed any significant retention on the AS11 column (Table 8). Both cyanide and iodide have retention times around 9.45 minutes, and the retention time of thiocyanate is 10.87 minutes. Perchlorate shows clear resolution from all of the tested anions, and is separated by three minutes from thiocyanate, which is the nearest eluting anion and which would not be typically found at high levels in drinking water or ground water.

**Table 8.** Retention times of common inorganic anions on the AS16 analytical separation column

| Anion | Anion retention time (min) | Perchlorate retention time (min) |
|---|---|---|
| arsenate | <4 | 12.72 |
| arsenite | <4 | 12.79 |
| bromate | <4 | 12.84 |
| bromide | <4 | 12.78 |
| carbonate | <4 | 12.78 |
| chlorate | <4 | 12.65 |
| chloride | <4 | 12.70 |
| chromate | <4 | 12.71 |
| cyanide | 9.32 | 12.71 |
| humic acid | <4 | 12.72 |
| iodate | <4 | 12.75 |
| iodide | 9.45 | 12.65 |
| molybdate | <4 | 12.50 |
| nitrate | <4 | 12.60 |
| nitrite | <4 | 12.50 |
| phosphate | <4 | 12.65 |
| phthalate | <4 | 12.65 |
| selenate | <4 | 12.55 |
| sulfate | <4 | 12.50 |
| sulfite | <4 | 12.50 |
| thiocyanate | 10.87 | 12.55 |
| thiosulfate | <4 | 12.55 |

In addition to those inorganic chemicals listed in Table 8, the first study also examined the possible interference associated with the organic chemicals listed in the experimental section. To examine the effects of the organics on perchlorate retention time and detector response, standards were prepared with each of the individual organic solvents at concentrations of 0, 100, and 1,000 ppb. Standards at each concentration level were then spiked with 25, 50, and 100 ppb perchlorate. Due to the volatile nature of the organics, samples containing the organic chemicals were carefully prepared in volumetric flasks and then transferred to vials with PTFE septa. Samples were analyzed within 24–48 hours of preparation. The organic interference study was performed in the absence of the EG40 Eluent Generator. A 35 mM NaOH mobile phase was used at a flow rate of 1.5 mL $min^{-1}$, rather than the 35 mM KOH.

With the exception of trichloroacetic acid (TCA), none of the organic chemicals were shown to affect either the chromatography or the detector response for perchlorate. Figure 8 shows the effect of 1 mM TCA on the chromatography and detector response for 50 ppb perchlorate-spiked samples. As illustrated in Figure 8 the 50 ppb perchlorate peak was expected to appear around 17.2 minutes. In the absence of perchlorate, 1 mM TCA appeared as a single peak around 14.0 min. However, samples with both perchlorate and TCA showed one large peak at 14.0 min, corresponding to the TCA, and one broadened perchlorate peak at 16.5 min. The presence of trichloroacetic acid affects the chromatography of perchlorate on the AS16 column by decreasing both the retention time and the detector response.

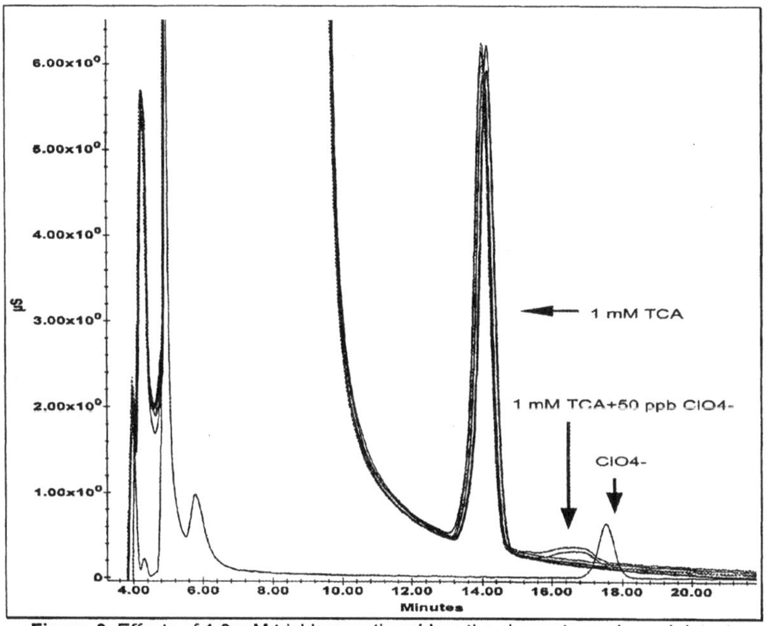


**Figure 8.** Effects of 1.0 mM trichloroacetic acid on the chromatography and detector response of 50 ppb perchlorate on an AS16 Anion Separation Column with 35 mM NaOH, at 1.25 mL $min^{-1}$.

The second study examined the effect of the presence of a few common interference anions at high ppm levels on perchlorate recovery. Perchlorate standards at 50 ppb were analyzed in the presence of 100, 500, 1,000, 2,500, 10,000, and 15,000 ppm chloride, carbonate, nitrate, and sulfate. As shown in Table 9, the presence of these common anions has no effect on the retention time. Furthermore, a decrease was not observed in the percent recovery of 50 ppb perchlorate spikes.

**Table 9.** Effect of ppm levels of carbonate, nitrate, chloride, and sulfate anions on 50 ppb perchlorate recovery on the Dionex IonPac AS16 column

| Common anion | Anion concentration (ppm) | Perchlorate retention time (min) | Perchlorate recovery (%) |
|---|---|---|---|
| carbonate | 100 | 12.75 | 96.6 |
| carbonate | 500 | 12.75 | 98.8 |
| carbonate | 1000 | 12.65 | 92.1 |
| carbonate | 2500 | 12.69 | 94.2 |
| carbonate | 10,000 | 12.63 | 95.5 |
| carbonate | 15,000 | 12.62 | 96.4 |
| nitrate | 100 | 12.66 | 99.1 |
| nitrate | 500 | 12.65 | 98.4 |
| nitrate | 1000 | 12.63 | 96.5 |
| nitrate | 2500 | 12.59 | 94.7 |
| nitrate | 10,000 | 12.55 | 95.9 |
| nitrate | 15,000 | 12.62 | 95.4 |
| chloride | 100 | 12.73 | 92.2 |
| chloride | 500 | 12.70 | 99.2 |
| chloride | 1000 | 12.66 | 98.7 |
| chloride | 2500 | 12.70 | 97.4 |
| chloride | 10,000 | 12.63 | 98.6 |
| chloride | 15,000 | 12.71 | 97.6 |
| sulfate | 100 | 12.59 | 94.4 |
| sulfate | 500 | 12.61 | 100.0 |
| sulfate | 1000 | 12.55 | 93.4 |
| sulfate | 2500 | 12.52 | 97.4 |
| sulfate | 10,000 | 12.44 | 96.9 |
| sulfate | 15,000 | 12.48 | 98.9 |

**Application of the AS16 Method**

In order to verify the accuracy of the AS16 method, and to determine the feasibility of its use on real water samples, the AS16 method was applied to the same water samples used by the ISPC collaborative study.[7] The results are shown in Table 10. The AS16 method showed results comparable to those of the accepted AS11 method. The two methods showed no more than 3.5% difference in perchlorate recovery.

**Table 10.** Analysis of IPSC Samples by AS16 Method

| Samples | Theoretical concentration (ppb) | AS16 concentration (ppb) | AS16 relative difference (%) |
|---|---|---|---|
| C1T1 | 0 | 0 | 0 |
| C1T2 | 0 | 0 | 0 |
| C1T3 | 0 | 0 | 0 |
| C2T1 | 5.8 | 5.6 | 3.5 |
| C2T2 | 5.8 | 6.0 | 3.4 |
| C2T3 | 5.8 | 6.0 | 3.4 |
| C3T1 | 17.9 | 18.4 | 2.8 |
| C3T2 | 17.9 | 18.7 | 4.5 |
| C3T3 | 17.9 | 18.1 | 1.0 |
| C4T1 | 35.4 | 36.5 | 3.2 |
| C4T2 | 36.1 | 36.2 | 2.8 |
| C4T3 | 35.5 | 36.1 | 1.7 |
| STO | 50.8 | 51.5 | 1.4 |
| QCS (50ppb) | 50.0 | 51.7 | 3.4 |
| 20ppb check | 20.0 | 20.2 | 1.0 |

## Effects of Sample Preparation Method on Perchlorate Retention Time and Recovery

The effect of the sample preparation method involving the use of Ag OnGuard, H-OnGuard, and Millex-$HV_{13}$ filtration cartridges and filters on perchlorate recovery was examined with the AS16 method for perchlorate analysis. Standards were prepared at 25 and 100 ppb perchlorate, and were then filtered either with the Millex-$HV_{13}$ cartridges only, or with the $Ag^+$, $H^+$, and Millex-$HV_{13}$ cartridges before analysis. The data presented in Table 11 shows that the Millex-$HV_{13}$ syringe filters have no significant effect on the percent recovery or retention time of perchlorate. Triplicate analysis of 25 and 100 ppb perchlorate standards filtered through the syringe filter showed better than 95 to 105% recoveries, as compared to the unfiltered standards. Table 12 shows the effect of the sample preparation method on the recovery of perchlorate from spiked water standards. The sample preparation method does not interfere with the quantitative determination of trace level perchlorate. In all samples, nearly 100% recovery was achieved with little or no effect on perchlorate retention time.

**Table 11.** Effects of Millipore Millex $HV_{13}$ Filter on 25 and 100 ppb Perchlorate Spike Recovery

| Perchlorate Concentrations (ppb) | Retention Time (min) | Area Count | Measured Concentration (ppb) | Percent Recovery (%) |
|---|---|---|---|---|
| 25 | 9.1 | 34034 | 24.0 | 96% |
| 25 | 9.1 | 34166 | 24.1 | 96% |
| 25 | 9.0 | 37053 | 26.1 | 104% |
| 100 | 8.8 | 146164 | 102.9 | 103% |
| 100 | 8.9 | 148607 | 104.7 | 105% |
| 100 | 8.8 | 146407 | 103.1 | 103% |

**Table 12.** Effects of Ag- and H-OnGuard Cartridges and Millipore Millex $HV_{13}$ Filter on 25 and 100 ppb Perchlorate Spike Recovery

| Perchlorate Concentrations (ppb) | Retention Time (min) | Area Count | Measured Concentration (ppb) | Percent Recovery (%) |
|---|---|---|---|---|
| 25 ppb | 9.1 | 33641 | 23.7 | 95% |
| 25 ppb | 9.0 | 36057 | 25.4 | 102% |
| 25 ppb | 9.1 | 35511 | 25.0 | 100% |
| 100 ppb | 8.9 | 143313 | 100.9 | 101% |
| 100 ppb | 8.8 | 148614 | 104.7 | 105% |

**Effect of Sample Preparation on Interference Anions**

The separation of bromate, chloride, chlorate, iodide, phosphate, nitrite, nitrate, sulfite, and sulfate was possible with the use of the AS16 gradient method. Mixed anion samples at 100, 500, and 1000 ppm were examined for the presence of the anions before and after the sample preparation method. The results for the mixed anion study are shown in Table 13. The actual mass (μg) of material was calculated by multiplying the measured concentration by the volume of the sample, where the sample volume was 2.0 mL.

Ion chromatography results indicate successful removal of the bromate, bromide, chloride, iodide, nitrite, and phosphate anions. Chlorate, nitrate, and sulfate were not removed by the sample preparation method. The amount of nitrate in the sample actually showed a slight (3.1%) increase after treatment with the sample preparation method, and the concentration of sulfate showed a 24% increase. Previous studies with the AS11 column showed bromide to be 100% removed from solution with the OnGuard cartridges (Table 13).

**Table 13.** Results for Mixed Anion Removal Study Using OnGuard Cartridges

| Anion | Expected Concentration (ppm) | Expected Amount (μg) | Measured Final Concentration (ppm) | Measured Final Amount (μg) | Percent Removal (%) |
|---|---|---|---|---|---|
| $NO_2^-$ | 5.65 | 11.3 | 1.13 | 2.26 | 80.0 |
| | 39.8 | 79.6 | 0.50 | 1.01 | 98.7 |
| | 87.6 | 175.3 | ND | ND | 100 |
| $NO_3^-$ | 7.81 | 15.6 | 6.38 | 12.8 | 18.2 |
| | 36.9 | 73.8 | 43.3 | 86.7 | –17.5 |
| | 80.2 | 160.4 | 88.1 | 176.3 | –9.91 |
| $PO_4^-$ | 4.79 | 9.57 | 3.76 | 7.53 | 21.4 |
| | 23.4 | 46.7 | 8.15 | 16.3 | 65.1 |
| | 51.9 | 103.9 | 11.6 | 23.3 | 77.6 |
| $SO_4^-$ | 4.46 | 8.92 | 7.74 | 15.5 | –73.5 |
| | 31.4 | 62.9 | 33.9 | 67.7 | –7.76 |
| | 71.8 | 143.7 | 65.2 | 130.3 | 9.26 |
| $BrO_3^-$ | 6.92 | 13.8 | 5.74 | 11.5 | 17.0 |
| | 40.9 | 81.8 | ND | ND | 100 |
| | 84.1 | 168.3 | ND | ND | 100 |
| $Cl^-$ | 7.18 | 14.4 | 3.09 | 6.18 | 56.9 |
| | 26.1 | 52.3 | 5.49 | 10.9 | 79.0 |
| | 53.5 | 107.2 | 9.49 | 19.0 | 82.3 |
| $ClO_3^-$ | 5.21 | 10.4 | 5.23 | 10.5 | –0.49 |
| | 41.3 | 82.5 | 41.2 | 82.5 | 0.09 |
| | 91.1 | 182.1 | 89.4 | 178.8 | 1.82 |
| $I^-$ | 5.18 | 10.4 | ND | ND | 100 |
| | 34.6 | 69.2 | ND | ND | 100 |
| | 85.3 | 170.6 | ND | ND | 100 |

A few individual anions were also examined at high TDS concentrations, in order to determine the capacity of the OnGuard cartridges, and to verify earlier results. Standards were prepared of chloride, chlorate, and bromide concentrations of 1000, 5000, and 10,000 ppm, and were analyzed with ion chromatography after treatment with the exchange cartridges. Table 14 shows the results obtained using the AS11 gradient method. Chloride and bromide were effectively removed from solution with the cartridges at concentrations as high as 10,000 ppm. Chlorate, as previously shown, is not removed with the sample preparation method.

**Table 14.** Results for individual anion removal study using OnGuard Cartridges

| Anion | Initial Concentration (ppm) | Final Concentration (ppm) | Percent Removal (%) |
|---|---|---|---|
| Bromide | 1000 | 1.34 | 99.9 |
| | 5000 | 3.83 | 99.9 |
| | 10000 | 14 | 99.8 |
| Chlorine | 1000 | 0.89 | 99.9 |
| | 5000 | 155 | 95.9 |
| | 10000 | 1619 | 83.8 |
| Chlorate | 1000 | 728 | 27 |
| | 5000 | 4956 | 0.8 |
| | 10000 | 9083 | 9 |

## DISCUSSION

A sensitive, selective, and robust ion chromatography method and an accompanied sample clean up procedure have been developed for perchlorate analysis with a Dionex AS16 Separation Column. The method detection limit and reporting limit for perchlorate with this method are 0.19 $\mu g\ L^{-1}$ and 1.9 $\mu g\ L^{-1}$, respectively. This method shows greater sensitivity than both the AS11 and AS5 methods, which have reported method detection limits of 0.5 $\mu g\ mL^{-1}$ and 1.0 $\mu g\ mL^{-1}$, respectively.

The proposed AS16 method has a higher capacity for applications such as industrial and wastewater samples, which contain high concentrations of contaminants. It was found from the interference study that high levels of inorganic anions that are commonly found in wastewater and other environmental samples do not interfere with the chromatographic separation and detection of perchlorate by the AS16 method. Of the inorganic anions tested, only three showed any appreciable retention on the AS16 column, and none interfered with either the retention time or the separation and recovery of perchlorate. It was also shown that 50 ppb perchlorate could be accurately quantified in the presence of up to 15,000 ppm carbonate, sulfate, chloride, or nitrate. Similar experiments performed at this laboratory with the AS11 column have shown that 50 ppb perchlorate spiked at up to approximately 7000 ppm of the anions could not be detected on an AS11 column.

Among the organic solvents tested, trichloroacetic acid was the only organic solvent reported to interfere with perchlorate analysis on the AS16 column. The presence of TCA in environmental samples not only distorts the perchlorate peak shape but also decreases both the retention time and recovery of perchlorate. TCA, DCA, and MCA are found in the environment typically as the byproducts of chlorinating surface water with humic acids. Humic acids are usually found from the natural degradation of plant materials. Hence, in the future, special consideration should be taken into account when developing an ion chromatography method specifically for analyzing perchlorate in the plant tissues. The other tested organic solvents did not interfere with the retention time or detector response.

Since some extraction techniques that are used in removing perchlorate from various natural settings involve the use of either HCl or NaOH in the sample preparation, the effect of sample pH on perchlorate recovery was examined. It was shown in this study, that with an increase in NaOH or HCl content in the sample, the perchlorate recovery was reduced by as much as 21% and 38%, respectively. When the pH of the perchlorate sample is varied from 7, the sensitivity of the AS16 method is diminished. Consequently, when an acid or base is required for perchlorate extraction, it is necessary to prepare the control standards in the same manner as the extracted samples. Failure to maintain controls and environmental samples at the same pH level would yield inaccurate results.

The sample preparation method is useful in allowing the analysis of trace level perchlorate in environmental matrices that contain high levels of contamination. It is possible, with the silver and hydrogen OnGuard cartridges, to remove many of the interference anions from solution without removing any perchlorate. The sample preparation method did not affect the recovery of perchlorate with the AS16 method. However, the cartridges effectively remove several of the anions commonly found in water, with almost 100% removal of chloride, bromide, iodide, bromate, nitrite, and phosphate.

The slight increase in nitrate concentration is probably due to the manufacturing process in which the silver cations are exchanged onto the resin. The most likely manufacturing procedure involves flushing the resin with silver nitrate, which could lead to some nitrate contamination in the resin. However, the presence of nitrate should not interfere with perchlorate analysis, as it is easily separated through chromatography. The increase in sulfate concentration may also be due to contamination by the resin. The resin contains sulfonyl groups, which act as exchange sites. It is possible that as the solution is pushed through the resin, some sulfate contamination is added to the sample. Again, the slight increase in sulfate should not interfere with perchlorate measurements, since the sulfate can be separated from perchlorate through ion chromatography.

In spite of the ease in handling and minimal preparation time, the advantages of using the commercially prepared silver and hydrogen OnGuard cartridges are counter-balanced by their relatively high cost. It is possible to prepare a silver-saturated exchange resin in-house by flushing Dowex AG50W-X8 cation exchange resin with a dilute silver nitrate solution. This resin was effective in the removal of the same interference anions that are removed with the cartridges. However, it is more likely to leave nitrate in the sample, and it has a greater potential for perchlorate loss than the cartridges due to difficulty in sample transfer.

## CONCLUSION

A new ion chromatography method was developed for perchlorate analysis using a Dionex AS16 Separation Column. The AS16 method proved to be more sensitive and selective than either the AS5 or the AS11 method, with a method detection limit for perchlorate of 0.19 $\mu g\ L^{-1}$. The higher capacity of the AS16 analytical column allows trace level perchlorate to be detected in samples that could not be analyzed with either of the previous ion chromatography methods, due to the high concentrations of inorganic and organic contaminants. The method was shown useful in analyzing environmental samples that contained up to 15,000 ppm total dissolved solids, as compared to the AS11 method, which could not detect perchlorate in samples containing 7,000 ppm total dissolved solids. A sample preparation method was also developed in order to remove interference ions, which can block the perchlorate signal to the detector. This preparation method reduces the background noise and increases the sensitivity of perchlorate analysis in environmental and industrial matrices.

## ACKNOWLEDGMENTS

The authors thank Dr. Peter Jackson (Dionex Corporation, Sunnyvale, California) and his department for his generous technical support. We also like to thank Dr. Sanwat Chaudhuri (Utah Department of Health, Division of Epidemiology and Laboratory Services) for providing the organic chemicals used in the interference study.

## REFERENCES

1. Manning, M. *Las Vegas Sun*. September 23, 1997.

2. Manning, M. *Las Vegas Sun*. September 24–25, 1997.

3. Manning, M. *Las Vegas Sun*. September 20, 1997.

4. Manning, M. *Las Vegas Sun*. September 8, 1997.

5. Manning, M. *Las Vegas Sun*. October 3, 1997.

6. Manning, M. *Las Vegas Sun*. January 27, 1998.

7 Chaudhuri, S.; Okamoto, H.; Pia, S.; Tsui, D. "Collaborative Study on AS5 and AS11 Methods." In *Interagency Perchlorate Steering Committee Analytical Subcommittee Report*. March 26, 1999.

8. Environmental Protection Agency, Office of Water website. "Standard Operating Procedure for Perchlorate." URL: http://www.epa.gov/OGWDW/ccl/perchlor/perchlo.html.

9. California Department of Health Services, Sanitation and Radiation Laboratories Branch. *Determination of Perchlorate by Ion Chromatography*. Rev. 0, June 3, 1997.

10. Dionex Application Note 121. Dionex Corporation: Sunnyvale, CA, 1998.

11. Wirt, K.; Laikhtman, M.; Rohrer, J.; Jackson, P.E. *American Environmental Laboratory* **1998**, *10*, 1,5.

12. Weast, R.C.; Astle, M.J.; Beyer, W.H. *CRC Handbook of Chemistry and Physics*. CRC Press, Inc: Boca Raton, FL, pp. B68–B146.

# Chapter 8
# SENSITIVITY AND SELECTIVITY ENHANCEMENT IN PERCHLORATE ANION QUANTITATION USING COMPLEXATION–ELECTROSPRAY IONIZATION–MASS SPECTROMETRY†⊗

Edward T. Urbansky* and Matthew L. Magnuson

United States Environmental Protection Agency, Office of Research and Development, National Risk Management Research Laboratory, Water Supply and Water Resources Division, 26 West Martin Luther King Drive, Cincinnati, Ohio 45268

## INTRODUCTION

The general nature and analytical chemistry of perchlorate have been reviewed elsewhere.[1-3] By its very existence this volume affirms the importance of having analytical chemistry tools capable of measuring low concentrations of perchlorate in water and other matrices. Other chapters delve into the toxicology[4] or regulatory and occurrence aspects.[5] The reader should refer to those for more information. The need for confirmatory techniques and methods of chemical analysis will exist whenever regulation occurs. Although we can expect ion chromatography to dominate drinking water monitoring for perchlorate, chromatographic and electrophoretic retention times are not unique to a particular analyte. Consequently, there is a need for robust, rugged methods using alternate techniques, such as mass spectrometry.

Because it is a small inorganic anion, perchlorate quantitation by mass spectrometry is most amenable to electrospray, thermospray, and similar means of ionization. For example, Barnett and Horlick used electrospray mass spectrometry and obtained a lower limit of detection of 0.050 μM (5 ng $mL^{-1}$) perchlorate.[6] However, natural and treated (finished

⊗This work was presented at the 218th national meeting of the American Chemical Society as part of the Environmental Division symposium Perchlorate in the Environment, held August 22–24, 1999, in New Orleans, Louisiana.



*Author to whom correspondence should be directed. Phone: 513-569-7655. Fax: 513-569-7658. Electronic mail: Urbansky.Edward@EPA.gov.

*Perchlorate in the Environment*, edited by Urbansky.
Kluwer Academic/Plenum Publishers, New York, 2000.

potable) water have a mixture of anions with mass below 300 u. Consequently, identification of perchlorate by mass ($m/z$) alone, can be challenging. Although the $^{35}Cl$:$^{37}Cl$ isotopic abundance ratios can be helpful, high background levels of ions can make it impossible to extract this information from the mass spectrum. It is therefore advantageous to complex the perchlorate anion ($ClO_4^-$) with another molecule that will raise its mass by a few hundred units out of the noisy region of the water mass spectrum. Clewell et al.[7] observed complexation (solvation) with acetic acid and we have reported on this phenomenon as well.[8–11]

## EXPERIMENTAL

Note: Mention of specific brand names and trademarks should not be construed to reflect an endorsement of products or manufacturers by the United States government.

### Instrumentation

A Finnigan (San Jose, Calif.) electrospray apparatus and Finnigan TSQ 700 quadrupole mass spectrometer were used throughout. Samples were introduced *via* a Rheodyne (Rohnert Park, Calif.) 7725 injector with a 200 µL sample loop. The liquid carrier (Fisher Optima® methanol) was supplied by a Waters (Milford, Mass.) 600-MS pump. Mass spectra were acquired in both positive ion and negative ion modes. ESI is a sufficiently soft ionization technique that fragmentation is generally not observed for these analytes. See references for additional details.[8–11]

### Complexing Reagents

Stock solutions of five minimally nucleophilic (2° or 3°) sterically hindered organic bases were prepared: 1,4-diazabicyclo[2.2.2]octane (DBO, Dabco™, triethylenediamine) *[280-57-9]* **1**, 1,5-diazabicyclo[4.3.0]nonene (DBN) *[3001-72-7]* **2**, 1,8-diazabicyclo[5.4.0]undec-7-ene, (DBU, 2,3,4,6,7,8,9,10-octahydropyrimido[1,2-*a*]azepine) *[6674-22-2]* **3**, 1,1′-hexamethylenebis[5-(4-chlorophenyl)-biguanide] (chlorhexidine) *[55-56-1]* **4**, 4,5-dihydro-2,4-diphenyl-5-(phenylimino)-1*H*-1,2,4-triazolium hydroxide, inner salt (nitron) *[2218-94-2]* (structure not shown), and triethanolamine (all from Aldrich). Concentrations of 0.010 M or 0.10 M were obtained by dissolving known masses of the commercial reagents into doubly deionized water. Acetic acid (HOAc, Mallinckrodt, Phillipsburg, NJ) at several concentrations was used to solubilize the chlorhexidine and the nitron.

1

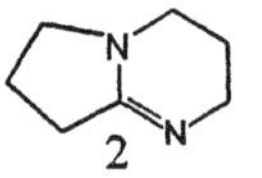

2

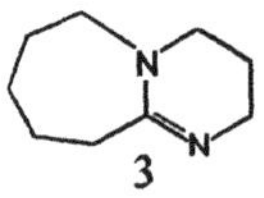

3

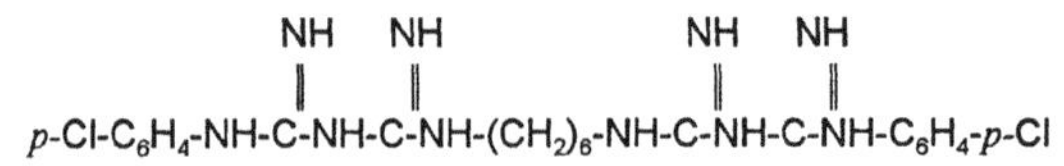

**4**

**Sample Preparation and Treatment**

**Test solutions with organic complexing agents.** Methanolic (high purity) solutions (containing 10-16% v/v water from dilution of stock reagents) were prepared containing perchlorate ranging from 0 to 50 μg $mL^{-1}$ and organic complexing agents (**1–4**) at concentrations from 1 μM to 10 mM (depending on the complexing agent). Methanolic solutions of ammonium perchlorate with no other reagents were used as a control group.

**Quaternary cations.** Given the behavior with the protonated complexing agents, we moved on to testing the possibility of extracting the perchlorate with a quaternary ammonium cationic surfactant of the form $C_nH_{2n+1}NMe_3^+$ where $n$ = 8, 10, 12, 14, 16. A volume of 500 mL of the aqueous sample, cationic surfactant, and 100 mL of the extraction solvent were shaken together vigorously. A separatory funnel was used to collect the organic phase. The organic phase was then evaporated through rotary evaporation to dryness at 60 °C (bath temp). The residue was redissolved in 5–7 mL of dichloromethane, transferred to a disposable test tube, and re-evaporated at 45–50 °C in a heater block. The localized residue was then reconstituted in 1.00 mL of the chosen solvent and transferred to a 1.8 mL glass vial. Injections of 50 μL of this solution were then analyzed by flow injection ESI-MS.

## RESULTS AND DISCUSSION

**Perchlorate Complexes with Organic Bases**

**Complexation.** The protonated organic bases, DBO, DBN, and DBU, form complexes with empirical formula $[(HB^+)(ClO_4^-)_2]$. The mass spectrum of the molecular ion for the DBU-perchlorate complex of this form is shown in Figure 1. As Figure 2 shows, chlorhexidine is capable of multiple protonations and up to three perchlorate complexations with molecular ions of the form [chlorhexidine•$ClO_4(HClO_4)_n]^-$; $n$ = 1 or 2.

Although we expected that a source of acid would be needed, we found that there was no effect on the peak areas when the acetic acid concentration was varied from 0 to 0.010 M with DBO, DBN, and DBU; at 0.10 M HOAc, the signal fell off dramatically, presumably due to a loss of electrospray efficiency induced by ionic strength. Apparently, the water present from the stock perchlorate solution (or the methanol itself) is sufficient to provide a source of protons. In the case of chlorhexidine, varying [HOAc]

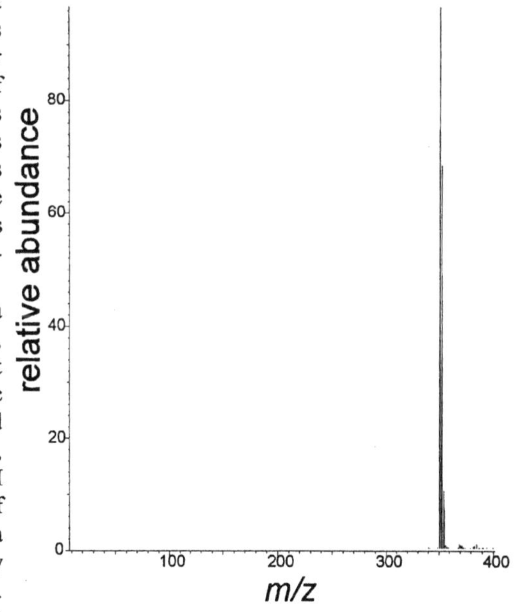


**Figure 1.** Negative ion ESI mass spectrum obtained for complex of 1,8-diazabicyclo[5.4.0]undec-7-ene (DBU) with perchlorate. Similar results were obtained for DBO and DBN. Reprinted from reference 8, *Journal of Analytical Atomic Spectrometry*, Royal Society of Chemistry.

from 0.01 to 10× (molar basis) the chlorhexidine concentration had no measurable effect on the peak areas.

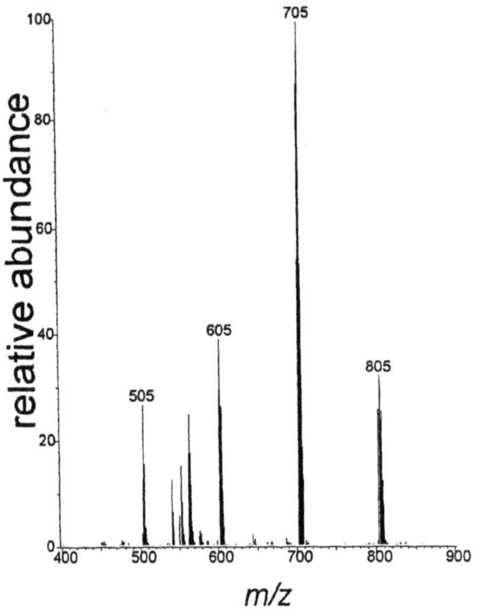


**Figure 2.** Negative ion ESI mass spectrum obtained for complexes of chlorhexidine with perchlorate. Note that multiple perchlorate ions can associate with a single chlorhexidine molecule. Complexes have the following empirical formulae: [chlorhexidine•$H(ClO_4)_n]^-$; $n$ = 1–3. Reprinted from reference 8, *Journal of Analytical Atomic Spectrometry*, Royal Society of Chemistry.

**Selectivity and sensitivity.** Of all the bases, chlorhexidine gave the greatest enhancement of selectivity Sensitivity must be gauged in terms of instrument response relative to a total concentration of complexing agent. The relative sensitivity for chlorhexidine is computed to be about 7 times greater. The chlorhexidine-perchlorate complex gave a signal equal to approximately 10% of the signal found for perchlorate alone at low concentrations. We determined a lower limit of detection ≤ 0.10 μM (10 ng $mL^{-1}$) for the perchlorate-chlorhexidine complex. This concentration gave a measurable signal distinct from the average noise by a factor of at least 2. With regard to capillary temperature, we eventually opted for 200 °C, which seemed to consistently give the greatest sensitivity and injection-to-injection reproducibility.

## Quaternary Ammonium Cation Extractions

**Extraction conditions.** The extraction solvent for these experiments was dichloromethane. Methyl isobutyl ketone (MIBK), ethyl acetate, and *t*-butyl methyl ether were also investigated. Dichloromethane was chosen because it resisted emulsification in the presence of the surfactant. MIBK also resisted emulsification but was not easily rotary-evaporated due to its high boiling point. Because of its efficiency, a single dichloromethane extraction of the aqueous solution was made.

Cationic surfactants differed in their sensitivity and their selectivity for perchlorate. Among common anions in a synthetic tap water solution, only nitrate, bromide, and chloride were detected. These complexes appear at different masses than the perchlorate, and do not present an interference in perchlorate detection. The choice of the cationic surfactant represents a compromise between selectivity for perchlorate and sensitivity for the extraction. The selectivity for perchlorate was investigated using the synthetic tap water. Nitrate, bromide, and chloride were the dominant species for which surfactant complexes were detected. Of the surfactants, C10 appeared to have the highest selectivity for perchlorate compared to nitrate or chloride, judging from the ratios of the peak areas of the respective complexes relative to the perchlorate complex.

The analytical sensitivity for the perchlorate extraction was investigated for several surfactants. Given the desirable ion-pairing abilities of C10, its high purity, and better sensitivity, C10 was chosen for future experiments. The concentration of surfactant was experimentally selected to provide a large excess of surfactant relative to perchlorate concentration. The amount of surfactant added to the aqueous solution resulted in a 1.0 mM surfactant concentration. The reconstitution solvent is important for sensitivity. Although C10 was selected for future experiments, all five cationic surfactants were investigated; interestingly, every combination of surfactant and solvent behaved differently.

**Electrospray Ionization-Mass Spectrometry Analysis**

Key parameters in the optimization of the ESI-MS systems are the carrier liquid flow rate, the pressure of the sheath (nebulizing) gas, the interface capillary temperature, and the applied electrospray voltage. The carrier flow represents a compromise between signal intensity/stability and analysis time. The signal intensity-stability dropped off when the carrier flow increased above approximately 0.4 mL $min^{-1}$. A flow rate of 0.3 mL $min^{-1}$ was chosen to optimize the peak height and minimize peak width. This carrier flow rate is sufficiently low enough that it does not require the auxiliary gas flow to assist the sheath gas, which was applied at 70 psi. The interface capillary temperature was investigated over the range 150-250 °C. The differences were small, and 200 °C was selected. The electrospray voltage was varied between 1 - 8 kV and underwent a shallow maximum at 4 kV. Removing the electrospray potential resulted in 80% signal reduction.

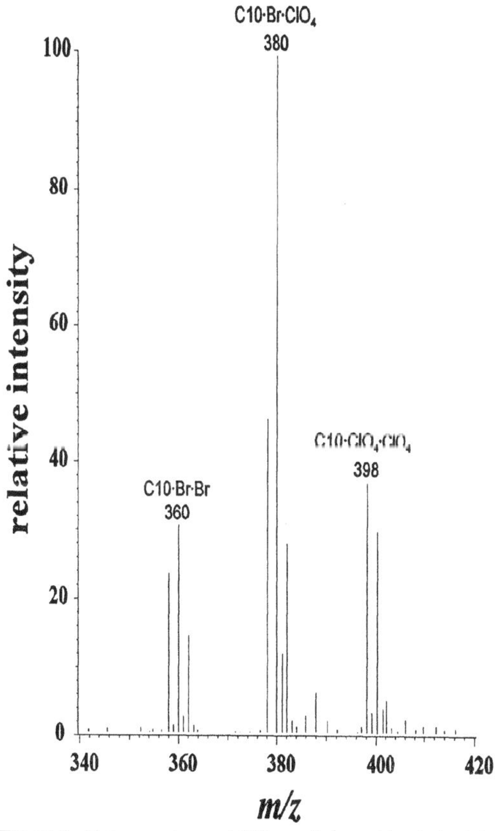


**Figure 3.** Mass spectrum of 100 ng $mL^{-1}$ perchlorate in deionized water extract. The ions with which the decyltrimethylammonium (C10) cationic surfactant is complexed are indicated. Injection volume was 50 µL. The remainder of the *m/z* range contains noise.

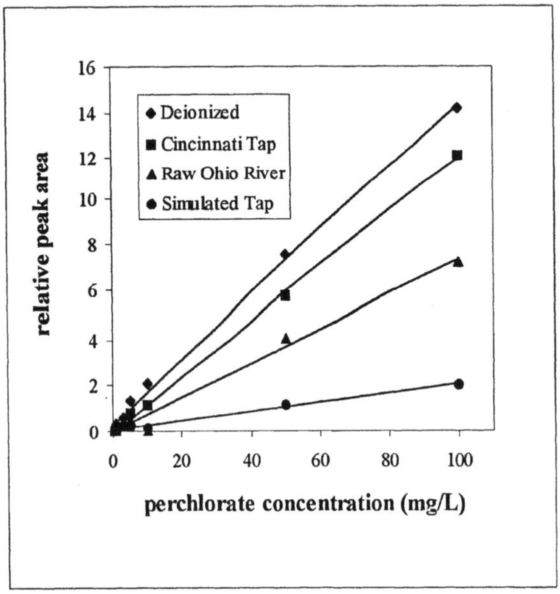


**Figure 4.** Peak area (*m/z* = 380 u) versus fortified perchlorate concentrations for four source waters. Least square regression lines are shown. Reprinted from reference 9, *Analytical Chemistry*, American Chemical Society.

Figure 3 is the mass spectrum of an extract from 100 μg $L^{-1}$ perchlorate in deionized water. The peak assignments are shown for the various C10 complexes. Because the C10 was used as bromide salt, three complexes are detected: (1) the complex with C10 and two bromide ions, (2) the complex with C10 and bromide and perchlorate, and (3) the complex with C10 and two perchlorate ions. It is more reliable to quantitate using the bromide-perchlorate-C10 complex rather than the perchlorate-perchlorate-C10 complex, probably due to the larger amount of bromide present as result of the C10 surfactant solution being prepared from the bromide salt.

As shown by Figure 4, calibration curves are affected by the matrix, but remain linear. The computation of perchlorate concentration is easily done using the method of standard additions. Most recently, we have moved to a microextraction procedure,[10] which consumes less solvent and sample. While it is possible to use MIBK as the solvent, we have found that the lower limit of detection decreases to 5 ng $mL^{-1}$, i.e., 1/10 of its value in $CH_2Cl_2$.[11] This is approximately the same as that for ion chromatography and thus would not affect use of ESI-MS for secondary confirmation.

## CONCLUSION

ESI-MS has been demonstrated to be sensitive and selective for the determination of perchlorate at trace levels. The use of the ion-pairing agent for extraction reduces the spectroscopic interference for samples not subjected to a separation prior to detection. The lower limit of detection is among the lowest reported in the literature. Results from this technique could be compared with those obtained by ion chromatography for the purpose secondary confirmation of concentrations determined by routine monitoring.

## REFERENCES


1. Urbansky, E.T. "Perchlorate chemistry: implications for analysis and remediation." *Bioremediation Journal* **1999**, *2*, 81–95, and references therein.

2. Urbansky, E.T.; Schock, M.R. "Issues in managing the risks associated with perchlorate in drinking water." *Journal of Environmental Management* **1999**, *56*, 79–95.

3. Schilt, A.A. *Perchloric Acid and Perchlorates.* GFS Chemicals, Inc.: Columbus, OH, 1979; p. 35, and references therein.

4. Clark, J.J.J. "Toxicology of perchlorate." THIS VOLUME: *Perchlorate in the Environment*, E.T. Urbansky, Ed. Kluwer/Plenum: New York, NY, 2000; Ch. 3, and references therein.

5. Pontius, F.W.; Damian, P.; Eaton, A.D. "Regulating perchlorate in drinking water." THIS VOLUME: *Perchlorate in the Environment*, E.T. Urbansky, Ed. Kluwer/Plenum: New York, 2000; Ch. 4, and references therein.

6. Barnett, D.A.; Horlick, G. *Journal of Analytical Atomic Spectrometry* **1997**, *12*, 497–501.

7. Clewell, R.A.; Chaudhuri, S.; Dickson, S.; Cassady, R.S.; Wallner, W.N.; Eldridge, J.E.; Tsui, D.T. "Analysis of trace level perchlorate in drinking water and ground water by electrospray mass spectrometry." THIS VOLUME: *Perchlorate in the Environment*, E.T. Urbansky, Ed. Kluwer/Plenum: New York, NY, 2000; Ch. 4, and references therein.

8. Urbansky, E.T.; Magnuson, M.L.; Freeman, D.; Jelks, C. "Quantitation of perchlorate ion by electrospray ionization mass spectrometry (ESI-MS) using stable association complexes with organic cations and bases to enhance selectivity." *Journal of Analytical Atomic Spectrometry* **1999**, *14*, 1861–1866.

9. Magnuson, M.L.; Urbansky, E.T.; Kelty, C.A. "Determination of perchlorate at trace levels in drinking water by ion-pair extraction with electrospray ionization mass spectrometry." *Analytical Chemistry* **2000**, *72*, in press.

10. Magnuson, M.L.; Urbansky, E.T.; Kelty, C.A. "Microscale extraction of perchlorate in drinking water with low level detection by electrospray–mass spectrometry." *Talanta*, submitted.

11. Urbansky, E.T.; Gu, B.; Magnuson, M.L.; Brown, G.M.; Kelty, C.A. "Quantitation of perchlorate ion in bottled water by electrospray ionization mass spectrometry (ESI-MS) and ion chromatography (IC)." *Journal of Environmental Monitoring*, submitted.

# Chapter 9

# REDUCTION OF PERCHLORATE ION BY TITANOUS IONS IN ETHANOLIC SOLUTION®

Joseph E. Earley, Sr.,* Daniel C. Tofan, and Giulio A. Amadei

Department of Chemistry, Georgetown University, Washington, DC 20057

## INTRODUCTION

It may seem odd that a chapter dealing with the mechanism of an inorganic reaction—a rather academic subject—is a part of a volume concerned with the environmental challenge raised by the detection of perchlorate ion in ground water. One reason that this chapter is included is to call attention to a second *perchlorate problem*—a purely scientific puzzle, quite unlike the practical difficulty posed by the presence of perchlorate ion in water supplies. Perchlorates are manufactured for their oxidizing power, useful in rocket propellants, pyrotechnics, and munitions. Since perchlorate ion is a strong oxidant, why do common *strong* reducing agents fail to react with that ion in aqueous media—while some *weaker* reductants do, in fact, reduce perchlorate? This is an certainly an interesting question in the area of inorganic reaction mechanism studies,[1] but, in addition, it is at least possible that the environmental threat of perchlorate contamination may not be fully contained until that scientific problem is resolved—until adequate understanding of the molecular-level basis of the unusual reactivity pattern exhibited by perchlorate ion is gained. The research reported here aims to clarify the mechanism of reduction of perchlorate by trivalent titanium ions, in order to contribute to understanding of the reasons why perchlorate ion displays the unusual reactivity pattern that it does. As a useful byproduct, this investigation may perhaps help identify a practical chemical means of destroying perchlorate ion in aqueous media.

®This work was presented at the 218th national meeting of the American Chemical Society as part of the Enviornmental Division symposium Perchlorate in the Environment, held August 22–24, 1999, in New Orleans, Louisiana.

*Author to whom correspondence should be directed. Phone: 703-532-5238. Fax: 202-687-6209. Electronic mail: earleyj@gusun.georgetown.edu

The powerful inorganic reductants, chromous ion ($Cr^{2+}$), stannous ion ($Sn^{2+}$), and uranous ion ($U^{3+}$) are all inert towards perchlorate ion in aqueous solutions. Vanadous ($V^{2+}$) and vanadic ($V^{3+}$) ions react with perchlorate,[2] but only at high temperatures and at low rates. Molybdate, tungstate, and niobate anions reportedly catalyze reduction of perchlorate ion by stannous ion.[3] $(CH_3)_2ReO_2$ reacts with perchlorate ion relatively rapidly.[4] Titanous ion ($Ti^{3+}$),[5] ruthenous ions ($Ru^{2+}$)[6] and some $Mo^{III}$ species[7] all reduce perchlorate ion in aqueous media, more or less sluggishly.

The rate of the oxidation-reduction (redox) reaction of $Ru^{2+}$ with perchlorate ion is quite comparable to the rates of the reactions of $Ru^{2+}$ with simple ions like $Cl^-$, $Br^-$, and $I^-$ to produce ordinary complexes of the $RuX^+$ type. This leads to the conclusion that the rate-limiting step in reaction of $Ru^{2+}$ with perchlorate ion is replacement of coordinated water by the anion to produce a perchlorato-complex. Subsequent redox steps are relatively rapid. Clearly, both ruthenium and rhenium species are able to reduce perchlorate ion rapidly—but both these elements are precious metals. Is it possible to locate a cheaper reducing agent that will rapidly reduce perchlorate, as these expensive complexes do?

Trivalent titanium seems to be a likely candidate for a practical perchlorate destroyer. Titanous ions reduce perchlorate, and have other desirable features. All forms of titanium are environmentally benign. $Ti^{III}$ when complexed with certain organic ligands is quite stable in aqueous solutions, even in the presence of dioxygen. Titanium compounds (particularly titanium dioxide) are abundant and cheap. The reaction of $Ti^{3+}$ with perchlorate ion is faster than that of other first-transition series reductants, but slower than replacement of water coordinated to that cation. The rate-limiting step must involve the redox change. Suitable ligands may vary such redox rates over many orders of magnitude.

There are some disadvantages to titanous ions as reductants for perchlorate. Reactions between $Ti^{III}$ complexes and perchlorate ion are generally slow (half-times of hours or days). Reaction of $Ti^{3+}$ (in excess) with perchlorate ions in acidic aqueous media is acid catalyzed; reduction of perchlorate ion by Ti(HEDTA) (HEDTA is the trianion of hydroxyethyl-ethylenedinitrilo-triacetic acid—*[150-39-0]*) gives evidence[8] of a *second-order* dependence on acid concentration. These reactions require high acid concentrations to achieve even modest rates. It would be useful to devise a way to increase the rate of reaction of $Ti^{III}$ with perchlorate ion, and to avoid complications caused by acid catalysis.

Other papers in this collection show that certain ion exchange resins effectively remove perchlorate from dilute aqueous solutions—but such resins do not destroy perchlorate ion, which still presents a disposal problem. Microbes can destroy perchlorate ion—but municipal water systems will not readily accept microbial treatment of drinking water. An inexpensive chemical agent able to destroy perchlorate ion rapidly might well be useful in water treatment. The research reported here aims to understand the mechanism of reduction of perchlorate by titanous ions—thereby contributing to understanding why some weak reductants react with perchlorate, while more powerful reductants do not react. Perhaps this research will also help identify a practical means of destroying perchlorate ion.

## PRELIMINARY OBSERVATIONS

Many redox reactions take place by transfer of an electron from a reductant ion or molecule to an oxidant molecule or ion, during a collision between those two particles. This

[*] As is usually done, coordinated water molecules will be omitted. E.g., $Cr(H_2O)_6^{2+}$ will be designated $Cr^{2+}$.

is called *direct electron transfer* or the *outer-sphere mechanism*. Perchlorate ion does not accept electrons directly from reductants—not even from the ultimate reductants, solvated electrons, either in water or in liquid ammonia. Reduction of perchlorate must involve formation of a perchlorato-complex of the reducing agent, and subsequent electronic rearrangement within that complex to yield products. This alternative type of process is called *bridged electron transfer* or the *inner-sphere mechanism*. When formation of a complex is involved as a preliminary step in a reaction, the equilibrium constant for the formation of the complex appears as a multiplicative factor in the observed rate constant. Perchlorate ions have a low tendency to form complexes with cations in aqueous solutions. Low values of the equilibrium constants for these complex-formation reactions contribute to the slow observed rates of perchlorate reactions.

But that effect is not sufficient to explain why some weak reductants ($Ru^{II}$, $Ti^{III}$) reduce perchlorate while strong reductants ($Sn^{II}$, $Cr^{II}$) do not. There must be some difference connected with the electron-transfer within the pre-assembled, bridged, intermediate complex. In order for an electron to move from one part of a molecule to another, there has to be a substantial interaction between the orbital that originally contains the electron, and the orbital that will hold that electron at the end of the process. That means the electron-donor orbital must overlap with the electron-acceptor orbital, and also that the two orbitals must be comparable in energy. It appears that the difference between reductants that do reduce perchlorate and those that do not is connected with the overlap requirement. What reductants that are active towards perchlorate have in common is that their electron-donor orbitals are *anti-symmetric* with respect to the line that connects the reductant atom, the bridging oxygen, and the chlorine center. That is, the electronic wave function of those orbitals changes sign (positive to negative, and vice-versa) on rotation about the bridge axis. (These are orbitals of *pi* symmetry—they are metal-centered $t_{2g}$ orbitals.) Reductants that do not reduce perchlorate have electron-donor orbitals that are *symmetric* with respect the bridge axis (*sigma* symmetry—$e_g$ metal orbitals). This suggests that the electron-acceptor orbital of perchlorate also is anti-symmetric with respect to the bridge axis—that it has *pi* symmetry with respect to that axis. Figure 1 gives a cartoon of how differences in symmetry of electron-donor orbitals determine the degree of overlap of redox (electron-donor and electron-acceptor) orbitals.

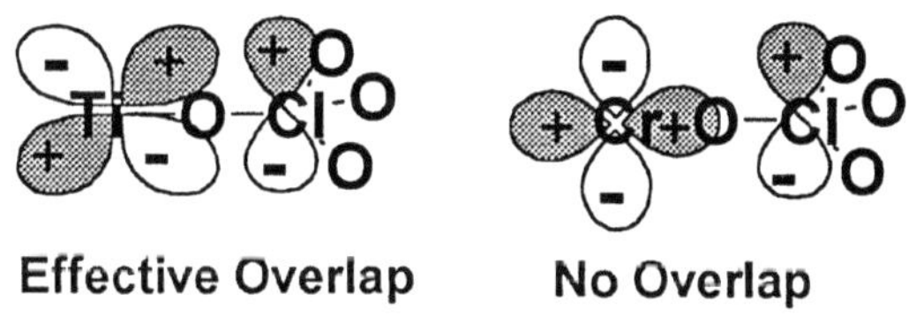


**Figure 1.** Cartoon of difference in effectiveness of overlap of an electron-acceptor orbital of *pi* symmetry on perchlorate ion with electron-donor orbitals that have *pi* symmetry (as in $Ti^{III}$) and *sigma* symmetry (as in $Cr^{II}$).

In addition to having appropriate symmetry for overlap, donor and acceptor orbitals must be of comparable energy or effective interaction will not take place. Tetrahedral perchlorate ion has a highly stable internal electronic structure. Low-energy bonding and non-bonding molecular orbitals are completely filled with electrons. A large energy gap separates the highest occupied molecular orbital (HOMO) and the lowest unoccupied molecular orbital (LUMO). Unfilled (electron-acceptor) orbitals of appropriate symmetry to overlap the electron-donor orbitals of $Ti^{III}$ exist on the tetrahedral perchlorate ion, but those orbitals are high in energy—much too high to interact with the metal-centered orbitals that contain the mobile electron. Distortion of perchlorate away from tetrahedral symmetry would lower the energy of such unfilled orbitals and to bring that energy into an appropriate range for effective interaction with the electron-donor orbitals on the reductant. It seems likely that distortion of perchlorate ion is a necessary precondition for its reduction.

Chlorate ion ($ClO_3^-$) is the first fairly stable reduction product of perchlorate. Therefore, perchlorate would be expected to function as a *two*-electron oxidant. The stable oxidation states of titanium are $Ti^{III}$ and $Ti^{IV}$. Compounds containing $Ti^{V}$ are rare and unstable. There is a mismatch between $Ti^{III}$ as a reductant and perchlorate as an oxidant, so far as number of electrcns is concerned. Each $Ti^{III}$ can donate *one* electron, but

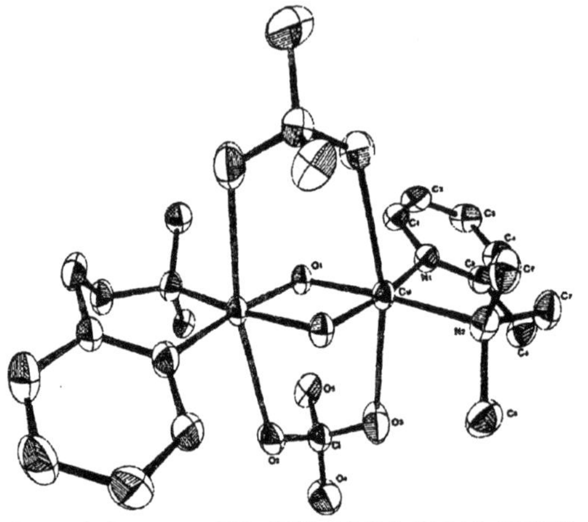

**Figure 2.** Structure of $[Cu(DMAEP)OH]_2(ClO_4)_2]$. DMAEP is 2-(N-dimethyl)-aminoethyl-pyridine. Reprinted with permission, from *Inorganic Chemistry* **1984**, *23*, 3418–3420. © 1984, American Chemical Society.

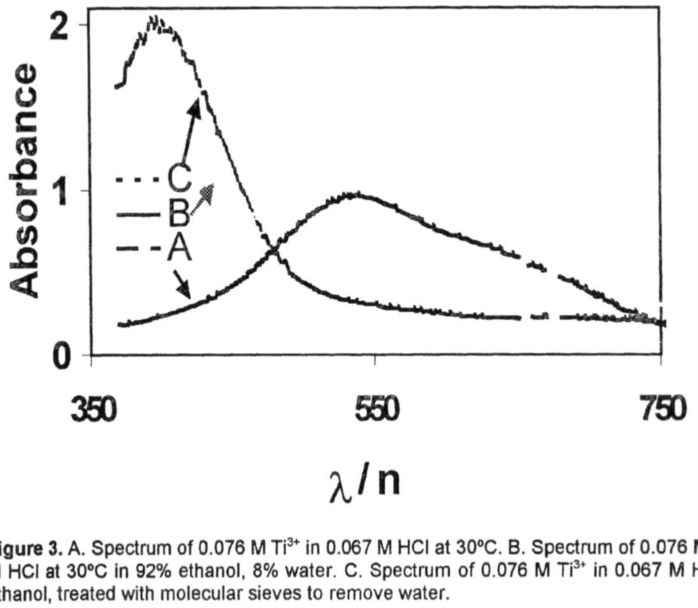


**Figure 3.** A. Spectrum of 0.076 M $Ti^{3+}$ in 0.067 M HCl at 30°C. B. Spectrum of 0.076 M $Ti^{3+}$ in 0.067 M HCl at 30°C in 92% ethanol, 8% water. C. Spectrum of 0.076 M $Ti^{3+}$ in 0.067 M HCl at 30°C in ethanol, treated with molecular sieves to remove water.

each perchlorate ion requires *two* electrons. One way to overcome this mismatch would be to involve *two* $Ti^{III}$ ions, and hence two electrons, in reducing each perchlorate ion. On the basis of these considerations, the mechanism for reduction of perchlorate ion by $Ti^{III}$ would be expected to involve a complex containing two (or more) $Ti^{III}$ ions and one perchlorate ion. It would also be expected that, in this complex, the coordinated perchlorate ion would be distorted away from its normal tetrahedral geometry. No stable $Ti^{III}$ perchlorato-complexes have been reported, but a complex containing two $Cu^{II}$ ions bridged by two perchlorate ions has been isolated from ethanol[9] (Figure 2).

In this structure, the bond-lengths within the coordinated perchlorate ions are severely distorted away from tetrahedral symmetry although angles are nearly tetrahedral. This structure suggested to us that it would be interesting to investigate the $Ti^{III}$–perchlorate reaction in ethanolic media.

## TITANOUS IONS IN ETHANOL

Quite concentrated titanium trichloride solutions can be prepared by dissolving titanium metal in hydrochloric acid. When such aqueous solutions of $TiCl_3$ are diluted with ethanol,[8] the absorbance peak at 500 nm (assigned to *d-d* transitions of $Ti^{3+}$) and its long-wavelength shoulder, persist—but an additional peak develops at 400 nm. The color of the solution changes from blue-violet to blue. When such ethanolic solutions are treated with molecular sieves to remove residual water, the original peaks are much reduced or eliminated and the 400nm peak is enhanced (Figure 3).

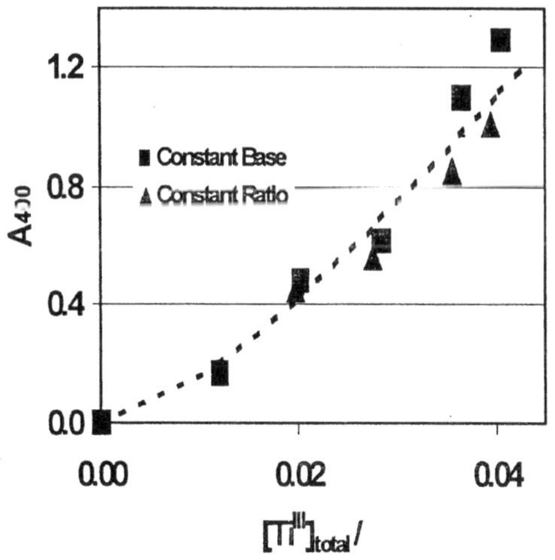


**Figure 4.** Variation of absorbance at 400 nm with total $Ti^{III}$ concentration, in ethanol at 30 °C, μ =1.0 M (LiCl). Perchlorate conc. 0.091 M, water 8% (1M).■ $[OH^-]_x$ = 0.02 M; ▲ $[OH^-]_x$ = 0.024 $[Ti^{III}]$. The dotted line assumes a dinuclear species : $K_f$= 25 $M^{-1}$ ; ε = 110 $M^{-1}$-$cm^{-1}$

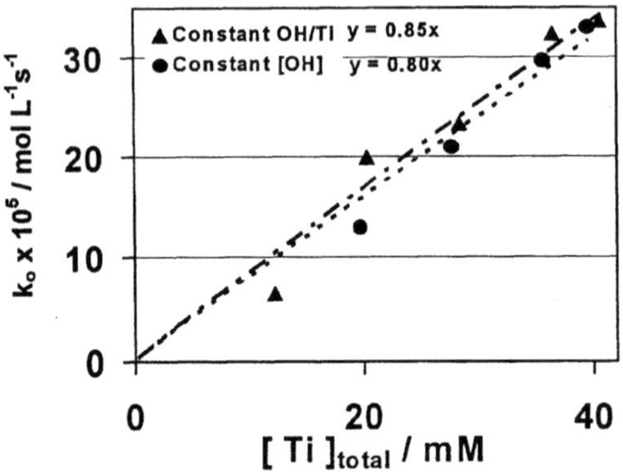


**Figure 5.** Consumption of perchlorate ion by $Ti^{III}$ when $Ti^{III}$ is in large excess, in ethanol at 30°C, measured at 400nm (diamonds) taking account of quadratic dependence of absorbance on $Ti^{III}$, and at 500 nm (triangles). Water 8% (1M), $Ti^{III}$ partially (~25%) deprotonated.

In ethanolic solutions containing 8% water ($[H_2O] \approx 1.0$ M), the absorbance at the 400 nm peak does not follow Beer's law (Figure 4) but is adequately fitted assuming formation of a low concentration of a dinuclear species that absorbs strongly at 400 nm. As dried ethanolic solutions of $TiCl_3$ stand in the absence of air for several days, the intensity of the 400 nm peak decreases, and that of the 500 nm peak increases. When a solution of LiOH in ethanol is added to a solution of $TiCl_3$, in aqueous (~8% water) ethanol, the absorbance peak at 400 nm increases and the 500 nm peak decreases, while the color of the solution changes from blue to green. (Precipitation of a green-black solid occurs when more than 1.8 ± 0.2 mol of base, in excess of the free acid originally present, have been added per mol of $Ti^{III}$). Increase of acid concentration, or addition of small amounts (~4%) of water, reduces the absorbance at 400 nm and tends to restore the original 500 nm peaks. The observations summarized above suggest that, in ethanolic solutions, two or more $Ti^{III}$ ions react with base to produce a polynuclear species that absorbs at 400 nm. Slow fading of these solutions suggests that the polynuclear species is unstable with respect to a form of $Ti^{III}$ (possibly mononuclear) that is fully equilibrated with solvent.

## REDUCTION OF PERCHLORATE ION BY TITANOUS ION IN ETHANOL

When ethanolic solutions containing perchlorate ion are reacted with solutions of $TiCl_3$ in ethanol ($\mu = 1$ M LiCl, water 8 ± 1%) the visible spectrum is bleached in a few minutes. When $Ti^{III}$ is in large (> 20-fold) excess, more than 4 mol $Ti^{III}$ are consumed for each 1 mol of perchlorate present—indicating that some chloride is produced. For somewhat higher perchlorate concentrations, after ten half-lives (~2 h), 2 mol $Ti^{III}$ are reduced for each 1 mol of perchlorate consumed—indicating that chlorate is the main initial product (Figure 5). Chloride is the eventual product after long times (~ 20 h), if $Ti^{III}$ is in excess.

The decrease of absorbance at 400 nm shows a brief induction period (seconds) followed by an essentially linear decrease of absorbance with time. Quasi zero-order behavior persists for about 60% of the absorbance-change. The slopes of the linear absorbance-decays measured

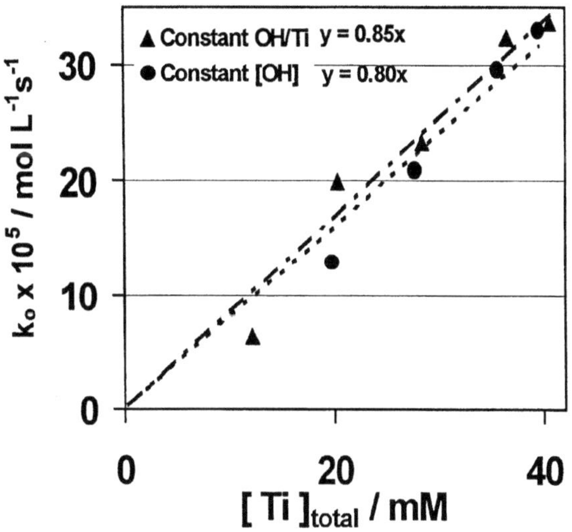


**Figure 6.** Dependence of zero-order rate constant measured at 400 nm on total $Ti^{III}$ concentration, in ethanol at 30°C, $\mu$ = 1.0 M (LiCl), $[ClO_4^-]$ = 0.091 M, water = 8% (~1M). ● $[OH^-]_x$ = 0.02 M; ▲ $[OH^-]_x = 0.024 \times [Ti^{III}]$.

at 400 nm increased linearly with $[Ti^{III}]$ (Figure 6). Rates were only slightly influenced by the concentration of water in the range, 0.8 M < $[H_2O]$ < 1.5 M. Rates remained constant on addition of base above the amount needed to neutralize free acid originally present, in the range 0.1 < (base added/$[Ti^{III}]_{tot}$) < 0.5.

Rates depend on perchlorate concentration (Figure 7) according to the phenomenological equation $k_{obs} = k[ClO_4^-]/(Q + [ClO_4^-])$, where $k$ is $2.9 \times 10^{-3}$ sec$^{-1}$ and $Q$ is 160 mM. The non-linear dependence on perchlorate concentration is probably the result of changes in medium brought about by replacement of more than 10% of the supporting electrolyte (LiCl) by $LiClO_4$. It is less likely that this curvature is the result of perchlorato-complex formation (mass law limitation).

The second-order rate constant determined from the best line through the low-perchlorate points for the reaction of present interest is 0.03 liters/mol-second. As noted previously, this value is independent of acid over a considerable range. Both of the previously studied reactions between $Ti^{III}$ species and perchlorate are acid-catalyzed. Acid-independent terms in the rate laws for the previously studied reactions have not been determined accurately, so precise quantitative comparisons cannot be made. On a semi-quantitative basis, it seems clear that the rate constant for the reaction of present interest is larger than the corresponding rate constant for the Ti(HEDTA) reduction of perchlorate, and much larger than reaction of aqueous $Ti^{III}$ with perchlorate ion.[10] Reactions that have a half-time of many hours in the absence of ethanol have a half-time of only a few minutes in ethanolic solution. Ethanolic media increase the rate of the titanous-perchlorate reaction by several orders of magnitude, relative to fully aqueous media.

## DISCUSSION

The basis for the relatively strong 400nm absorption is not yet clear. An oxo-bridged dinuclear $Ti^{III}$ species in which electrons are weakly anti-ferromagnetically coupled[10] does not show such absorbance, nor do simple oxo-ions.[11] It is possible that di-μ-hydroxy $Ti^{III}$ dinuclear complexes may have strongly coupled electrons and therefore exhibit high-energy electronic transitions that correspond to absorbance at 400 nm.

To convert perchlorate to chlorate, *two* electrons must be added to the chlorine center, and one oxide ion removed. The highly symmetrical perchlorate ion is resistant to acceptance of electrons, even from solvated electrons in water or in liquid ammonia—and also is a poor oxide ion donor. Although perchlorate readily complexes α-cyclodextrin,[12] it is a poor ligand for cations. Solids that contain perchlorate ions coordinated to two metal ions at once have been obtained from ethanolic media (Figure 2). In such complexes, coordinated perchlorate ion is distorted. Our present observations are consistent with the mechanistic model that, in ethanolic solution, a polymeric form of $Ti^{III}$ binds perchlorate ion between two cations, and distorts the perchlorate ion away from tetrahedral symmetry. Electronic rearrangement within that species then results in transfer of two electrons from a pair of $Ti^{III}$ ions to the chlorine center, and concomitant transfer of an oxide ion to the titanium centers.

That hypothetical mechanism involves two steps that would be expected to be partially rate-determining: formation of the dinuclear complex, and the redox reaction of that species with perchlorate ion. Simulation of this mechanism using the dynamic system modeling software package STELLA 5 generates a time-series fitting the unusual absorbance-time curves quite well. A fully adequate fit would require explicitly including a rate constant for

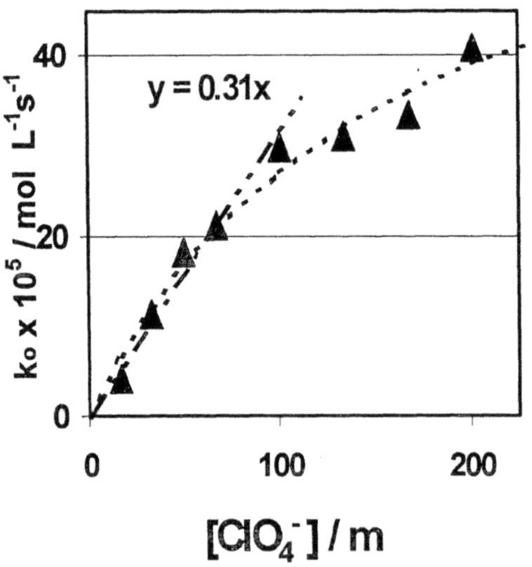


**Figure 7.** Dependence of zero-order rate constant measured at 400 nm on perchlorate concentration in ethanol at 30°C, water = 8% (~1 M), $[Ti^{III}]$ = 0.030M. The curved (dotted) line is calculated as described in the text.

formation of a perchlorato-titanous complex. Substitution on $Ti^{III}$ is sufficiently slow that formation of that complex would be expected to be partially rate-determining.

These preliminary data are consistent with a mechanism in which distortion of a perchlorate ion—concomitant with coordination into a species containing two $Ti^{III}$ centers lowers the energy of a chlorine-centered orbital of appropriate symmetry to accept an electron from $Ti^{III}$. Rapid transfer of a second electron from a second titanium center—to complete reduction of perchlorate to chlorate—follows this initial slower process.[13] Similar mechanisms operate in other perchlorate reductions. We are currently investigating the influence of several ligands on perchlorate reduction by $Ti^{III}$. The effectiveness of ethanol in speeding up the $Ti^{III}$-perchlorate reaction is probably due to fostering formation of perchlorato complexes in a slightly less polar environment than water. We are working with other sources of such environment in order to develop systems that effectively destroy perchlorate in aqueous solutions.

## ACKNOWLEDGMENT

We are grateful for helpful conversations with Professor H.B. Gray.

## REFERENCES

1. Taube, H. "Observations on atom-transfer reactions." *Mechanistic Aspects of Inorganic Reactions. ACS Symposium Series* **1982**, *198*, 151–173.

2. King, W.R.; Garner, C.S. "The kinetics of oxidation of V(II) and V(III) ions by perchlorate ion." *Journal of Physical Chemistry* **1954**, *58*, 29–33.

3. Haight, G.P. Jr. "Mechanism of the tungstate catalyzed reduction of perchlorate by stannous chloride." *Journal of the American Chemical Society* **1954**, *76*, 4718–4721.

4. Abu-Omar, M. M.; Espenson, J. H. "Facile abstraction of successive oxygen atoms from perchlorate ions by methylrhenium dioxide." *Inorganic Chemistry* **1995**, *34*, 6239–6240.

5. Duke, F.R.; Quinney, P.D. "The kinetics of reduction of perchlorate ion by Ti(III) in dilute solution." *Journal of the American Chemical Society,* **1954**, *76*, 3800–3803

6. Kallen, T.W.; Earley, J.E. "Reduction of the perchlorate ion by aquoruthenium(II)." *Inorganic Chemistry* **1971**, *10*, 1152–1155.

7. Hills, E.F.; Sharp, C.; Sykes, A.G. "Kinetic studies on the perchlorate oxidation of $Mo^{3+}$ and $Mo_2^{III}$." *Inorganic Chemistry* **1986**, *25*, 2566–2569.

8. Liu, B.; Wagner, P.A.; Earley, J.E. "Reduction of perchlorate ion by (N-hydroxyethyl)-ethylenediaminetriacetatoaquotitanium(III)." *Inorganic Chemistry* **1984**, *23*, 3418–3420.

9. Lewis, D.L.; Hatfield, W.E.; Hodgson, D. "Crystral and molecular structure of α-di-μ-hydroxo-bis[2-(2-dimethylamino-ethyl)pyidine]di-copper(II) perchlorate." *Inorganic Chemistry* **1974**, *13*, 147–152.

10. Jeske, P.; Wieghardt, K.; Nuber, B. "New mono- and dinuclear titanium(III) complexes: the crystal structure of and exchange coupling in $[\{LTi^{III}(NCO)_2\}_2(\mu\text{-}O)]$ (L=1,4,7-trimethyl-1,4,7-triazacyclononane)." *Inorganic Chemistry* **1994**, *33*, 47–53.

11. Miskowski, V.M.; Gray, H.B.; Hopkins, M.D, "Electronic structures of metal-oxo complexes." *Advances in Transition Metal Coordination Chemistry* **1996**, *1*, 159–186.

12. Cramer, F.; Saenger, W.; Spatz, H.-Ch.. "Inclusion compounds. XIX. The formation of inclusion compounds of α-cyclodextrin in aqueous solutions thermodynamics and kinetics." *Journal of the American Chemical Society* **1967**, *89*, 14–20.

13. Dunn, B.C.; Ochrymowycz, L.A.; Rohrabacher, D.B. "Conformational constraints on gated electron transfer kinetics. 2. Copper (II/I) complexes with phenyl-substituted [14]ane-$S_4$ ligands in acetonitrile." *Inorganic Chemistry* **1997**, *36*, 3253–3257.

# Chapter 10

# INVESTIGATION OF PERCHLORATE REMOVAL IN DRINKING WATER SOURCES BY CHEMICAL METHODS[⊗]

Mirat D. Gurol* and Kyehee Kim

Department of Civil and Environmental Engineering, San Diego State University
San Diego, California 92182-1324

## INTRODUCTION

Soon after a new IC method that achieved a method detection limit of approximately 1 ppb was developed in 1997, perchlorate has been detected in many drinking water wells and surface and groundwater in the western states, including Colorado River water. Perchlorate is a chemical of health concern due to its interference with the activity of the thyroid gland, and therefore its removal from drinking water sources is very desirable.

A few promising technologies are being developed for removal of perchlorate, however, the stability of perchlorate makes treatment technologies difficult, especially at low concentration levels. There is no doubt that perchlorate can be removed from water by separation processes, such as ion exchange and reverse osmosis. Many researchers have been also investigating biological techniques to reduce perchlorate by organic chemicals, e.g., ethanol, acetate, and lactate under anaerobic conditions. A recent literature review indicates that many metals, including Ti(III), V(II), Cr(II), Mo(III) are capable of reducing perchlorate to chloride or chlorate.[1] However, perchlorate reduction by common reducing agents, e.g., $Fe^0$, $S_2O_3^{2-}$, and $SO_3^{2-}$, is believed to be too sluggish to be practically useful.

The present study was designed to evaluate the interaction of perchlorate with metallic iron and an iron oxide mineral. The objectives were to investigate the feasibility of (1)

[⊗]This work was presented at the 218th national meeting of the American Chemical Society as the part of the Environmental Division symposium Perchlorate in the Environment, held August 22–24, 1999, in New Orleans, Louisiana.

*Author to whom correspondence should be directed. Phone: 619-594-0391. Fax: 619-594-8078. Electronic mail: mgurol@mail.sdsu.edu.

perchlorate reduction by metallic iron ($Fe^0$) using ultraviolet light to promote the reaction, and (2) adsorption of perchlorate on the surfaces of metallic iron and goethite crystal ($\alpha$–Fe(O)OH) under various conditions.

## BACKGROUND

Perchlorate ion is the conjugate base of perchloric acid with a $pK_a$ value of –7, existing in ground and surface waters as the salts of ammonium, potassium, magnesium, or sodium. The chlorine atom in perchlorate has an oxidation valence of +7, and its reduction to chloride or chlorate is thermodynamically very favorable, as shown in Table 1.[2]

**Table 1.** Reduction of perchlorate

| Reduction Half-Reactions | $E^\circ_{red}$ (V) | $\Delta G^\circ$ (kcal mol$^{-1}$) |
|---|---|---|
| $ClO_4^- + H_2O + 2e^- \rightleftharpoons ClO_3^- + 2OH^-$ | 0.36 | –16.62 |
| $ClO_4^- + 2H^+ + 2e^- \rightleftharpoons ClO_3^- + H_2O$ | 1.189 | –54.88 |
| $ClO_4^- + 8H^+ + 8e^- \rightleftharpoons Cl^- + 4H_2O$ | 1.389 | –256.47 |
| $ClO_4^- + 8H^+ + 7e^- \rightleftharpoons 1/2Cl_2 + 4H_2O$ | 1.389 | –224.41 |

However, the reduction reactions of perchlorate are kinetically controlled by their large activation energies. An input of energy as heat or light or the presence of a catalyst would be needed to accelerate the reaction rate.

Metallic ($Fe^0$) iron was used as the reducing agent of choice in our study, because zero-valent metals, mainly Fe, Sn, and Zn, have been effective in enhancing the rate of removal of a wide range of heavy metals and halogenated compounds under anoxic conditions.[3–5] These reactions involve the dissolution of ions, e.g., $Fe^{2+}$, from the metal surface coupled with release of electrons, which provide reduction and dehalogenation of the targeted chemicals. However, the reaction rates are generally quite low, and depend upon such parameters as pH, temperature and surface area of the metal under consideration. In fact, reduction of perchlorate by metallic iron has been tried earlier by Yarmoff and Amrhein,[6] who reported no observable change in perchlorate concentration. In the present study, ultraviolet (UV) light was used as a catalyst since perchlorate is known to absorb UV in the wavelength range shorter than 185 nm.[7] Furthermore, radiolytic decomposition of perchlorate by the action of X-rays and $\gamma$-rays on the alkali metal and alkaline earth perchlorates has been reported by Prince and Johnson.[8] Chemically identifiable fragments, including $O_2^-$, $ClO_3^-$, $ClO_2$, $ClO_2^-$, $ClO^-$, $Cl^-$, $O^-$, ClOO, $ClO_3$ and $ClO_4$ have been detected in radiated samples.

## EXPERIMENTAL APPROACH

The metallic iron used in the experiments was an electrolytically-produced 100 mesh particles (certified grade, 95%; Fisher) with a surface area of 0.74 $m^2\ g^{-1}$.[9] The iron oxide ($\alpha$-Fe(O)OH) was a hydrated, 30-50 mesh catalyst grade particles (Aldrich Chemical) with a surface area of 120 $m^2\ g^{-1}$, as determined by a BET surface analyzer.[10] The pH effect was studied by using phosphoric, hydrochloric and sulfuric acids, all ACS reagent grade. The effect of UV radiation was studied by using a Rayonet® Photochemical Chamber equipped with a 800 ml quartz reactor and low pressure mercury lamps emitting 99% of light at 254 nm and 1% at 185 nm. The chamber accommodated up to sixteen lamps with a total emitted light intensity of 26 W, and an incident light intensity of 3.4 W, as measured by actinometry.[11] A Masterflex® hydraulic pump with the speed of 60–600 rpm (Cole-Palmer) was used to continuously circulate the reactor contents. The samples were filtered through a 0.45 μm filter (Micron Separation Inc.) for separation of the iron and iron oxide particles.

The experiments using $Fe^0$ were conducted under anoxic conditions. Deionized water was degassed by vacuum pump and purged with $N_2$ gas before preparing perchlorate solutions. The concentration of $Fe^0$ was investigated in the range of 10–100 g $L^{-1}$. The reactor was a quartz container of 0.8 L that was operated as batch; however, the contents were continuously circulated to provide complete mixing. The contents were exposed to UV light under neutral pH.

Similar experiments without exposure to UV light were conducted using $Fe^0$ or Fe(O)OH in order to investigate the adsorption of perchlorate on the surfaces in the presence of $H_3PO_4$, HCl or $H_2SO_4$. The pH of the reactor contents were maintained in the range of 2.0–4.5.

## ANALYTICAL METHODS

An ion chromatograph (Dionex, Sunnyvale, CA, DX-500 IC) was used to identify and quantitate perchlorate and other chlorine oxyanion species in this study. The components of the IC included a 100 μL sample loop, an ED40 conductivity detector, AS11 and AG11 columns, and a GP40 gradient pump. Because the method developed by the California Department of Health Services in 1997 could not separate the peaks of $Cl^-$, $ClO_2^-$, and $ClO_3^-$ from each other, a modified method with varying eluent concentrations was employed in this study. The modified method involved the use of 10 mM of NaOH solution for the first 7 min as the eluent, followed by application of 100 mM of NaOH solution through the 20 minute. This did not change the peak heights and peak areas, but enabled the separation of the peaks. The calibration of concentrations of the chlorine oxyanions was made in the range of 10 to 1000 μg $L^{-1}$. Method detection limit (MDL) was calculated thusly: MDL = $t \times s_{area} \times [ClO_4^-]/\mu_{area}$, where Student's $t$ = 3.14 for seven replicates (6 degrees of freedom) at the 99% confidence level, $s_{area}$ is the estimated standard deviation, and $\mu_{area}$ is the average peak area obtained. The MDL for perchlorate achieved here was 2.74 μg $L^{-1}$. The results of calibration studies of chlorine oxyanions are presented in Table 2.

**Table 2.** Results of Calibration Study of Chlorine Oxyanions

| Chlorine Oxyanion | Retention time (min) M1* | M2** | MDL (μg $L^{-1}$) | Equation $y$: area $x$: concentration in μg $L^{-1}$ | $R^2$ |
|---|---|---|---|---|---|
| $Cl^-$ | 1.85 | 2.70 | 3.53 | $y = 1938.4x + 57208$ | 0.9998 |
| $OCl^-$ | 1.85 | 2.70 | 3.02 | $y = 410.7x + 76284$ | 0.9965 |
| $ClO_2^-$ | 1.80 | 1.95 | 3.03 | $y = 1326.1x + 36915$ | 0.9968 |
| $ClO_3^-$ | 2.08 | 4.80 | 5.72 | $y = 750.37x$ | 0.9993 |
| $ClO_4^-$ | 7.52 | 16.50 | 2.74 | $y = 547.2x$ | 0.9997 |

*M1: Isocratic eluent mode (100 mM NaOH) for entire run.
**M2: Gradient eluent mode (10 mM NaOH for 0–7 min, and 100 mM NaOH for 7–20 minutes).

## RESULTS AND DISCUSSION

### Perchlorate Removal in $Fe^0$-UV System

The results of perchlorate removal by $Fe^0$ without UV exposure are presented in Figure 1. While no appreciable reduction was observed in the presence of 10 g $L^{-1}$ $Fe^0$, about 37% of 1000 μg $L^{-1}$ of perchlorate was removed within 3 h, indicating the feasibility of the following reactions:

$$4Fe^0 + ClO_4^- + 8H^+ \rightarrow Cl^- + 4Fe^{2+} + 4H_2O \quad (1)$$

$$Fe^0 + ClO_4^- + 2H^+ \rightarrow ClO_3^- + Fe^{2+} + H_2O \quad (2)$$

The direct role of $Fe^0$ as a reactant implies the involvement of reactive sites on the metal and, therefore, the quantity and the condition of the metal surface is expected to strongly influence the rate of perchlorate reduction.

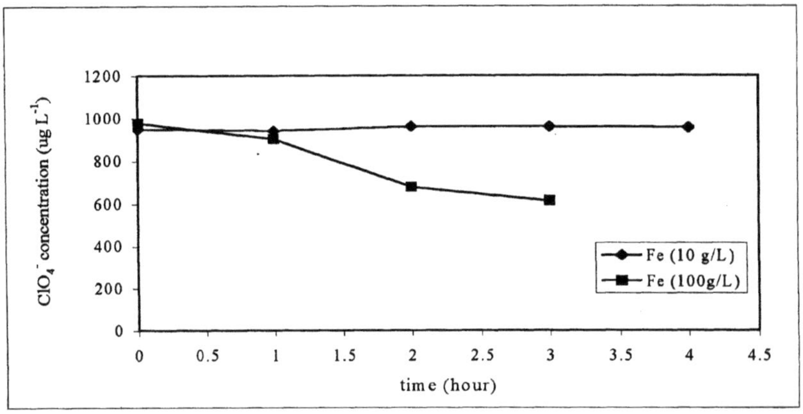


**Figure 1.** Change in perchlorate concentration by metallic iron.

Irradiation by UV light has been tried to accelerate removal of perchlorate by $Fe^0$. The result is shown in Figure 2 that while UV light without any $Fe^0$ could not reduce the perchlorate concentration, simultaneous application of $Fe^0$ and UV light was very effective on perchlorate removal. Furthermore, increasing the $Fe^0$ concentration for a constant light intensity increased the perchlorate removal rate. In the presence of UV light, perchlorate was reduced by 77% by 100 g $L^{-1}$ of $Fe^0$ in 3 h, wheras the removal was only 37% without UV. These experiments were conducted in unbuffered neutral solutions (pH about 6.6) where the pH has increased by up to 2.0 units at the end of the experiments.

The results of additional experiments showed that removal rate of perchlorate was a function of the UV intensity. For example, 77% removal of perchlorate was achieved with 100 g $L^{-1}$ of $Fe^0$ and a total UV intensity of 0.9 W $cm^{-2}$, while only 40% of perchlorate was

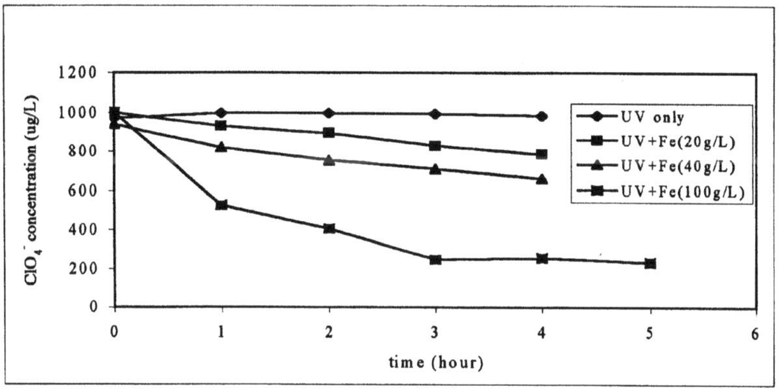


**Figure 2.** Change in perchlorate concentration by metallic iron and UV light (UV intensity = 0.9 W $cm^{-2}$).

removed using the same concentration of $Fe^0$ , but a total UV intensity of 0.6 W $cm^{-2}$. Furthermore, ion chromatograms of the treated samples showed significant increase in peaks that belong to $Cl^-$ and $ClO_3^-$. Conversion of perchlorate to these ions was complete, with more than 99 % of perchlorate reduced to $Cl^-$ while less than 1% converted to $ClO_3^-$.

Hence, it is apparent from these results that 1) UV light acts as a catalyst, 2) both the concentration of $Fe^0$ and dosage of UV affect the reaction rate significantly, and 3) perchlorate is reduced quantitatively to $Cl^-$, with less than 1% reduced to $ClO_3^-$. Mechanistically, it is conceivable that perchlorate ion first adsorbs on the surface of $Fe^0$, and then undergoes an electron transfer process that is facilitated by UV excitation.

**Perchlorate Removal in $Fe^0$-$H_3PO_4$ and FeOOH-$H_3PO_4$ Systems**

The metallic iron in the presence of phosphoric acid was capable of removing large amounts of perchlorate from water, as shown in Figure 3. However, within about 5 min of contact, perchlorate concentration started to increase, indicating that the removal was due to a reversible adsorption on the metallic surface. The initially low pH of about 1, which was provided by phosphoric acid, started to increase in parallel with perchlorate desorption, which is coupled with pH increase. The pH increase might be due to the dissolution of metallic iron to ferrous ion and the reduction of water. It should be noted that no removal of perchlorate was observed when sulfuric or hydrochloric acids were used in amounts to also reduce the pH to less than 2.

The same process process was repeated for goethite. As presented in Figure 4, goethite during the initial phase of contact. Orthophosphate is known to have high affinity towards iron and iron oxide surfaces. Hence, it is conceivable that the removal of perchlorate is due to a complexation between perchlorate and phosphoric acid followed by adsorption of the complex on the surfaces of the particles. However, as in the case for metallic iron, proved

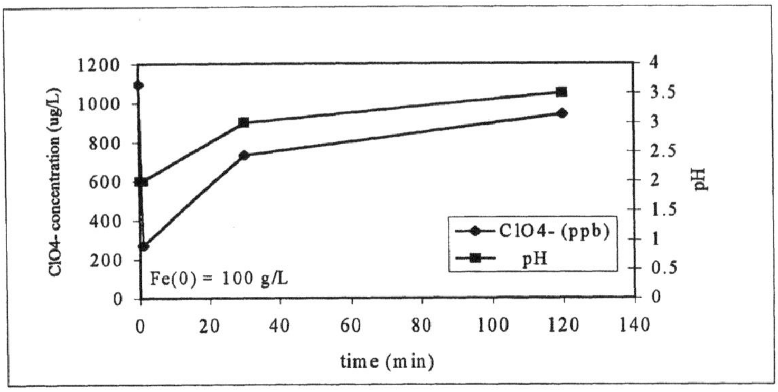


**Figure 3.** The change of perchlorate concentration and solution pH in the presence of metallic iron and phosphoric acid.

to be very effective in removing perchlorate in the presence of phosphoric acid perchlorate concentration stared to increase upon prolonged contact. The desorption rate was however very slow for goethite for the same initial pH of 2.5, and particle concentration of 12 g $L^{-1}$. This could be due to stronger binding of the complex on goethite compared to metallic iron, or more likely goethite particles having mostly internal surface provided by the porous structure, which provides resistance to back diffusion of the complex.

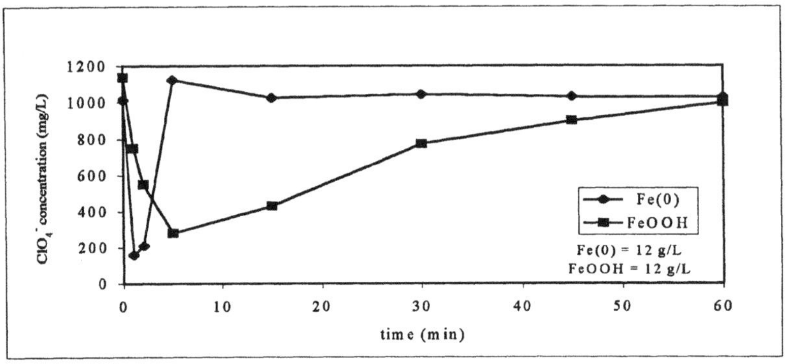


**Figure 4.** Change of perchlorate concentration in the presence of metallic iron and phosphoric acid, and goethite and phosphoric acid; $pH_0$=2.5 for both adsorbents.

Perchlorate removal was observed only at acidic pH values. In fact, the desorption during the contact is very likely due to pH of the suspension increasing from initial pH values of about 2 to more than 3 during prolonged contact, for both metallic iron and goethite. This pH dependence might be explained in the context of speciation of the surface and phosphate.

The surface charge of goethite, changes with the pH change of the solution as shown in equations 3 and 4.[12]

$$\equiv FeOH_2^+ \rightleftharpoons H^+ + \equiv FeOH \qquad pK_1^s = 6 \qquad (3)$$

$$\equiv FeOH \rightleftharpoons H^+ + \equiv FeO^- \qquad pK_2^s = 8.8 \qquad (4)$$

Equations 5-7 summarize the chemical speciation of phosphate: [13]

$$H_3PO_4 \rightleftharpoons H^+ + H_2PO_4^- \qquad pK_{a1} = 2.1 \qquad (5)$$

$$H_2PO_4^- \rightleftharpoons H^+ + HPO_4^{2-} \qquad pK_{a2} = 7.2 \qquad (6)$$

$$HPO_4^{2-} \rightleftharpoons H^+ + PO_4^{3-} \qquad pK_{a3} = 12.3 \qquad (7)$$

Hence, when the pH is about 2, the surface becomes positively charged due to the dominance of $\equiv FeOH_2^+$ sites, whereas the dominant phosphate species is phosphoric acid, which is neutral. In the case of complexation between $H_3PO_4$ and $ClO_4^-$, the complex will have a net charge of −1, and therefore will exhibit an electrostatic attraction towards the positively charged surface. However, the negatively charged phosphate species that become dominant at elevated pH values may not necessarily form a complex with $ClO_4^-$, and there should be no removal of perchlorate at higher pH. It should be noted that $ClO_4^-$ did not adsorb well on the positively charged surface in the absence of phosphoric acid, as checked by reducing the pH to 2 with HCl and $H_2SO_4$ acids.

Additional experiments were conducted using FeOOH-$H_3PO_4$ system for lower initial perchlorate concentrations of 200 and 500 μg $L^{-1}$, keeping all other conditions the same. The removal of perchlorate as $C/C_0$, where $C_0$ is the initial concentration, is presented in Figure 5 for three different initial concentrations of perchlorate. About 70-75% of perchlorate was removed for all three cases within the first few minutes. However, it was released back to the solution upon prolonged contact, although more slowly for the lowest perchlorate concentration. The percent removal of perchlorate was independent of its initial concentration.

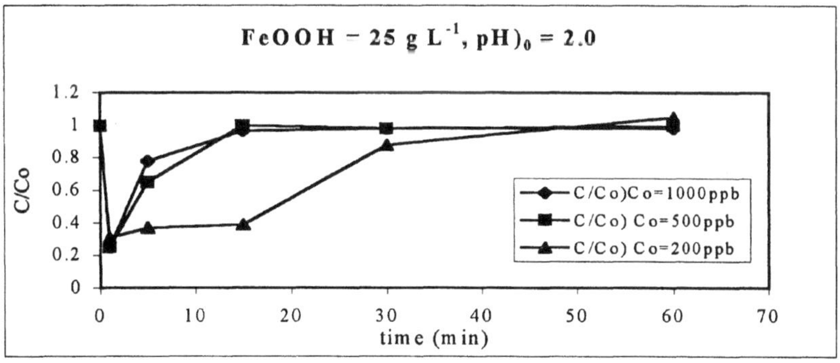


**Figure 5.** Perchlorate removal in FeOOH-$H_3PO_4$ system for various initial perchlorate concentrations.

## CONCLUSIONS

Two innovative chemical processes were investigated to determine the feasibility of perchlorate removal from water. The first process involves the exposure of perchlorate simultaneously to metallic iron and UV light under anoxic conditions. Despite the concerns of many researchers regarding the high kinetic inertness of perchlorate, it was shown that perchlorate can be reduced by metallic iron, and furthermore that UV light can accelerate the reaction rate to levels that could make the process viable for practical applications.* The results can be summarized as follows: (1) UV light promotes the reaction, while metallic iron provides electrons for reduction of perchlorate, (2) both the concentration of metallic iron and dosage of UV affect the reaction rate significantly, and (3) more than 99% of perchlorate is reduced to $Cl^-$, with less than 1% reduced to $ClO_3^-$. It is believed that perchlorate ion is adsorbed on the surface of metallic iron, and then undergoes an electron transfer process that is facilitated by UV excitation. It should be noted that $ClO_4^-$ absorbs light at wavelengths shorter than 185 nm; however, the low pressure mercury lamps used in this study generate light primarily at 254 nm (99%), with only 1% emitted at 185 nm. Thus, these lamps are not efficient for $ClO_4^-$ excitement. Better results can be obtained by using lamps that emit primarily at lower wavelength.

The second process involves the contact of perchlorate with the surfaces of metallic iron or an iron oxide mineral (goethite) in the presence of phosphoric acid. The experimental results suggest that perchlorate can be removed up to almost 100% during the initial phases of the contact in the pH ranges of 2.0–2.5. This removal is believed to be due to formation of a complex between perchlorate and phosphoric acid that subsequently adsorbs to particle surfaces. At higher pH values very little removal of perchlorate can be observed. However, even at acidic pH, continuous contact with the surface—coupled with agitation and pH rise—seems to release the perchlorate back to the solution. It is obvious that the particles must be separated from solution before desorption of perchlorate if this is to be used as a treatment method. Unfortunately, the requirement of very acidic conditions and subsequent neutralization for pH restoration might make this process relatively expensive for typical applications.

* Patent application date: August 1999.

## REFERENCES


1. Urbansky, E.T. "Perchlorate chemistry: implications for analysis and remediation." *Bioremediation Journal* **1998**, *2*, 81–85.

2. Lide, D.R., Ed. *CRC Handbook of Chemistry and Physics,* 75th ed. Chemical Rubber Company: Boca Raton, FL, 1995; Ch. 8.

3. Gotpagar, J.; Grulke, E.; Tsang, T.; Bhattachayya, D. "Reductive dehalogenation of trichloroethylene using zero-valent iron." *Environmental Progress* **1997**, *16*, 137–143.

4. Gillham, R.W.; O'Hannesin, S.F. "Enhanced degradation of halogenated aliphatics by zero-valent iron." *Ground Water* **1994**, *32*, 958–967.

5. Blowes, D.W.; Ptacek, C.J.; Jambor, J.L. " In-situ remediation of Cr(VI)-contaminated groundwater using permeable reactive walls: laboratory studies." *Environmental Science and Technology* **1997**, *31*, 3348–3357.

6. Yarmoff, J.A.; Amrhein, C. "Fundamental studies of the removal of contaminants from ground and waste waters via reduction by zero-valent metals." Progress Report. DE-FG07-96ER14707. URL: http://www.doe.gov/em52/55061.html/.

7. Bailer, J.C.; Emeleus, H.J.; Nyholm, R.; Trotman-Dickenson, A.F. *Comprehensive Inorganic Chemistry*. Pergamon: Elmsford, NY, 1973; Vol. 2, Ch. 7.

8. Price, L.A.; Johnson, E.R. "The radiation-induced decomposition of the alkali and alkaline earth perchlorates." *Journal of Physical Chemistry* **1965**, *62*, 359–377.

9. Matheson, L. J.; Tratnyek, P. G. "Reductive dehalogenation of chlorinated methanes by iron metal." *Environmental Science and Technology* **1994**, *28*, 2045–2053.

10. Lin, S. S.; Gurol, M. D. "Catalytic decomposition of hydrogen peroxide on iron oxide: kinetics, mechanism, and implications." *Environmental Science and Technology* **1998**, *32*, 1417–1423.

11. Gurol, M.D.; Akata, A. "Kinetics of ozone photolysis in aqueous solution." *AICHE Journal* **1996**, *42*, 3283.

12. Stumm, W.; Morgan, J.J. *Aquatic Chemistry*. Wiley: New York, NY, 1995; Ch. 1.

13. Snoeyink, V.L.; Jenkins, D. *Water Chemistry*. Wiley: New York, NY, 1980; Ch. 4.

# Chapter 11

# MODELING THE FORMATION OF ION PAIRS IN ION EXCHANGE RESINS AND EFFECTS ON PERCHLORATE TREATMENT CHEMISTRY[⊗]

Gerald A. Guter*

Guter Consulting, 215 Monte Vista, San Clemente, California 92672

## INTRODUCTION

### Background

**Need for study**. This study develops information regarding perchlorate ion exchange chemistry by studying theoretical aspects of the process, applies the information to treatment processes and evaluates resin performance by projecting process costs and product water quality. Perchlorate is difficult to treat. It has a very high affinity for most ion exchange resins, making regeneration difficult and costly as well as producing large wastewater discharges to the environment.[1] Treatment complexities arise also when ion exchange is coupled with biological or chemical processes.[1] Perchlorate ion exchange with iodide is proposed to be the basis of thyroid disorders when it is present in drinking waters.[2] It is, therefore, important to develop information on perchlorate ion exchange.

**Importance of affinity of interfering ions.** The affinity between perchlorate ion and resins is unusually high and has been demonstrated[3] to be dependent on resin loading. Different resins will show different perchlorate affinity. In the ion exchange process, not only is the affinity of a single ion important but the affinities of all other ions in the untreated water are equally important because they will compete with perchlorate for resin sites and influence over all product water quality. For example, ion exchange resins also

[⊗]This work was presented at the 218th national meeting of the American Chemical Society as part of the Environmental Division symposium Perchlorate in the Environment, held August 22–24, 1999, in New Orleans, Louisiana.

*Phone: 949-487-7767. Fax: 949-487-0958. Electronic mail: ionexchange@earthlink.net.

remove bicarbonate ion which influences pH of the treated water and the treatment necessary to follow the Lead and Copper Rule. This paper develops data not only for perchlorate but for other competing ions present in groundwater as well.

**Perchlorate anomalies.** Anomalous selectivity behavior of perchlorate has been observed. Gregor[3] reported that the selectivity coefficient approximately tripled as the amount of perchlorate on the resin increased. This behavior prompted speculation that the formation of *ion pairs* within resins accounted for the increased and changing affinity. No definition was given. Diamond[4] has also reported the existence of perchlorate ion pairs in concentrated aqueous solutions. Again, no structural definition was offered.

### Objectives

Using the above background information, two separate objectives of this study become necessary in order to make engineering projections of treatment costs.

**Theoretical objective:** The first objective is to develop a model from which selectivity coefficients can be predicted from resin and ion molecular structures.

**Treatment objective:** The second objective is to determine how the predicted values of selectivity coefficients will influence bottom line treatment costs and waste quantities and product water quality.

### Approach

In this paper, computer models of various ion pairs are constructed from which their bond energies are calculated using the PM3 semi-empirical method. Linear correlations are developed which allow estimation of selectivity coefficients for perchlorate and other ions. Useful and reasonable assumptions regarding limits and trends in selectivity for various ions and resin types are made. The practical significance of these predictions is evaluated using ion exchange process simulation software that accounts for variability in water composition and resin characteristics.

### Summary of Significant Results

**Theoretical study.** The study of ion pair formation using PM3 computational modeling shows that strong attraction of perchlorate for resins can be attributed to stable forms of diperchlorate ion pairs bonded to the functional sites within the resin. A method was developed to estimate selectivity coefficients for perchlorate and several other ions using an empirically observed linear correlation between the quantum mechanical energy for a model ion exchange reaction and the log of the selectivity coefficient. A perchlorate selectivity coefficient of 26,000 was estimated for an SR6 type resin (a hydrophobic resin); whereas, a value of 5 was estimated for a hydrophilic resin.

**Treatment simulation study.** The selectivity coefficients obtained for four resins differing in hydrophobic character in the theoretical study were used in the treatment simulation of water from the San Gabriel Ground Water Basin. The lowest treatment costs, lowest waste production, and least impact on pH change was found for the hydrophobic resins. Use of the SR6 as a throwaway resin, requiring no on-site regeneration or wastewater production was the most attractive process. The chemical cost of such a process is estimated as $60 for each million gallons of water produced.

## PART I. USE OF COMPUTATIONAL CHEMISTRY

To achieve the theoretical objective expressed above, a review of the various kinds of theoretical models was made to find one that appeared to be adaptable to the ion exchange system. Computational chemistry is a rapidly developing field. This development is at least partly due to the explosive development in computer speed and availability over the last few years, [6] as well as continued improvement in software design. It is the objective of all models to compute various molecular properties for which the model is most suitably constructed. There are three general types of computational models available to and employed by the bench chemists today that do not require in depth mastery of quantum mechanical mathematics or theory.[7] Chemical knowledge of the system under study is the primary pre-requisite. The software for each model can compute the properties of a starting structure or make iterative calculations to determine a structure having a minimum energy, but cannot supply the experience of the bench chemist. The three general types are as follows:

**Molecular mechanical models.** The oldest and simplest is the **mechanical model (MM)**. These treat a molecule as a system of weighted balls and springs to account for atomic masses, atomic diameters, and bond energies. The MM model is used to produce a starting structure for use by other computational models.

***Ab initio* models**. The most sophisticated models are referred to as *ab initio*; i.e., they make their computations using fundamental parameters. The only inputs required are (1) a good starting structure, (2) the velocity of light, (3) Planck's constant, (4) the quantum numbers of each atom. The mathematics is based on wave mechanics to obtain the properties of atomic and molecular orbitals at various points throughout the molecule as well as a total energy value. The major disadvantage to the bench chemist is the time required to make the computations. This limits the method to studies of simple molecules of approximately ten atoms (other than hydrogen).

**Semi-empirical models.** The semi-empirical models solve the quantum mechanical wave equations to obtain electronic energies using simplified mathematics by injecting empirical parameters for each element into the procedure. Clark,[8] has reviewed the AM1 model (Austin model 1) and the PM3 model that is based on AM1 but contains empirical parameters for more of the elements related to contaminants in drinking water. Because of the greater speed and the ability to produce useful molecular energy data, the PM3 and AM1 are the methods used in the ion exchange studies reported here.

**Background and procedures.** Using the above computational models over the past several years, we have studied molecular models of ions interacting with the functional groups of ion exchange resins. The use of the PM3 method required that all computations be made on gas phase systems. This restriction is suitable for the resin phase species but ignores any ion hydration effects in the aqueous phase.

Our model of the resin-ion pair does not include the supporting polymer structure for the following reason: As shown in structure **1**, the sum of the number of atoms in the inorganic ion X, plus the number of atoms in the functional part of the molecule, plus the number of carbon atoms in the polymer is large, and computation time was found to exceed 1 h. Consequently, the supporting polymer was deleted from the model and a hydrogen atom was added to the methylene group connecting the nitrogen of the functional group to the polymer chain to terminate the model. The resulting ion pair model **2** became a tetramethylammonium ion paired with the inorganic ion. This is the **ion pair model** structure referred to in the remainder of this paper.

$[-CH_2\text{-}(\text{-}CH)\text{-}C_6H_4\text{-}CH_2\text{-}N(CH_3)_3 \text{ --- } X\ ]_{n \text{ repeating units}}$ $CH_3\text{-}N(CH_3)_3 \text{ --- } X$

**1**. (ResinX) **2**. Ion Pair Model

The procedure used here to find minimum energy of the ion pair model structure involved the following steps:

1. Select small molecules to represent the functional group of resins. Quaternary ammonium ions were selected as described above.
2. Construct the first guess structure of the ion pair model.
3. Use a molecular mechanics model to obtain a starting structure.
4. Submit the model to PM3 minimization computations.
5. Change the computed structure to a reasonably more stable structure, if any, and repeat steps 4 and 5 until reasonable structural possibilities are exhausted. Designate the computed structure having the lowest energy as the ion pair model structure.

**Model reactions**. Following the above steps for commonly occurring and representative inorganic ions and the tetra methyl ammonium ion, the chemical reactions involving these species had to be defined and related. If X has a single charge, the ion exchange reaction that defines the *selectivity coefficient* (a measure of affinity between ion and resin) is represented by eq 1 (for singly charged ions) and eq 2 (for doubly charged ions. Ionic species are aqueous phase and resin species are resin phase.

$$X^{1-} + (\text{Resin})Cl \rightleftharpoons (\text{Resin})X + Cl^- + \text{Energy of Reaction} \quad (1)$$

$$\tfrac{1}{2}X^{2-} + (\text{Resin})Cl \rightleftharpoons (\text{Resin})X + Cl^- + \text{Energy of Reaction} \quad (2)$$

where (Resin) represents the resin phase species including polymer and functional group.

From the Gibbs relationship it may be written that:

$$E \propto -\log K_X \quad (3)$$

The PM3 method did not allow direct calculations of hydrated ions, consequently, gas phase reactions were adopted as **Model Ion Exchange Reactions**, equation 4 and 5:

$$HX + RCl \rightleftharpoons RX + HCl + \text{Energy of Reaction} \quad (4)$$

$$\tfrac{1}{2}H_2X + RCl \rightleftharpoons \tfrac{1}{2}R_2X + HCl + \text{Energy of Reaction} \quad (5)$$

The reaction energies are designated as $E_M$ for these reactions and were calculated from the gas phase PM3 computed minimum energies of the reactants and products of eq 4 (for singly charged ions) or eq 5 (for doubly charged ions). $E_M$ was calculated for several anions, X as discussed below. In addition to the ion pair models, RX and $R_2X$, the energies for dimerized forms $(RX)_2$ were computed as well when the dimerization energy was significantly greater than that of the monomer RX.

## RESULTS AND DISCUSSION OF THEORETICAL STUDY

### Data for Ion Pairs and Model Reactions

**Table 1.** The data computed (PM3) for various ions and their ion pairs is listed in Table 1 and is compared to the published selectivity coefficients for type 1 resin.[9] In Table 1, the value of *E*, RX is the bonding energy for the ion pair. The calculated value of $E_M$ is the difference between the bonding energy of the products less the bonding energy for the reactants for the model ion exchange reaction (eqs 4 and 5). Fourteen different ions are listed including regulated contaminants, perchlorate, nitrate, arsenate, fluoride, and common ground water and other familiar ions. Different dimerized forms of the perchlorate and nitrate ion pairs were shown to be stable in the gas phase and are also listed. The lower and upper limits to the range of values of the selectivity coefficients for perchlorate as reported by Gregor are listed in line numbers 12 and 14. The value of 400 in line 13 is an extrapolated value of a selectivity coefficient based on the computed value of $E_M$ for the most stable dimer of the perchlorate ion pair model as is described in the following sections. This high value represents an upper limit to the selectivity coefficient for perchlorate and a type 1 resin.

**Structural data.** Another set of computed data is represented in Figure 1. It represents an ion exchange site that consists of dimerized perchlorate ion paired with two tetramethyl-ammonium ions (no. 14 in Table 1). This is a model structure for the ion pairs of a perchlorate loaded type 1 strong base resin and shows the structure (three molecules in lower left of Figure 1) and charge distribution that results from the energy calculations by the PM3 method. The O–O bonds between the two perchlorate ions are covalent type bonds. The perchlorate dimer portion was found to exist in two stable forms each in a six membered ring configuration: a *chair* conformation and a stabler *twisted chair* conformer. Other forms may also be possible, but these were not investigated.

**Table 1.** Energy data for various ion pairs and the model reaction

| No. | Ion Pair, RX | *E*, RX | *E*, HX | $E_M$ | $K_X$ | log $K_X$ |
|---|---|---|---|---|---|---|
| 1 | $(CH_3)_4N{\cdot}F$ | –1458.37 | –133.76 | 35.25 | 0.09 | –1.05 |
| 2 | $0.5{\cdot}\{[(CH_3)_4N]_2{\cdot}HAsO_4\}$ | –1650.74 | –331.825 | 26.685 | 0.14 | –0.85 |
| 3 | $(CH_3)_4N{\cdot}H_2PO_4$ | –2080 | –740.44 | 20.3 | 0.2 | –0.70 |
| 4 | $(CH_3)_4N{\cdot}CH_3CO_2$ | –2119.21 | –771.38 | 12.03 | 0.25 | –0.60 |
| 5 | $0.5{\cdot}\{[(CH_3)_4N]_2{\cdot}SO_4\}$ | –1643.46 | –298.515 | 14.915 | 0.18 | –0.74 |
| 6 | $(CH_3)_4N{\cdot}HCO_3$ | –1955.7 | –593.79 | –2.05 | 0.3 | –0.52 |
| 7 | $(CH_3)_4N{\cdot}Cl$ | –1461.43 | –101.57 | 0 | 1 | 0.00 |
| 8 | $(CH_3)_4N{\cdot}Br$ | –1450.48 | –73.54 | –17.08 | 2 | 0.30 |
| 9 | $(CH_3)_4N{\cdot}NO_3$ | –1748.45 | –381.86 | –6.73 | 3 | 0.48 |
| 10 | $2{\cdot}[(CH_3)_4N{\cdot}NO_3]$ | –1755.66 | –381.86 | –13.94 | 3 | 0.48 |
| 11 | $(CH_3)_4N{\cdot}I$ | –1442.59 | –48.83 | –33.9 | 7 | 0.85 |
| 12 | $(CH_3)_4N{\cdot}ClO_4$ | –1740.91 | –345.88 | –35.17 | 10 | 1.00 |
| 13 | $[(CH_3)_4N{\cdot}(ClO_4)]_2$ | –1792.7 | –345.88 | –86.96 | 400 | 2.60 |
| 14 | $[(CH_3)_4N]_2{\cdot}(ClO_4ClO_4)$ | –1762.17 | –345.88 | –56.43 | 50 | 1.70 |

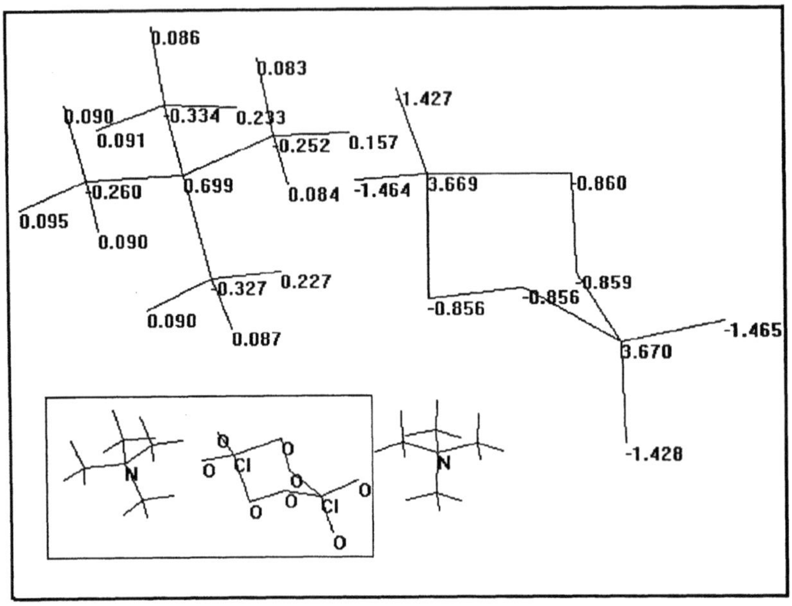


**Figure 1.** Structure of diperchlorate ion paired with two tetramethylammonium ions is shown by the three molecules in lower left center.

Large area in Figure 1, shows PM3 computed charge distributions for molecular portion shown in framed insert. Diperchlorate ion is shown in its twisted boat conformation.

**Linear energy correlation.** By inspection of the data of Table 1, it is apparent that log $K_X$ and $E_M$ are linearly related. For the purpose of this study, the correlation is represented by equation 6. The correlation is shown in Figure 2. The correlation coefficient (excluding the extrapolated value of 400 for perchlorate) between the data is 0.98. A linear equation can be written as:

$$\log K_X = E_M \times R_C \qquad (6)$$

Where log $K_X$ is the selectivity coefficient (as defined by eq 1 or 2) for the ion X and type 1 resin, $E_M$ is the ion pair model energy assigned to the ion, and $R_C$ is the slope of the linear relationship. This linear relationship infers parallel reaction chemistry between the ion exchange reaction and the model reaction. Although some speculative rationale can be stated for the observed linearity, the author asks that it be accepted for its purely empirical nature at this time and the study proceed to use the findings to reach the second objective of this study.

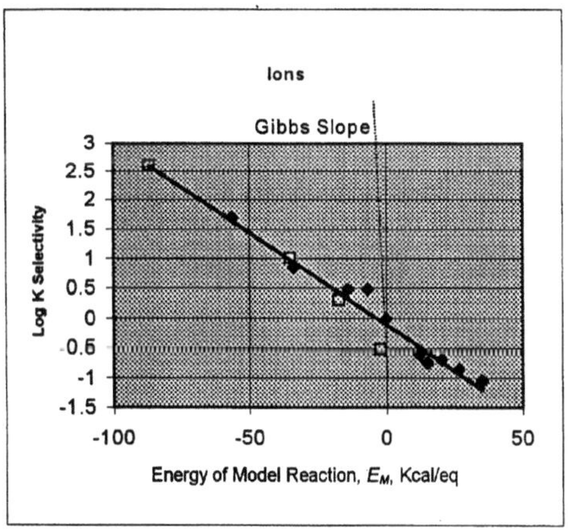


**Figure 2.** Correlation between model reaction energy $E_M$ and selectivity coefficient for type 1 SDVB resin and various ions.

Equation 6 can be used to estimate selectivity coefficients for differing ions by building a computer model and determining the $E_M$ parameter associated with the ion. The slope value for type 1 SDVB resin is 0.031. The value of the selectivity coefficient for any of the ions can be easily found. Diperchlorate is an example. The highest point in Figure 2 was calculated to obtain the selectivity coefficient on line 14 Table 1.

The observed linearity gave rise to speculation regarding the generality of the relationship and the significance of the parameters that comprise it. For example, the following questions come up: Is the relationship specific for type 1 resins? Does the value of $E_M$ for an ion hold if the functional group is changed? What is the significance of the slope value of $R_C$?

**Application to other kinds of resins.** The above values for $E_M$ were computed for different inorganic ions using the model tetramethylammonium group. It was easily determined by making computations with other starting structures that multi-carbon alkyl groups surrounding the nitrogen atom had little or no effect on the $E_M$ value for any of the ions. **As far as this study goes, $E_M$ values are specific for an ion paired with quaternary ammonium ions, this infers linearity between $E_M$ and log $K_X$ for all tetraalkyl quaternary SDVB resins.** The difference then between type 1 resin and a nitrate selective resin (if the linear relation holds for nitrate selective resins) is the slope value $R_C$.

Although there are very few selectivity coefficient data available on nitrate selective resins which have quaternary ammonium functionality, the author and associates have collected some data on these resins which indicates the linear trends will hold for these resins as well. Nitrate selective resins are characterized by having higher than usual nitrate to chloride selectivity and lower than normal affinity for sulfate ions. In Figure 2, the points representing nitrate (no. 10 in Table 1) would be shifted vertically upward and the point representing sulfate (no. 5 in Figure 1) would be shifted vertically downward. The net effect is that the line of Figure 2 would shift *clockwise* around the origin (0,0 for X = $Cl^-$). Nitrate selective resins are less hydrophilic than type 1 resins. The clockwise movement of the line of Figure 2 is consistent with this property, i.e., as the resin becomes more hydrophobic, its attraction for hydrophobic ions increases and for hydrophilic ions, decreases. The slope value $R_C$ for a resin more hydrophobic than type 1 SDVB such as a tributylbenzylammonium SDVB resin (Sybron SR6) is estimated to be about 0.080.

Resins of hydrophilic nature would likewise be represented by a linear relationship with the correlation line rotated *counterclockwise*; i.e., the bonding between the hydrophobic ions would decrease while the bond to hydrophilic ions would increase. We assign an arbitrary slope of 0.015 to a resin more hydrophilic than a type 1 SDVB resin for our comparison studies. We also assign an arbitrary slope of 0.070 to a resin less hydrophobic than SR6 resin and more hydrophobic than type 1 SDVB resin for our comparison studies.

**Estimation of selectivity coefficients**. The selectivity coefficients for the resins discussed above were calculated using eqs 7–10 and the ionic $E_M$ values (Table 1) and the resin $R_C$ values assigned above.

For the type 1 SDVB resin: $\log K_X = E_M \times 0.031$ (7)

For the SR6 resin: $\log K_X = E_M \times 0.080$ (8)

For the hydrophobic resin: $\log K_X = E_M \times 0.070$ (9)

For the hydrophilic resin: $\log K_X = E_M \times 0.015$ (10)

The results are shown in Table 2.

**Table 2**. Calculated values of selectivity coefficients

| | Observed type 1 SDVB | Calculated values for SDVB resins | | | |
|---|---|---|---|---|---|
| | | Type 1 SDVB | SR6 type | Hydro-phobic | Hydro-philic |
| $SO_4^{2-}$ | 0.20 | 0.08 | 0.004 | 0.005 | 0.23 |
| $HCO_3^-$ | 0.30 | 0.55 | 0.30 | 0.34 | 0.66 |
| $Cl^-$ | 1.0 | 1.0 | 1 | 1 | 1.0 |
| $NO_3^-$ | 3.0 | 2.1 | 10 | 7.5 | 1.3 |
| $ClO_4^-$ | 50 | 45 | 26000 | 7081 | 5.6 |

The ions listed in Table 2 are frequently encountered in groundwater. *It is especially important to include bicarbonate because of the effect of pH changes that may occur and impact the Lead and Copper Rule.* The selectivity coefficients for the sulfate ion were obtained by squaring the value obtained from eqs 7–10 to obtain the selectivity coefficient for the following reaction:

$$X^{2-} + 2\text{ResinCl} \rightleftharpoons \text{Resin}_2X + 2Cl^- \quad (11)$$

This form of the selectivity coefficient for doubly charged ions is used in the treatment simulation software discussed below in this paper.

It can be noted that the trends in the selectivity coefficients as shown in the last three columns is what might be expected when compared to the type 1 SDVB resin. For the hydrophobic ions, nitrate and perchlorate, large values of the selectivity coefficients appear for the hydrophobic (SR6) resin and are less for the hydrophilic resin. Conversely for the hydrophilic ions, sulfate and bicarbonate, the values are larger for the hydrophilic resin and smaller for the hydrophobic resin. It can also be noted that for the hydrophilic resin, the selectivity coefficients are all small, indicating that all ions will compete for the ion exchange sites. Whereas for the hydrophobic resins, the perchlorate greatly outweighs the other ions, consequently, there will less competition from the other ions.

With the above-demonstrated ability to compute selectivity coefficients, not only for perchlorate but other ions as well, the first objective of this study was reached. The next section evaluates the effect of the values of the selectivity coefficients determined above on the treatment costs and waste production.

## CONCLUSIONS OF THEORETICAL STUDIES

1. PM3 derived energies for ion exchange model reactions for quaternary ammonium SDVB resins were found to have a linear energy correlation with known selectivity coefficients. From these correlations, empirical formulae are developed to compute selectivity coefficients for various ions and the types of resins stated.

2. Selectivity coefficients for perchlorate are estimated to range as high as 400 for a type 1 SDVB resin. Nitrate selective resins may show even higher selectivity for perchlorate, the selectivity coefficient for SR6 resin was estimated to be 26000.

3. The high affinity of perchlorate for ion exchange resins can be attributed to strong bond formation between perchlorate ions within the ion exchange sites to form dimerized forms of perchlorate ion bonded to the quaternary ammonium functional group as is demonstrated by computational model programs.

4. By definition of an ion pair model and a model ion exchange reaction, an ionic parameter $E_M$ is derived using quantum mechanical computations. A second parameter $R_C$ relates the $E_M$ value to the selectivity coefficient for the ion and is a property of the resin.

5. Preliminary studies indicate that $E_M$ appears to have a specific and constant value for a given ion when the resin is a quaternary ammonium SDVB resin.

6. Preliminary studies indicate that $R_C$ appears to be higher if the resin is hydrophobic and lower if hydrophilic for the quaternary ammonium SDVB resins.

## PART 2. TREATMENT SIMULATION MODELING

### Description of Treatment Simulation Models

**Background**. The testing of ion exchange processes to determine best resin and operating conditions has been done in the past solely through laboratory or pilot scale studies. This method is quite time consuming as a laboratory column operation can consume as much as 100–500 person-hours of labor for lab work and as much as $10,000–20,000 per month of operation for a pilot study using portable equipment. However, the currently widespread availability of computer power has stimulated use of simulations of the ion exchange process and replace the expensive manned experimental field work. The chemical equations and related mathematics have been known for a long time and have been frequently used to make limited but reliable predictions of performance.[10] The simulation of column performance descriptive of full-scale plant operations, however, requires repetitive and tedious chemical equilibrium calculations of complex equations and is done conveniently with modern personal computers.

**IX WINDOWS™ PRO 300**. A detailed description of the use of such computer based simulations and the algorithm is given in a study comparing the performances of nitrate selective resins and optimizes the process design and operation for full scale nitrate plants.[11] The simulation software used here is commercially available as IX WINDOWS™ PRO 300 version which operates under an MS Windows™ operating system on IBM compatible PCs.[12] This program uses both chemical equilibrium and diffusion kinetics models to simulate a variety of column designs and experimental or operating conditions. A free sample program CD is available.[12]

The required **raw data inputs** are as follows: untreated water composition (≤ 5 major ions), selectivity coefficients for each ion, starting resin composition and total ionic capacity, regenerant strength and composition. Optional operating parameters can be chosen for fixed or moving bed design (carousel), co-current or counter current regeneration, resin declassification, flow rates, Harries mass transfer computation, bead diameter, column dimensions, number of repeated runs, and various program control parameters.

Numerous output data from a single or multiple run are obtained. For this study the underlined outputs are the most useful: Regenerant **quantity and cost**, treated water **composition**, breakthrough curves, **wastewater quantity and composition**, regeneration rinse curves, final resin composition at various bed depths, data snapshots at various run times, plant design, full text report and animated graphics.

**Treatment Simulation Computations**

**Resins other than SR6**. Except for the SR6 resin which has a relatively low capacity and showed very high selectivity for perchlorate, treatment simulation runs were made for identical processes (fixed bed, countercurrent regeneration) using the other three resins listed in Table 2 to compare the performances of each one to the other.

**Simulation of SR6 performance.** The selectivity coefficient for SR6 was so high that repeated cycles of loading and regeneration were not possible to simulate. Over 100,000 bed volumes could be treated in a single loading. This, however, with repeated regenerations, provided enough information to give the performance data needed. However, the SR6 usage as a "throw away" or "tow away" resin is the most likely usage. This process is described below as an alternate process.

## RESULTS OF SIMULATION COMPUTATIONS

**Comparative Performance Projected for Three Resins**

The results of the simulation runs for the three resins (other than SR6) are given in Table 3. They clearly indicate that the SDVB resin that is more hydrophobic than the type 1 resin is projected to have a superior performance to the other two resins.

**Table 3**. Comparison of simulated performances for SDVB resins

| Parameter | Type 1 | Hydrophobic | Hydrophilic |
|---|---|---|---|
| $ClO_4^-$ Leakage, ppb | 0.9 | 2.5 | 1.2 |
| Bed Run, BV | 1000 | 33,000 | 400 |
| NaCl, lbs/ft$^3$ | 98.5 | 1167 | 36.5 |
| NaCl Cost, $/MG | 329 | 118 | 304 |
| Waste Water, % | 0.80 | 0.13 | 1.25 |
| $HCO_3^-$ Reduction,% | 7 | 0.2 | 32 |

This resin would be similar to a triethylbenzyl SDVB resin or one that is mildly nitrate selective. The projected performance is more attractive than the other two because of the low wastewater production, the lowest salt cost and the small impact on bicarbonate (Lead and Copper Rule). The effect of the low perchlorate selectivity coefficient to make the hydrophilic resin more easily regenerable is offset by the strong competition of the other ions for the exchange sites. The net result is a shorter bed run (400 BV) and nearly much more wastewater production.

**Projected Performance of SR6 Resin**

The performance of the SR6 resin was projected on the basis that it would be used in a *throw away operation* because of the large amount of water treatable by a resin column. The throw away operation consists of disposing the resin directly to an acceptable disposal site. The perchlorate is so tightly bound that leaching should not be a problem. The spent resin could also be returned either to the manufacturer or other resin service to chemically remove the perchlorate in an industrial operation rather than within the operation of public water system. The operating and performance characteristics of the SR6 resin are given in Table 4.

**Table 4.** Projected performance of SR6 perchlorate removal

| Operational parameter | Projected value |
|---|---|
| Resin volume required for plant at $10^6$ gal day$^{-1}$, ft$^3$ | 200 |
| Maximum amount treated per cycle, bed volumes | 650,000 |
| Million gallons treated per cycle *(days of operation per cycle)* | 972 *(972)* |
| Cost of resin, \$ ft$^{-3}$ *(per plant)* | 300 *(60,000)* |
| Cost of treatment/million gallons, \$ | 62 |
| Solid waste resin from plant, ft$^3$day$^{-1}$ | 0.21 |
| Bicarbonate removal, g dL$^{-1}$ (%) | Near 0 |
| Waste water produced, % of treated | 0 |

The values in Table 4 suggest an extremely attractive mode of operation giving very low operating costs and virtually no environmental discharges. The plant could employ two beds of resin in series, using fresh resin in the downstream bed. This is a desirable configuration to prevent perchlorate breakthrough.

## CONCLUSIONS OF PROCESS SIMULATIONS

1. The full-scale performances of perchlorate ion exchange plants were simulated using IX WINDOWS™ PRO 300 software. Four different resins were evaluated and compared. The selectivity coefficients for the major ions in the San Gabriel Ground Water Basin for each resin was estimated by a method described in Part I of this paper.

2. The resins evaluated all had a SDVB supporting structure and a quaternary ammonium functionality. The resins differed by a theoretical parameter $R_C$ (derived in Part I). $R_C$ is proposed to be a measure of hydrophobic character of the resin.

3. The resins which were compared are a type 1 SDVB resin and one hypothetical resin more hydrophobic and one hypothetical resin less hydrophobic. An SR6 resin was also compared.

4. The best resin for treating perchlorate was the most hydrophobic of the three resins whose operation was simulated using a conventional ion exchange process with counter current regeneration. The least amount of salt was used and the least amount of waste was produced. The least hydrophobic resin allowed sites to be filled with

other competing ions and therefore operated at a reduced efficiency. The strongly hydrophobic resin was able to adsorb the most perchlorate per cycle and could be regenerated sufficiently to allow resin cycling.

5. The most hydrophobic resins also had the least impact on the bicarbonate removal. This allows a more stable pH. This is an important consideration regarding the Lead and Copper Rule.

6. SR6 resin performance was also simulated and gave the best projected results when its use as a throw away resin was simulated. Up to 650,000 bed volumes are projected per cycle. In this case the resin is not regenerated on site but is wasted or returned to an industrial service to strip the perchlorate. The projected cost of using the resin in this way is $62 per million gallons of treated water with no salt cost or wastewater disposal necessary.

7. The formation of dimerized perchlorate ion bonded to the quaternary functional groups of resins explains perchlorate anomalies of varying selectivity coefficient with resin loading and the very large selectivity coefficients for hydrophobic type resins. The existence of dimerized perchlorate ion is suggested in the literature, however, no experimental proof that dimers exist was found in literature searches. Studies on electrospray mass spectrometry of perchlorate have, however, detected existence of perchlorate ion adducts with organic molecules and quaternary ammonium groups.[13]

## ACKNOWLEDGMENTS

The author acknowledges the assistance and inspiration of the late Professor Robert Taft of University of California–Irvine Chemistry Department and presents this paper as a personal dedication to him. Robert Taft was enthusiastic about applying computational chemistry methods to studies relating molecular structure of resins and ions to chemical reactivity. He was engaged in these studies until his untimely death in early 1996. Assistance from Professor Warren Hehre on the use of SPARTAN™ computational programs is also acknowledged.

## REFERENCES

1. Venkatesh, K.R. "Removal and destruction of perchlorate and other anions from ground water using *ISEP+* system." Presented at the Perchlorate Conference Sponsored by East Valley Water District, March 19, 1999, Pomona, CA; Paper No. 21.

2. McCracken, C. "Stakeholder involvement on perchlorate issues", presented at the Perchlorate Conference Sponsored by East Valley Water District, March 19, 1999, Pomona, CA; Paper no. 22.

3. Gregor, H., Belle, J., Marcus, R. "Studies on ion-exchange resins. XIII. Selectivity coefficients of quaternary base anion-exchange resins toward univalent anions." *Journal of the American Chemical Society* **1955**, *77*, 2713–2719.

4. Diamond, R. and Whitney, D.C. "Resin selectivity in dilute to concentrated aqueous solutions." In *Ion Exchange: A Series of Advances,* J.A. Marinsky, Ed. Dekker: NewYork, 1966; Vol. 1. See also *Journal of Physical Chemistry* **1963**, 67, 2513–2514.

5. Reichenberg, D. "Ion-exchange selectivity." In *Ion Exchange: A Series of Advances,* J.A. Marinsky, Ed. Dekker: New York, 1966; Vol. 1.

6. Jurs, P.C. *Computer Software Applications in Chemistry*, 2nd ed. Wiley: New York, 1996; Ch. 1.

7. Hehre, W.J.; Yu, J.; Klunzinger, P.E. *A Brief Guide to Molecular Mechanics and Quantum Chemical Calculations.* Wavefunction, Inc.: Irvine, CA, 1998, passim.

8. Clark, T. *A Handbook of Computational Chemistry.* Wiley: New York, 1986. Ch. 4.

9. Wheaton, R.M; Bauman, W.C. "Properties of strongly basic anion exchange resins." *Industrial and Engineering Chemistry* **1951**, *43*, 1088–1093.

10. Anderson, R.E. "Estimation of ion exchange process limits by selectivity calculations." In *Adsorption and Ion Exchange*, I.M. Abrams, I. Zwiebel, N.H. Sweed, Eds. AIChE Symposium Series: New York, 1975, Vol. 152, 236–242.

11. Guter, G.A. "Nitrate removal from contaminated ground water by anion exchange."In *Ion Exchange Technology Advances in Pollution Control.* A.K. Sengupta, Ed. Technomic: Lancaster, PA, 1995; Ch. 2.

12. Cathedral Peak Software. 215 Monte Vista, San Clemente, CA 92672. Fax: 949-487-0958.

13. Tsui, D.T. "Electrospray mass spectrometric determination of perchlorate in drinking water." Presented at the Perchlorate Conference sponsored by East Valley Water District, March 19, 1999, Pomona, CA; Paper no. 6.

12. Urbansky, E.T.; Magnuson, M.L. "Sensitivity and selectivity enhancement in perchlorate anion quantitation using complexation–electrospray ionization–mass spectrometry." THIS VOLUME: *Perchlorate in the Environment*, E.T. Urbansky, Ed. Kluwer/ Plenum: New York, NY, 2000; Ch. 8.

# Chapter 12

# THE TREATABILITY OF PERCHLORATE IN GROUNDWATER USING ION-EXCHANGE TECHNOLOGY[⊗]

Anthony R. Tripp and Dennis A. Clifford*

Civil and Environmental Engineering Department, University of Houston,
4800 Calhoun Road, Houston, Texas 77204

## INTRODUCTION

The study of the interaction of ion-exchange resins with the perchlorate ion is not a recent development. There are numerous reports in the ion-exchange literature of the 1950s related to the behavior of various anions and the new synthetic ion-exchange resins of that time period.[1–6] The perchlorate ion was among many different anions tested for the construction of affinity sequences in an attempt to determine the basis for selectivity. From these tests is was shown that the affinity for the perchlorate ion by these early resins was very high. The nonpolar, hydrophobic matrix of these resins provided an environment that the nonpolar, hydrophobic perchlorate ion found much preferable to that of the aquatic environment.

## EXPERIMENTAL

### Resin Characteristics

Binary isotherms were constructed for seventeen commercially available strong-base anion exchange resins with variable compositions. These resin variables were polymer matrix,

[⊗]This work was presented at the 218th national meeting of the American Chemical Society as part of the Environmental Division symposium Perchlorate in the Environment, held August 22–24, 1999, in New Orleans, Louisiana.

*Author to whom correspondence should be directed. Phone: 713-743-4266. Fax: 713-743-4260. Electronic mail: DAClifford@uh.edu.

functional group, percent cross-linking, and porosity. Three polymer matrices tested were: polystyrene, polyacrylic, and polyvinylpyridine. See Figures 1 through 3.

Figure 1 is a standard type I strong-base anion-exchange resin with a quaternary ammonium trimethylamine functional group attached to the aromatic ring of styrene or divinyl-benzene, the cross-linking agent. This resin is in the chloride form, i.e., the quaternary amine cation charge is balanced by a chloride anion, which is exchanged for other anions during the process of ion exchange. The polystyrene divinylbenzene matrix is considered to be nonpolar compared with matrices containing oxygen molecules.

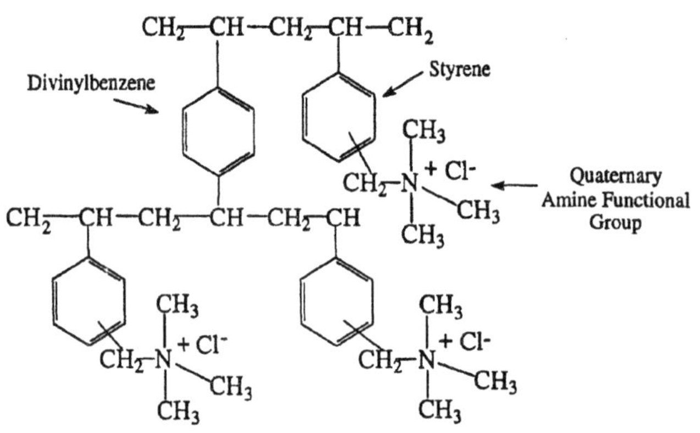


**Figure 1.** Diagram of a polystyrene type I trimethyl quaternary amine, strong base anion-exchange resin in the chloride form.

Figure 2 describes a typical polyacrylic anion-exchange resin. Here the acrylic monomer has replaced the styrene monomer in the polymer chain resulting in an ion-exchange resin that is relatively hydrophilic. As with the polystyrene resin shown in Figure 1, the functional group in Figure 2 is a trimethyl quaternary amine and the cross-linking is accomplished with divinylbenzene.

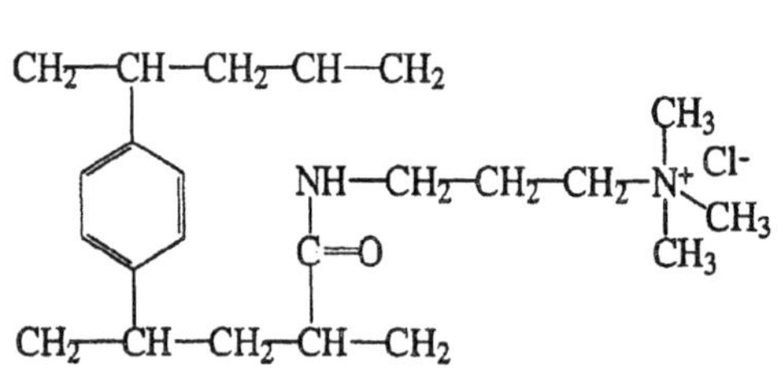

**Figure 2.** Diagram of a polyacrylic, trimethyl quaternary amine, strong base anion-exchange resin in the chloride form.

Figure 3 is a representation of the polyvinylpyridine anion-exchange resin in which the styrene monomer is replaced with a vinylpyridine unit in the polymer chain. This once again results in a resin that is hydrophobic. The pyridine resins differ from the polystyrene and polyacrylic resins in that the quaternary methylated nitrogen is within the aromatic ring rather than pendant to the ring.

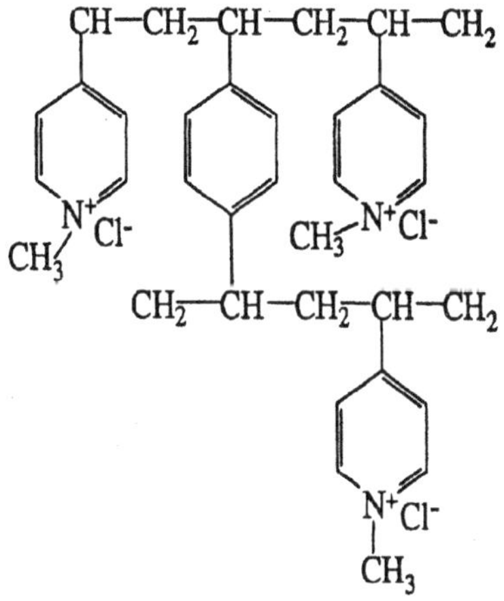


**Figure 3.** Diagram of a chloride-form polyvinyl-pyridine, strong-base anion-exchange resin with methylpyridine quaternary amine functional groups.

The different functional groups present on the resins studied are represented in Figure 4. The basic trimethyl quaternary amine functional group is the standard found on type I strong-base anion exchange resins. The type II dimethylethanolamine functional group is formed by the replacement of one of the methyl groups with an ethanol group resulting in a slightly hydrophilic resin with lower basic strength. Increasing the length of the alkyl group from methyl to ethyl to propyl results in resins that are relatively more hydrophobic and have increased charge-separation distances. As the alkyl length increases, the selec-tivity for divalent ions decreases.[7] For example, the type I and II resins have a greater affinity for sulfate (a divalent ion) relative to nitrate (a monovalent ion). In fact, triethyl and longer functional group resins prefer nitrate over sulfate and for this reason are referred to as nitrate-selective resins.

As mentioned earlier, the pyridine resin is different in that the quaternary nitrogen atom is contained within the pyridine molecule. As a result, the attachment of a single methyl group is all that is required to achieve a quaternary amine functionality. As a result of this, comparisons between the vinylpyridine resin and styrene resins can be difficult.

There are also differences in the microstructure of the resins. The amount of cross-linking between the polymer chains can be varied to produce resins of differing characteristics. The amount of cross-linking usually varies between 2% to 20% with 8% cross-linking being typical. Resin flexibility decreases as percentage cross-linking increases. Low percentage cross-linking leads to flexibility, which allows the resin to swell upon immersion in water to a greater degree than the resins with larger percentages.

The internal porosity of the resins can also be altered by the inclusion of nonreactive solvents during the polymerization and cross-linking process. The pore diameter for resins ranges from essentially zero for microporous gels to several hundred angstroms for macroporous resins.

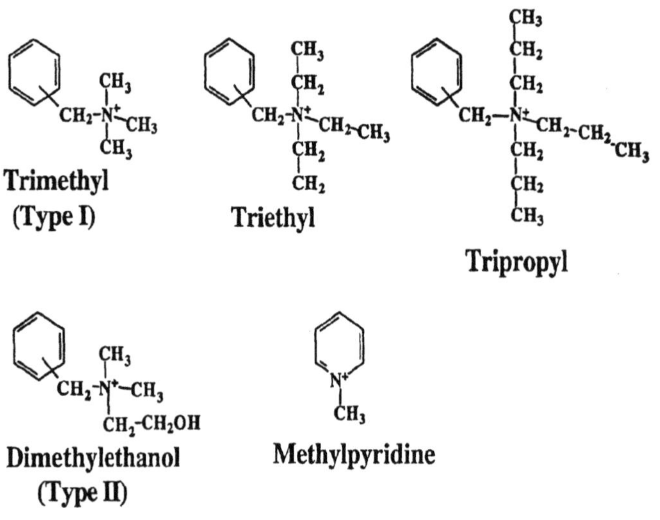


**Figure 4.** Quaternary amine functional groups studied in this research.

### Isotherm Procedure

During the equilibrium batch isotherm procedure, a mass of air-dried chloride-form resin (measured to the nearest 0.1 mg) was added to 100 mL of solution containing 5 μeq $L^{-1}$ perchlorate (500 μg $L^{-1}$) and 4.995 meq $L^{-1}$ chloride (177 mg $L^{-1}$). This mixture was shaken in a water bath maintained at ±1 °C of the target temperature for 24 h. At this time a sample was taken for the measurement of both perchlorate and chloride. This procedure was repeated for varying masses of resin resulting in different equilibrium concentrations. At least five different equilibrium concentrations spread over the range 5–50 μg $L^{-1}$ were used in the construction of isotherms in this study. The isotherm was constructed by plotting $y_{ClO_4^-}$ (equivalent fraction of $ClO_4^-$ on the resin) vs $x_{ClO_4^-}$ (equivalent fraction $ClO_4^-$ in solution) in the equivalent fraction range (0–0.014) of interest in this study.

The perchlorate and chloride ions were added as the sodium salt, the perchlorate being obtained from a NIST-traceable stock solution from GFS Chemicals (Columbus, OH). Perchlorate and chloride were analyzed by ion chromatography using the perchlorate procedure described in Application Note 121 from Dionex using AS5/AG5 analytical/guard columns,[8] and the chloride procedure described in EPA method 300.1.

An amount of air-dried chloride-form resin was also taken and dried at 105 °C for over 24 h to determine the amount of moisture present so that corrections could be made for the dry mass of resin added.

In addition to the 24-h isotherm tests, a seven-day equilibration comparison was performed to determine if equilibrium was achieved in 24 h. For this process three identical sets of samples were tumbled at room temperature for either 1, 3 or 7 days, and the respective isotherms were constructed to determine if the separation factors increased with increasing equilibration time.

## RESULTS

### Effect of Cross-Linking on Perchlorate Selectivity

The trace-concentration-range $ClO_4^-$–$Cl^-$ isotherms for three type I polystyrene resins are shown in Figure 5. These resins are identical except for the degree of cross-linking, with two different resins (Amberlite™ IRA-402 and IRA-404, Rohm & Haas, Philadelphia, PA) having 4% cross-linking and one (Amberlite IRA-400) with 8% cross-linking. In the figure, the line with the greatest slope has the larger separation factor ($\alpha$). Here it is the resin with the 8% cross-linking that has higher perchlorate affinity and the larger separation factor ($\alpha = 125$) compared with the two resins with 4% cross-linking ($\alpha = 100$). The lower perchlorate affinity for the lower cross-linked resin is attributed to its ability to swell and hydrate to a greater extent, creating a slightly less hydrophobic environment within the resin matrix.

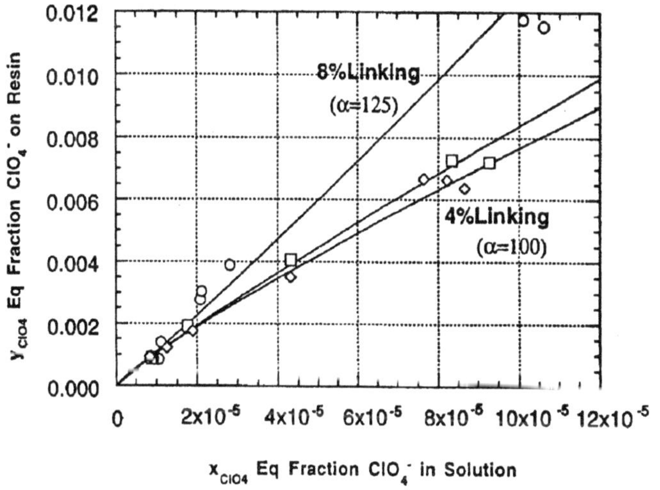


**Figure 5.** $ClO_4^-$–$Cl^-$ isotherms for resins that differ only in their percent cross-linking. As the cross-linking increases, the per-chlorate affinity (separation factor) increases due to decreased ability to swell upon hydration.

### Effect of Porosity on Perchlorate Selectivity

The $ClO_4^-$–$Cl^-$ isotherms for two polyacrylic resins (Amberlite IRA-458 and IRA-958) with differing porosity are shown in Figure 6. The increase in porosity for the macroporous resin appears to result in an increase in the separation factor relative to the microporous resin. To introduce porosity into a resin, the resin matrix is made stiffer by in-creasing the percent cross-linking. As shown previously, increasing the cross-linking also increases the separation

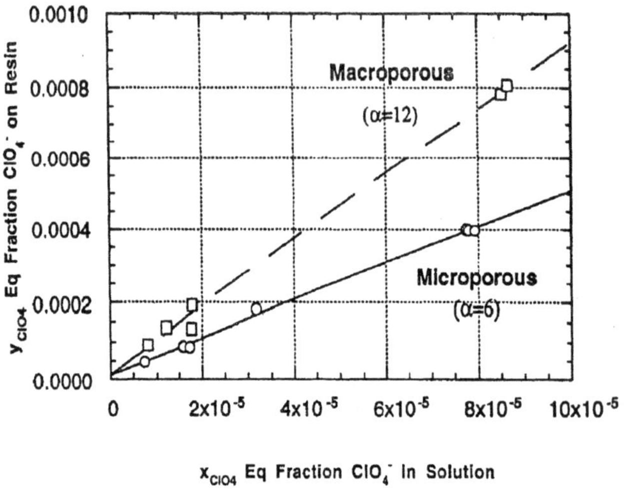


**Figure 6.** $ClO_4^-$–$Cl^-$ isotherms for macroporous versus microporous resins. Macroporous resins have a higher per-chlorate affinity because they are highly crosslinked and less able to swell upon hydration.

factor. Therefore, the increase in the separation factor seen for the macroporous resin may be due to increased cross-linking or to a combination of increased cross-linking and porosity.

**Effect of Matrix**

Figure 7 shows the $ClO_4^-$–$Cl^-$ isotherms for the three different resin matrices studied. The functional groups on the resins are similar. The polystyrene (Amberlite IRA-400) and polyacrylic (Amberlite IRA-458) resins have trimethyl quaternary amine functional groups and the polyvinylpyridine (Reillex™ HPQ, Reilly Industries, Indianapolis, IN) resin has methylpyridine quaternary amine func-tional groups. The hydrophilic, acrylic resin exhibited a much lower separation factor ($\alpha = 6$) than the hydrophobic styrene and vinylpyridine resins ($\alpha = 125$ and $\alpha = 275$, respectively). So it appears that, in general, the more hydrophobic the resin matrix, the greater the perchlorate separation factor. Although there was a dramatic difference in perchlorate affinity when comparing the acrylic and polystyrene or polyvinylpyridine resins, there also appears to be a significant difference between the polyvinylpyridine ($\alpha = 275$) and poly-styrene ($\alpha = 125$) resins. This latter difference is attributed to the relatively more hydro-phobic nature of the polyvinylpyridine resin compared with the polystyrene resin. This greater hydrophobicity was noted earlier when studying the removal of nitrate (also a hydrophobic ion) from water by these resins.[9]

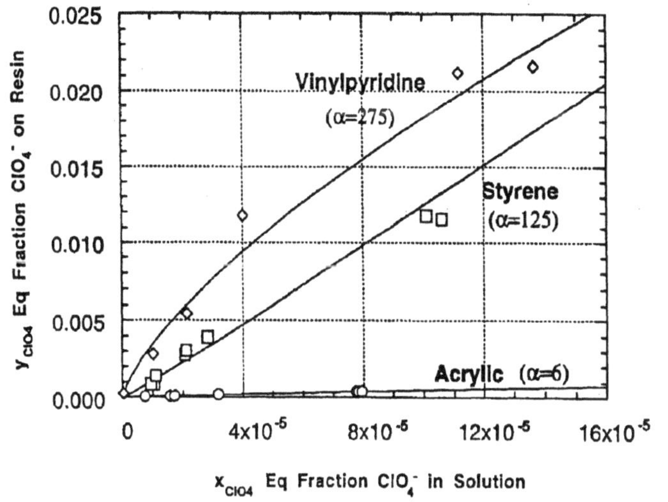


**Figure 7.** $ClO_4^-$–$Cl^-$ isotherms for the three resin matrices studied. The separation factors for the hydrophobic poly-styrene and polyvinylpyridine resins are much greater than the hydrophilic polyacrylic resins.

## Effect of Functional Group

All three of the resins in Figure 8 have the same matrix (polystyrene), but different functional groups attached to that matrix. As the length of the alkyl group on the quaternary amine increases from methyl (Amberlite IRA-400) to ethyl (Amberlite IRA-996) to propyl (Sybron™ SR-7, Sybron Chemicals, Birmingham, NJ), the perchlorate affinity increases from $\alpha = 125$ to $\alpha > 1100$ to $\alpha > 1500$ respectively. Thus, the selectivity for perchlorate is highly dependent upon the functional group. Note that for the longer chain (triethyl and tripropyl) alkyl groups, greater-than values (>1100, >1500) were given for the separation factors since these resins did not reach equilibrium within the 24 h period of the tests.

## Kinetics

Figure 9 is a plot of perchlorate affinity (separation factor) versus equilibration time for 1, 3, and 7 day tests. As can be seen, there is no effect of time upon the polystyrene trimethyl quaternary amine resin (Sybron ASB-1). In contrast, the polystyrene tripropyl quaternary amine resin (Sybron SR-7) has not reached equilibrium even after seven days of mixing.

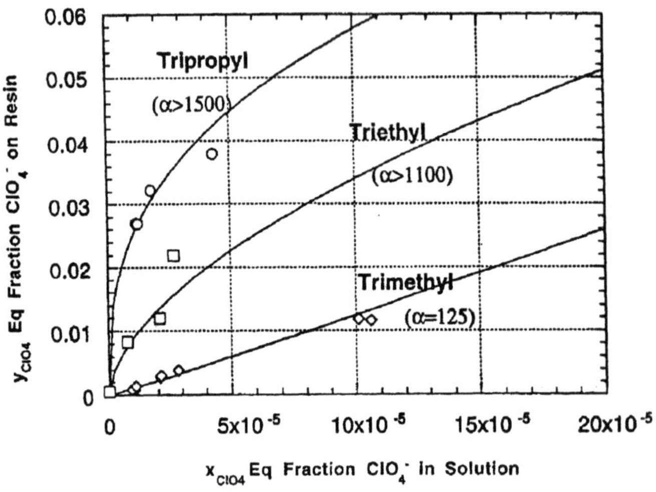


**Figure 8.** $ClO_4^-$–$Cl^-$ isotherms for polystyrene resins with different functional groups. Effect of functional groups (alkyl chain length) on perchlorate selectivity.

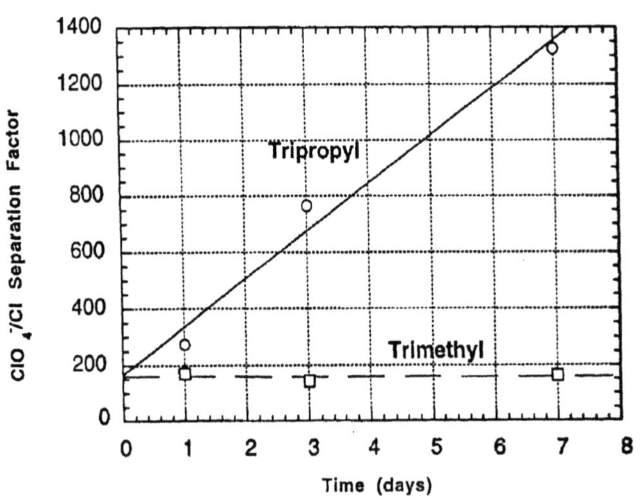


**Figure 9.** The effect of equilibration time upon the $ClO_4^-$–$Cl^-$ separation factor. After seven days, the perchlorate separation factor is still increasing for the tripropyl resin, which has not reached equilibrium.

Apparently, these resins with very high separation factors are so hydrophobic that the transport of water becomes the limiting step in achieving equilibrium. It is possible that resins with very high separation factors may not perform as well as resins with lower separation factors due to this transport problem.

**Effect of Temperature**

The $ClO_4^-$–$Cl^-$ isotherms for a polystyrene type I resin (Sybron ASB-1) at 20, 40 and 60 °C are shown in Figure 10. As the equilibration temperature was increased from 20 to 40 to 60 °C, the separation factors decreased from $\alpha = 170$ to $\alpha = 90$ to $\alpha = 60$, respec-tively. This trend indicates that the ion exchange reaction is exothermic and, as temperature increases, the equilibrium constant (separation factor) decreases. This was not the case for the polyacrylic resin (Amberlite IRA-458) shown in Figure 11. For this resin there was no significant difference in the separation factor for the three temperatures tested.

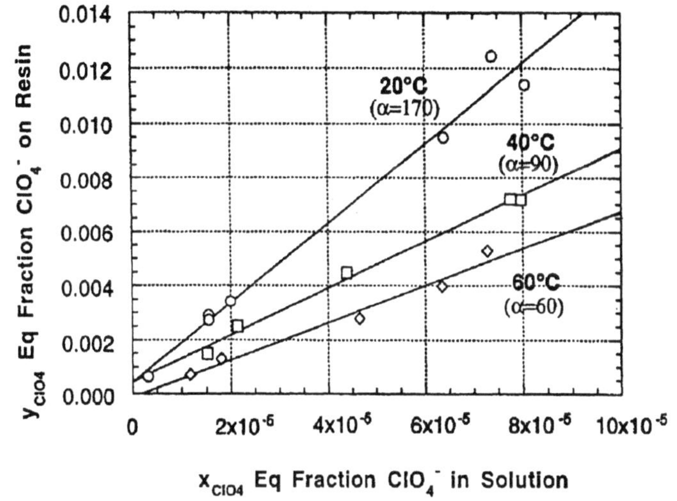


**Figure 10.** $ClO_4^-$–$Cl^-$ isotherm for a polystyrene ion-exchange resin at 20, 40, and 60 °C. With increasing temperature the $ClO_4^-$–$Cl^-$ separation factor decreases for this polystyrene resin, indicative of an exothermic reaction, which was typical for the hydrophobic resins.

The percent change of the $ClO_4^-$–$Cl^-$ separation factor with temperature for all seven-teen resins tested is shown in Figure 12. Resins 1 and 2 have polyacrylic type matrices while all others have either polystyrene or polyvinylpyridine. The two acrylic resins show little change with temperature while for the remaining resins there is an approximately 30% decrease in the separation factor for both the 20–40 °C and the 40–60 °C temperature changes with an overall decrease of 60% for the 20–60 °C change.

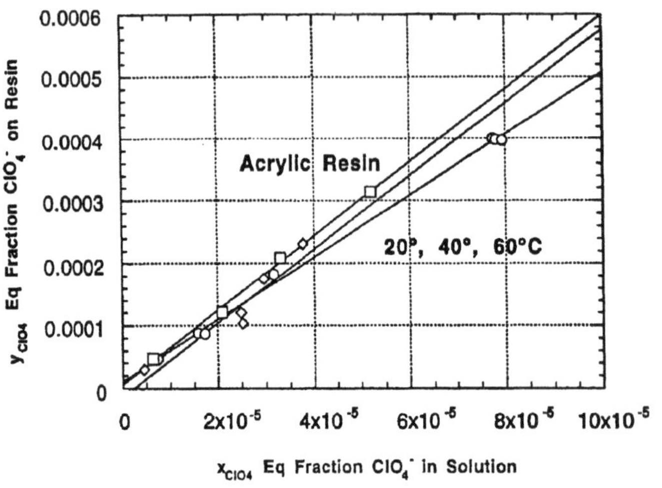


**Figure 11.** $ClO_4^-$–$Cl^-$ isotherm for a polyacrylic ion-exchange resin at 20, 40, and 60 °C. Very little change was seen in the separation factor for this polyacrylic resin, typical for the hydrophilic resins.

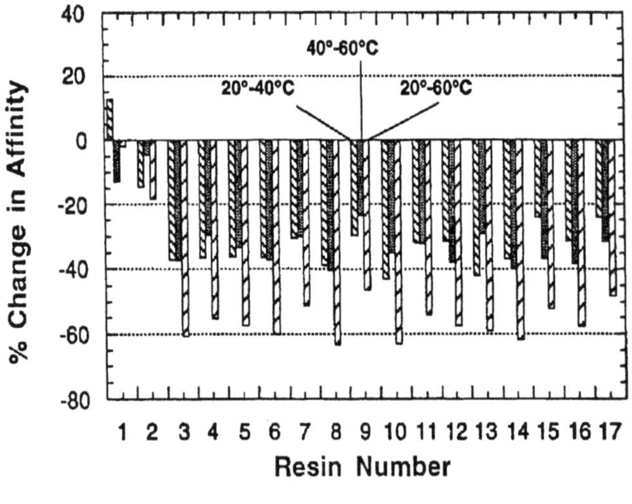


**Figure 12.** Percent change in the $ClO_4^-$–$Cl^-$ separation factor with temperature for the seventeen resins tested. The hydrophilic resins (1,2) showed little change while hydrophobic resins (3-17) exhibited approximately 30% decrease with each 20 °C increase in temperature.

## CONCLUSIONS

Among the resins tested, the overall foundation for perchlorate selectivity was the hydrophobic nature of the resin matrix (polystyrene, polyvinylpyridine, polyacrylic), i.e., the more hydrophobic resins exhibited higher separation factors. The functional group attached to the resin matrix was also important in determining the ultimate degree of perchlorate selectivity. The highest separation factors were for the polystyrene resins with triethyl and tripropyl functional groups (Amberlite IRA-996 and Sybron SR-7, respectively). Unfortunately, these high separation factors were associated with slow kinetics. It requires several days, if not weeks, for these resins to reach equilibrium.

The effects of temperature, porosity and percent cross-linking upon the separation factor were not as significant as that of matrix composition or functional group, but reductions in kinetic limitations may be achieved by the judicious selection of these three variables. For example, the equilibrium time for the polystyrene tripropyl functional group resin was decreased from greater than 7 days to less than 1 day when the temperature was increased to 60 °C.

## ACKNOWLEDGMENTS

Funding for this research has been provided by the American Water Works Association Research Foundation. The information contained herein is based upon intellectual property which is jointly owned by the University of Houston and the Foundation. The Foundation retains its right to publish or produce the jointly owned intellectual property in part or in its entirety. The comments and views detailed herein may not necessarily reflect the views of the American Water Works Association Research Foundation, its officers, directors, affiliates or agents, or the views of the U.S. government.

## REFERENCES

1. Gregor, H. P.; Belle, J.; Marcus, R.A. "Studies on ion exchange resins. IX. Capacity and specific volumes of quaternary base anion exchange resins." *Journal of the American Chemical Society* **1954,** *76,* 1984–1987.

2. Gregor, H.P.; Belle, J.; Marcus, R.A. "Studies on ion exchange resins. XIII. Selectivity coefficients of quaternary base anion-exchange resins toward univalent anions." *Journal of the American Chemical Society* **1955,** *77,* 2713–2719.

3. Aveston, J.; Everest, D.A.; Wells, R.A. "Adsorption of gold from cyanide solutions by anionic resins." *Chemical Society Journal (London)* **1958**, 231–239.

4. Chu, B.T. *Factors Governing the Selectivity of Anion Exchangers.* Ph.D. Dissertation. Cornell University: Ithaca, NY, 1959.

5. Freeman, D.H. "Thermodynamics of binary ion-exchange systems." *Journal of Chemical Physics* **1961,** *35,* 189–191.

6. Reichenberg, D. "Ion-Exchange Selectivity." In *Ion Exchange: A Series of Advances.* J.A. Marinsky, Ed. Dekker: New York, NY, 1966; pp. 227–276.

7. Subramonian, S.; Clifford, D.A. "Monovalent/divalent selectivity and the charge separation concept." *Reactive Polymers* **1988**, *9*, 195–209.

8. *Analysis of Low Concentrations of Perchlorate in Drinking Water and Ground Water by Ion Chromatography*. Application Note 121. Dionex: Sunnyvale, CA, 1998.

9. Huang, X. *Laboratory Tests for In-situ Immobilization of Uranium in Contaminated Groundwater*. Master's Thesis. University of Houston: Houston, TX, 1996.

# Chapter 13

# THE REMOVAL OF PERCHLORATE FROM WATERS USING ION-EXCHANGE RESINS®

Jacimaria R. Batista,* Frank X. McGarvey,② and Adriano R. Vieira①

① Department of Civil and Environmental Engineering, University of Nevada–Las Vegas 4505 Maryland Parkway, Las Vegas, Nevada 89154-4015
② 39 Woodbury Drive, Cherry Hill, New Jersey 08003

## INTRODUCTION

The recent discovery of perchlorate ($ClO_4^-$) in several groundwater wells in Nevada, California, and Utah, has generated considerable interest in potential treatment technologies to remove the contaminant from water supplies. Biological and physico-chemical treatment technologies are currently under investigation for their potential to economically remove perchlorate from waters. In November 1998, several researchers were awarded grants from the American Water Works Association Research Foundation (AWWARF)[1] to investigate the potential of ion-exchange, biodegradation, membrane filtration, and ozone/granular activated carbon systems to remove perchlorate from waters. Strong-base ion-exchange resins have proven to be very effective in removing perchlorate from waters to very low levels.[2,3,4] There are, however, two issues that deserve further consideration before ion-exchange can be used to economically remove perchlorate from waters. The first issue is resin regeneration. Regeneration of perchlorate-laden resins has been proven to be very difficult since perchlorate attaches very strongly to the resins. Several bed volumes of 12% sodium chloride (NaCl) solution were able to remove only a small portion of the perchlorate loaded to styrenic type strong-base resins[2] and heating perchlorate-laden strong-base resins during regeneration has been investigated with some degree of success.[3] The regeneration of acrylic type strong-base resins, however, has proven to be quite effective.[2] The second issue is the final disposal of regenerant brines containing high concentrations of perchlorate. Any potential technology to

®This work was presented at the 218th national meeting of the American Chemical Society as part of the Environmental Division symposium Perchlorate in the Environment, held August 22–24, 1999, in New Orleans, Louisiana.

*Author to whom correspondence should be directed. Phone: 702-895-1585. Fax: 702-895-3936. Electronic mail: jaci@ce.unlv.edu.

*Perchlorate in the Environment*, edited by Urbansky.
Kluwer Academic/Plenum Publishers, New York, 2000.

remove perchlorate, based on separation, will have to address the final disposal of highly concentrated perchlorate solutions. So far, this issue has not been given much attention.

The authors believe that a combination of ion-exchange resins with biological reduction constitute a very effective treatment technology to completely removed perchlorate from waters. In this case, the ion-exchange resin is used to remove the perchlorate from the water and biodegradation is used to treat the solutions resulting from the resin regeneration.

The removal of perchlorate from waters by strong-base anion exchange resins (in the chloride form) can be described by the following reactions:

$$\text{Resin-Cl}^- + ClO_4^- \rightleftharpoons \text{Resin-}ClO_4^- + Cl^- \quad (1)$$

When the breakthrough point, the point at which a specific amount of the influent is detected in the effluent, is reached the resin can potentially be regenerated with sodium chloride (NaCl) according to the reaction below:

$$\text{Resin-}ClO_4^- + NaCl \rightleftharpoons \text{Resin-Cl}^- + Na^+ + ClO_4^- \quad (2)$$

In the regeneration process, NaCl is used in excess, thus the regenerant solution is very saline and contains high concentrations of perchlorate.

Biodegradation may hold a great potential to treat ion-exchange regenerant solutions. Perchlorate is easily biodegradable and has been shown to be effective in treating water with high concentrations of perchlorate.[5–7] Romenenko,[8] Attaway,[5] and Wallace[6] have reported the anaerobic biodegradation of perchlorate in suspended growth with the addition of nutrients, trace minerals, and a carbon source. Several strains of microorganisms have been shown to reduce perchlorate to chlorite under anaerobic conditions. Hackenthal[9] demonstrated that several species of heterotrophic bacteria are able to reduce perchlorate to chloride. Wallace[6] showed that *Wolinella succinogenes*, an obligate anaerobe is capable of reducing >7000 mg $L^{-1}$ perchlorate to chloride. Korenkov[10] also showed that *Vibrio dechlorticans* Cuznesove B-1168 was able to reduce perchlorate concentrations of 300 mg $L^{-1}$ to chloride. The greatest challenge in treating ion-exchange regenerant solutions biologically is the high salinity of these solutions. Biodegradation of perchlorate-containing regenerant brines would require the isolation and acclimation of salt tolerant (halophiles) microorganisms which are able to use perchlorate as an electron acceptor (Figure 1). To the best of the authors knowledge, biodegradation of perchlorate on highly saline environments have not been reported to date.

Instigated by the need for technologies to treat ion-exchange regenerant solutions from perchlorate removal systems and by the work performed by McGarvey[11] and Kunin[12] on deionization of waters, the authors explored the feasibility of using weak-anion exchange resins to remove perchlorate from waters.

The primary advantage of such a system is that the resin can potentially be regenerated using ammonium hydroxide instead of sodium chloride. In this case, instead of sodium chloride, the regenerant will contain ammonium hydroxide, which can be used as nutrient by microbes. This process involves carbonating the weak-base anion-exchange resin by passing water saturate with $CO_2$ over the basic form of the resin. Based on the work of Kunin[12] and McGarvey,[11] the authors hypothesized the following steps are needed to remove perchlorate from waters using a weak-anion-exchange resin:

**Carbonation Step:**

$$[R_3N:] + H_2O + CO_2 \rightleftharpoons [R_3NH^+]HCO_3 \quad (3)$$

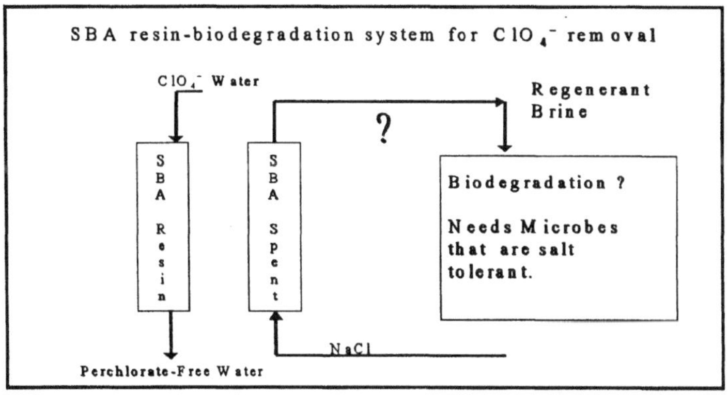


**Figure 1.** Potential combination of strong-base anion exchange resin and biodegradation to remove perchlorate from waters. The regenerant brine is very saline and biodegradation may not be possible, unless salt tolerant microbes capable of perchlorate biodegradation are isolated.

**Perchlorate Exchange Step:**

$$[R_3NH^+]HCO_3 + NaClO_4 \rightleftharpoons [R_3NH^+]ClO_4 + NaHCO_3 \quad (4)$$

**Regeneration Step:**

$$[R_3NH^+]ClO_4 + NH_4OH \rightleftharpoons [R_3N:] + NH_4ClO_4 + H_2O \quad (5)$$

Carbonation of the weak resin can be performed in two ways:

1. by adding carbon dioxide ($CO_2$) to the feed solution, or

2. by generating bicarbonate ($HCO_3^-$) from a strong-acid cationic resin fed a sodium bicarbonate solution (Figure 2).

## EXPERIMENTAL

Several strong and weak base anion-exchange resins were tested for their ability to remove perchlorate from waters. Resin samples were provided by Sybron Chemicals[13] and Purolite.[14] The tests were performed in 15.0 and 25.4-mm I.D. fixed-bed glass columns with bed height of approximately one foot. The empty bed contact time (EBCT) for all tests run was between 4–6 minutes. Perchlorate solutions containing 40-50 mg $L^{-1}$ were fed to the columns. The resins were supported by glass beads. Peristaltic pumps were used to feed and backwash the columns. Solutions of sodium perchlorate (99% purity, Aldrich chemicals) in DI water were fed to the columns in a downflow mode (Figure 3).

The effluent from the columns was collected at determined time intervals and analyzed for perchlorate. Breakthrough concentration was chosen as 20 ppb. Perchlorate analysis was performed using a Dionex-120 ion chromatograph.[15] Sodium chloride was used to regenerate strong-base anionic resins, while varying concentrations of ammonium hydroxide, sodium chloride and sodium hydroxide were used to backwash the weak-base anionic resins. The regeneration was performed in a upflow mode.

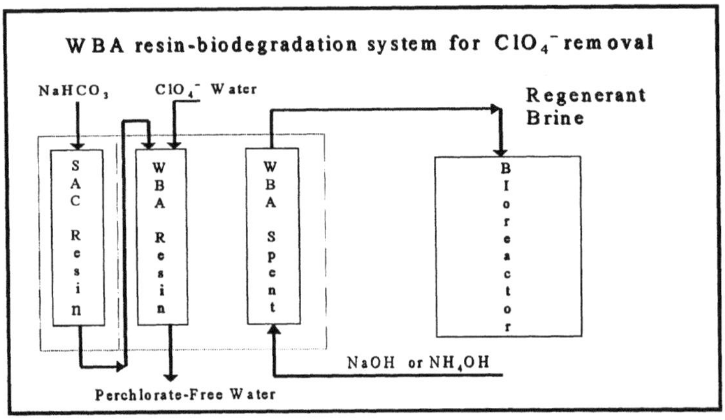


**Figure 2.** Potential combination of weak-anion exchange resin and biodegradation to remove perchlorate from waters. When caustic solutions such as $NH_4OH$ or NaOH are used, the regenerant solution is not a brine and biodegradation is potentially more favorable.

**Figure 3.** Experimental set-up for perchlorate removal testing by ion-exchange.

## RESULTS AND DISCUSSION

Several strong and weak base anion exchangers were investigated for their capacity to remove perchlorate from waters and for their regeneration potential.

### Strong Base Anion Exchange Resins

Figure 4 shows the loading and regeneration of a strong-base styrenic type I resin (Sybron Chemical–ASB-1) with a trimethyl amine functional group. Notice that about 600 bed volumes

(BV) were processed before breakthrough was reached. However, the column utilization was only 21% of the total column capacity. The breakthrough curve for this resin shows an apparent kinetic effect that may be due to the slow diffusion rate inside the resin beads. This indicates that high capacity type I anion exchange resins may present a loading problem if full loading of perchlorate is expected. A porous strong base resin is expected to load more efficiently. Studies are needed to confirm the role of resin bead size on the loading of perchlorate to strong base resins. Regeneration with a low concentration of sodium chloride (0.5%) was very ineffective and capable of removing only 5.2% of the loaded perchlorate. A low concentration of sodium chloride was used as an attempt to generate a brine with less salinity, more favorable to biodegradation.

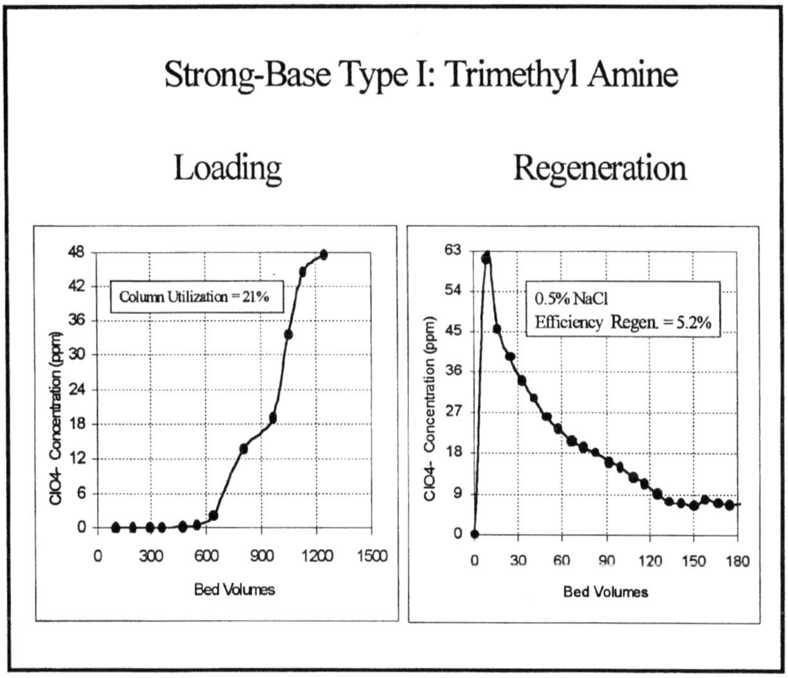


**Figure 4.** Loading and regeneration of ASB-1 resin column for perchlorate removal.

The theory that hydrophobic styrenic resins in combination with the hydrophilic perchlorate anion hinders the approximation of the chloride ion during regeneration is supported by these results.

Figure 5 shows loading and regeneration data for another styrenic strong-base resin (Sybron Chemicals SR-7). In this case, the functional group is tripropyl. For this resin, column utilization was very satisfactory (60%) and more than 1300 BV were processed before breakthrough was reached. Regeneration with 12% NaCl was very ineffective and only 17.3 % of the loaded perchlorate could be stripped out of the resin.

The loading and regeneration data for an acrylic strong base resin with a quaternary amine group (Sybron Chemicals Macro-T) is shown in Figure 6. More than 55 % of the resin bed was utilized and regeneration with 12% NaCl was able to remove 96% of the loaded perchlorate. More than 1800 BV could be processed before breakthrough. The above results indicate that acrylic strong-base resins load and regenerate better than styrenic type.

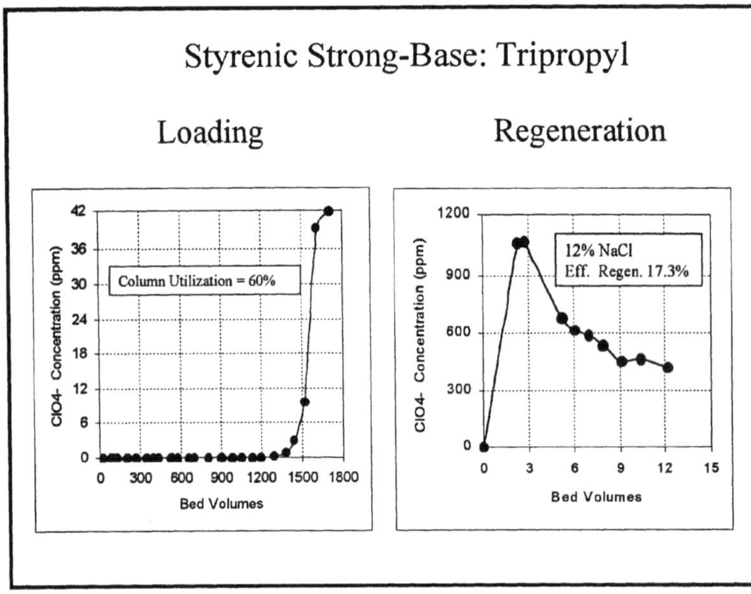


**Figure 5.** Loading and regeneration of SR-7 resin column for perchlorate removal.

**Weak-Base Anion Exchange Resins**

Figure 7 shows the result of duplicate column testing using a styrenic weak-anion exchange resin (Sybron Chemicals AFP-329) after carbonation, as described by Eq 3. Carbonation was performed by passing a $NaHCO_3$ solution trough a bed of strong-acid cationic resin (Sybron C-267). Notice that column utilization for this resin was very poor (13.2%). About 500 BV of solution could be processed before breakthrough. Regeneration of one of the columns using 1% NaOH was able to remove only 15.3 % of the perchlorate loaded. However, regeneration with 12% NaCl was able to remove 77% of the loaded perchlorate. These results were unexpected, since the perchlorate loaded in the weak-anion exchange should be removed by NaOH. Further investigation of the issue indicated that the weak anionic resin used in this test, contains 15-20% strong-base groups. Calculations of the perchlorate loading indicated that the amount of perchlorate removed by the resin is roughly equivalent to the capacity of the strong-base groups present. Therefore, it is possible that the strong-base groups played a role on perchlorate removal by this resin. Further investigation is needed to evaluate the influence of these strong-base groups on perchlorate removal.

Figure 8 shows the removal of perchlorate by an acrylic weak-base resin (Purolite A-830), with tertiary amine functional groups. More than 2700 BV of perchlorate solution were processed before breakthrough was reached and column utilization was 35.5%. Regeneration with 1% NaOH was able to remove more than 76.5%. These results show that acrylic weak base resins have high column capacity and regenerate easily with caustic solution. These types of resin can be used to give regenerant solutions, which are amenable to biodegradation.

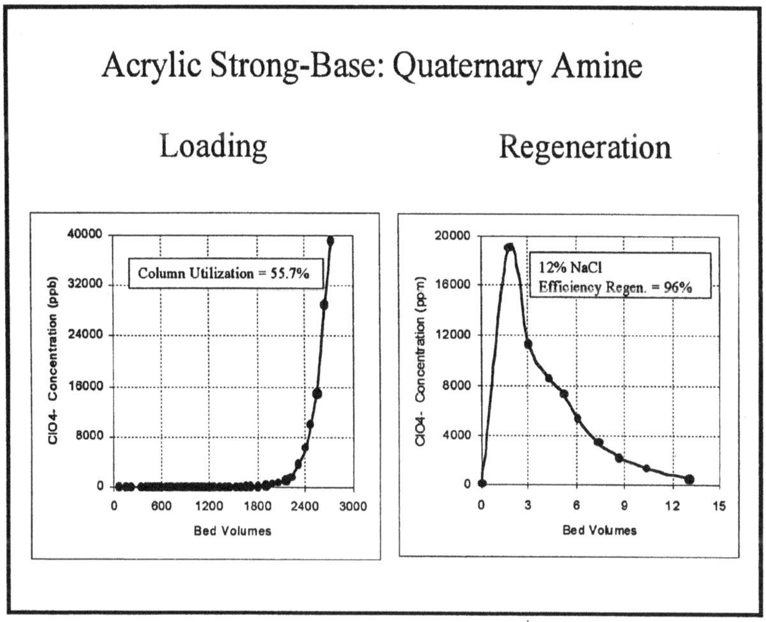


**Figure 6.** Loading and regeneration of Macro-T resin column for perchlorate removal.

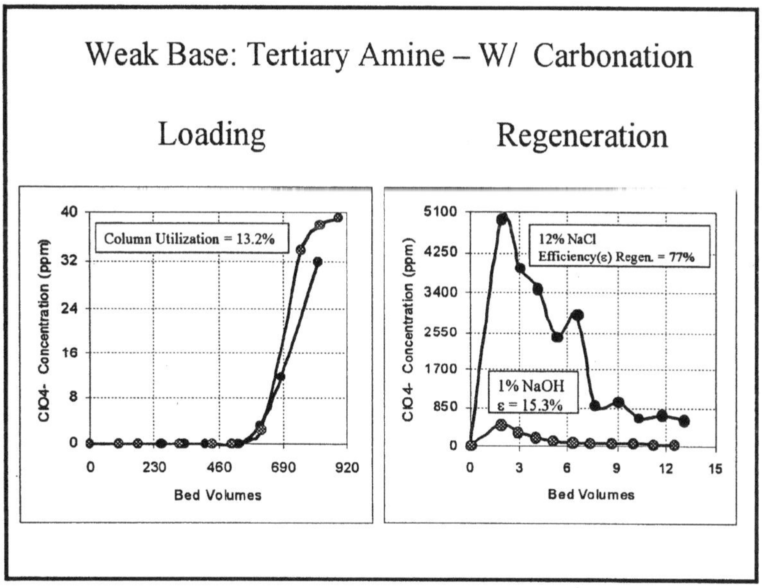


**Figure 7.** Loading and regeneration of AFP-329 resin for perchlorate removal.

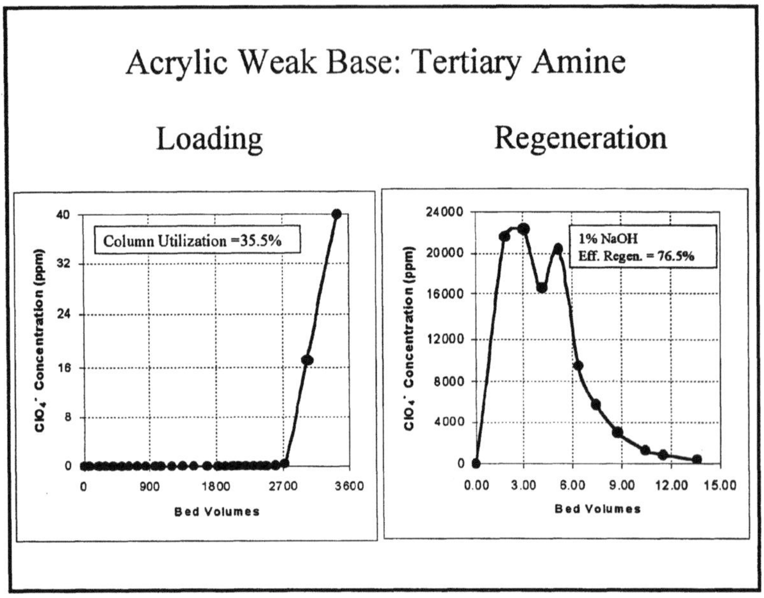


**Figure 8.** Loading and regeneration of A-830 resin for perchlorate removal.

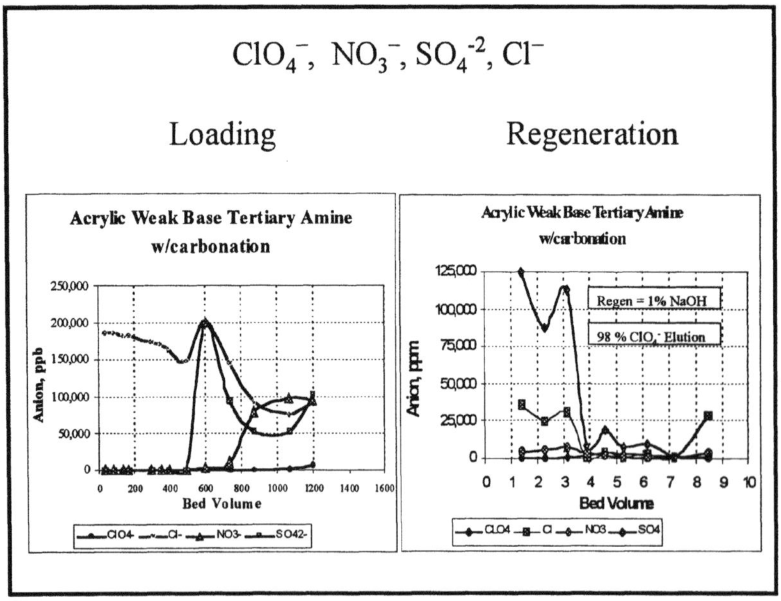


**Figure 9.** Loading and regeneration of several anions, including perchlorate, to A-830 resin.

The removal of 10 mg $L^{-1}$ perchlorate by the A-830 acrylic resin in the presence of 100 mg $L^{-1}$ sulfate, 100 mg $L^{-1}$ nitrate, and chloride is shown in the Figure 9. The loading and regeneration of perchlorate for the same column in question is shown in Figure 10. Notice in Figures 9 and 10 that more than 800 BV were processed before breakthrough and 98% of the perchlorate loaded could be removed by regenerating with 1% NaOH. Sulfate breakthrough occurred after about 450 BV followed by nitrate breakthrough at 760 BV. Chloride concentrations in the effluent varied greatly because the weak resin used in this test was in the chloride form.

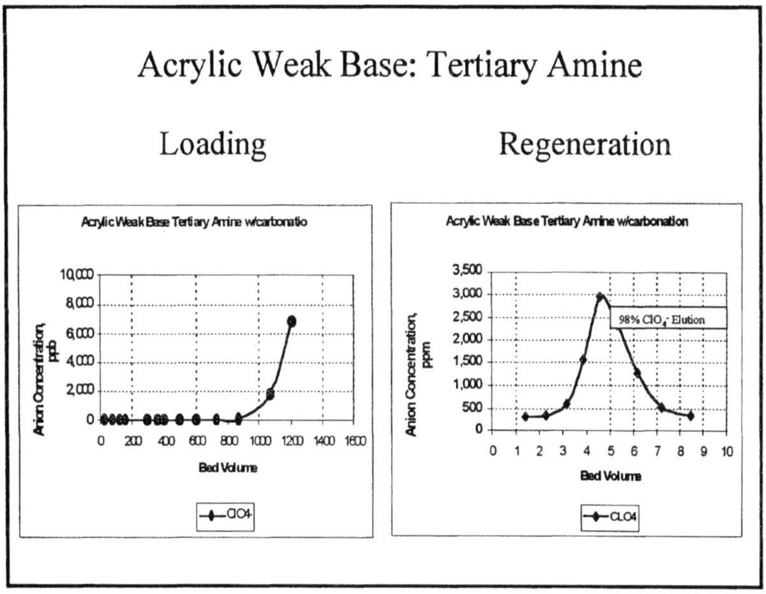


**Figure 10.** Loading and regeneration of $ClO_4^-$ from resin A-830.

## CONCLUSIONS

The following preliminary conclusions can be made about the data that have been obtained to date:

1. Both acrylic and styrenic strong-base resins remove perchlorate from waters very effectively. However, styrenic type resins do not regenerate.

2. The poor column utilization of styrenic type resins seems to be the result of a kinetic effect due to slow diffusion rate inside the resin bead. Studies are needed to further investigate this issue.

3. Acrylic type resins have high capacity for perchlorate and regeneration efficiency is much higher than that for styrenic strong base resins.

4. Column utilization for acrylic type resins is much higher than that of styrenic type resins. The styrenic resin SR-7 is an exception; it has 60% column utilization.

5. Styrenic-weak base resins have a high capacity for perchlorate, but regeneration with caustic solutions was initially ineffective due to the strong-base groups present in this resin. Additional work is required to determine the best way to use this type of resin.

6. Acrylic weak base resins have high column utilization and regeneration with caustic is very effective. These types of resins could be utilized to give regenerant solutions which can be fed to biological processes.

The mechanisms of perchlorate removal from waters by weak-base resins are not yet well understood. The mechanism proposed by the authors may be reasonable, but more studies are needed to evaluate the hypothesized reactions.

## REFERENCES

1. AWWARF, 1998, RFPs on perchlorate. URL: http://www.awwarf.com/research/ guides.

2. Batista, J.R.; McGarvey, F.X.; Vieira, A. Unpublished data.

3. Clifford, D.A. Perchlorate Conference, Ontario, CA, March 18–19, 1999.

4. Calgon Carbon Corporation. Big Dalton Perchlorate Removal Pilot Study. October 30, 1998.

5. Attaway, H.; Smith, M. "Reduction of perchlorate by an anaerobic enrichment culture." *Journal of Industrial Microbiology* **1993**, 12, 408–412.

6. Wallace, W.; Ward, T; Breen, A.; Attaway, H. "Identification of an anaerobic bacterium which reduces perchlorate and chlorate as *Wollinella succinogenes*." *Journal of Industrial Microbiology* **1996**, 16, 68–72.

7. Logan, B.E. "A review of chlorate- and perchlorate-respiring microoganisms" *Bioremediation Journal* **1998**, *2*, 69–79.

8. Romanenko, V.I.; Korenkov, V.N.; Kuznetsov S.I. "Bacterial decomposition of ammonium perchlorate." *Mikrobiologiya* **1976**, *45*, 204–209.

9. (a) Hackenthal, E. "Reduction of perchlorate by bacteria I: intact cells." *Biochemical Pharmacology* **1963**, *13*, 195–208. In German. (b) Hackenthal, E. "Reduction of perchlorate by bacteria II: identity between the nitrate reductase and the perchlorate-reducing enzyme in *Bacillus cereus*." *Biochemical Pharmacology* **1965**, *14*, 1314–1324. In German.

10. Korenkov, V.N.; Ivanovich, V.; Kuznetsov, S.I.; Vorenov, J.V. "Process for purification of industrial wastewaters from perchlorates and chlorates." U.S. Patent No. 3,943,055, March 9, 1976.

11. McGarvey, F.X. "Ion-Exchange Development and Applications." In *Ion Exchange Developments and Applications: Proceedings of IEX '96.* (International Ion Exchange Conference), J.A. Greig, Ed. Cambridge, England (UK): Royal Society of Chemistry Information Services, 1996; p. 82.

12. Kunin, R.; Vassilou, B.I. "New deionization techniques based upon weak electrolyte ion exchange resins." *Industrial and Engineering Chemistry Process Design and Development* **1964**, *3*, 404–409.

13. Sybron Chemicals, Birmingham Road, P.O. Box 66, Birmingham, NJ 98011.

14. The Purolite Company, 150 Monument Road, Bala Cynwyd, PA 19004.

15. Dionex Corporation, 1228 Titan Way, Sunnyvale, CA 94088.

# Chapter 14

# REMOVAL AND DESTRUCTION OF PERCHLORATE AND OTHER ANIONS FROM GROUND WATER USING THE *ISEP+*™ SYSTEM[⊗]

K. Raman Venkatesh,* Scott M. Klara, Dale L. Jennings, and Norman J. Wagner

Calgon Carbon Corporation, 500 Calgon Carbon Drive,
Pittsburgh, Pennsylvania 15205

## INTRODUCTION

Recently, there has been increased regulatory attention towards the presence of perchlorate ($ClO_4^-$) in many ground water aquifers in California and several other states. At least 11 states in the U.S. are believed to have sites contaminated with perchlorate in their aquifers or waterways.[1] Perchlorate in drinking water is believed to cause potential problems for infants and patients with hypothyroidism by interfering with the ability of the thyroid gland to process iodine.[2] A provisional action level (PAL) of 18 $\mu g\ L^{-1}$ perchlorate in drinking water has been established by the Department of Health Services in California.

Perchlorate and other anionic contaminants in ground water are effectively removed by ion-exchange, a process where contaminant anions are exchanged and replaced by an innocuous anion, typically chloride. Ion-exchange is one of the most effective methods for most ground water treatment applications due to its efficiency in removing contaminants present in varying concentrations at relatively low costs. Most of the ion-exchange resins manufactured are used for water treatment, and ion-exchange resins have been treating drinking water for several years.[3] Although ion-exchange technology is well-known, the effectiveness of an ion-exchange process depends, among other factors, on the operational configuration of the process. Key parameters that determine the efficiency and impact the

[⊗]This work was presented at the 218th national meeting of the American Chemical Society as part of the Environmental Division symposium, Perchlorate in the Environment, held August 22–24, 1999, in New Orleans, Louisiana.

*Author to whom correspondence should be directed. Phone: 412-787-6674. Fax: 412-787-6607. Electronic mail: venkatesh@calgoncarbon.com.

economics of an ion-exchange process are treatment ratio and regeneration waste. Treatment ratio refers to the volume of feed water that can be treated before breakthrough of the contaminant(s) is obtained. Regeneration waste refers to the volume of waste generated by the ion-exchange process while regenerating the ion-exchange resin saturated with contaminants. An effective ion-exchange process is one that achieves high treatment ratios while producing low regeneration waste. Calgon Carbon's ISEP® system utilizes an effective ion-exchange process configuration that achieves high perchlorate treatment ratios while producing minimal waste.

Calgon Carbon utilized its patented multi-port ISEP® valve in developing an ion-exchange process for the removal of perchlorate and other anionic contaminants from ground water. The system bears similarity to Calgon Carbon's ISEP®-based system used commercially for the treatment of nitrate from drinking water. The two fundamental advantages of the ISEP® system are better utilization of the mass transfer zone[#] and continuous split-flow or counter-current regeneration, which lead to high treatment ratios and low regeneration waste as compared to conventional ion-exchange processes using fixed bed systems. The ISEP® system involves sequential segmentation of the mass transfer zone where on a continuous basis, the loaded segment of the resin is removed from the top of the mass transfer zone and the regenerated resin is added back at the bottom of the mass transfer zone, leading to better utilization of the ion-exchange resin during the adsorption cycle.

In the regeneration cycle, the technology utilizes a split-flow regeneration scheme where the regenerant flow is split equally and pumped into each column in the regeneration zone in parallel. This allows the fresh regenerant to be available to each column in the regeneration zone that results in efficient removal of perchlorate from the loaded resin compared to conventional fixed bed systems where split-flow regeneration is not feasible. This enables a highly efficient use of the regenerant thereby producing significantly lower wastes than the fixed-bed system. In another variation, regeneration of the resin in the ISEP® system may be accomplished by a staged counter-current mass transfer approach wherein the entire regeneration flow is passed through each column in series. In this configuration, the concentration of contaminants in the regenerant stream progressively increases as it traverses through each column in the regeneration zone. The resin columns, traveling in a direction counter-current (opposite) to the flow of regenerant, get progressively regenerated. This scheme can also result in an effective utilization of the regenerant, depending on the levels of contaminant loading on the resin. The ability to vary the configuration of adsorption and regeneration zones independently to operate at their optimum efficiencies on a continuous basis makes the ISEP®-based design effective, versatile and economical for ground water treatment applications involving ion-exchange.

A total treatment technology for the removal and destruction of perchlorate and other contaminant ions (nitrate and sulfate) from ground water has been successfully piloted by Calgon Carbon Corporation. The technology (*ISEP+™*) incorporates a continuous counter-current ion separation (ISEP®) system for the removal of perchlorate, nitrate and sulfate from the water. The spent brine, after regenerating the resin in our ISEP® system, is treated by our perchlorate and nitrate destruction module (PNDM) that reduces the perchlorate and nitrate to chloride and nitrogen, respectively, and also consists of a nanofiltration unit to remove the sulfate present in the brine. The purified brine is recycled back into our ISEP® system as regeneration feed. The entire process results in a very small waste stream free of perchlorate and nitrate that can be easily disposed. A schematic of the *ISEP+™* treatment system is shown in Figure 1.

[#] The portion of the bed where ion exchange is taking place, sometimes called the "wavefront."
'Counter-current' in ISEP® refers to the fact that the direction of stream flow is opposite to the direction of rotation of the ISEP® columns.

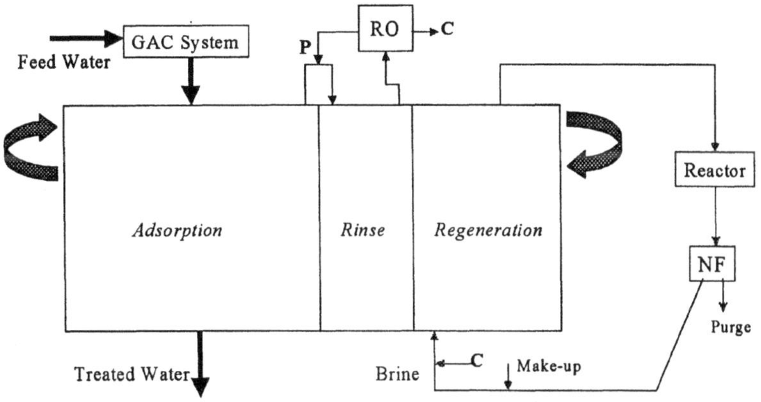


**Figure 1.** Schematic of the ISEP+™ perchlorate removal and destruction system.

## EXPERIMENTAL/PILOT METHODOLOGY

Static equilibrium tests and dynamic column tests were conducted in our laboratory using several candidate anion-exchange resins to select the resin for use in the ISEP® system. A key objective was to choose a resin that offered the best compromise between selectivity and capacity for perchlorate removal versus ease of regeneration. Based on evaluation of selectivity, capacity and adsorption vs. regeneration profiles, a strong base anion-exchange resin was selected to be further evaluated in the ISEP® system at the pilot stage.

Pilot-scale testing was conducted with the selected resin at two different field sites for a total period exceeding six months. Testing was conducted with the pilot-scale ISEP® unit designed to treat feed water at 16.2 L $min^{-1}$ (4.28 gal $min^{-1}$). Regeneration was conducted using a 7 w/w NaCl solution. The pilot testing covered various objectives ranging from waste minimization to performance validation at varying influent perchlorate concentrations. Over the course of the two pilot studies, the ISEP® system successfully treated over 750,000 gallons of contaminated ground water. The brine effluent from the ISEP® system was treated by the catalytic reactor followed by a nanofiltration (NF) system for removal of sulfate. A slight stoichiometric excess of ethanol, which gets oxidized to carbon dioxide and water during the reaction, was used as a reductant in the catalytic reactor. The samples generated were analyzed by ion chromatography (EPA Method 300.0 *modified*)[4] for perchlorate and other anions. The minimum analytical detection limit for perchlorate was reported to be 4 μg $L^{-1}$ for water samples and 125 μg $L^{-1}$ for the brine samples.

## RESULTS AND DISCUSSION

Removal of perchlorate by anion exchange can be represented by simple ion-exchange equilibria as shown below, where R represents the anion-exchange resin containing an appropriate functional group:

$$ClO_4^- + R\text{-}Cl \longrightarrow R\text{-}ClO_4 + Cl^- \quad \text{(Adsorption)} \quad (1)$$

$$R\text{-}ClO_4 + NaCl \longrightarrow R\text{-}Cl + NaClO_4 \quad \text{(Regeneration)} \quad (2)$$

Other anions present in groundwater (such as nitrate and sulfate) are removed by equilibria similar to that represented above for perchlorate. An efficient ion-exchange process is one where the resin has reasonably high selectivity for the target ion while being regenerated with relatively small amounts of regenerant. Reduced usage of regenerant (such as brine) understandably leads to lower operating and waste disposal costs.

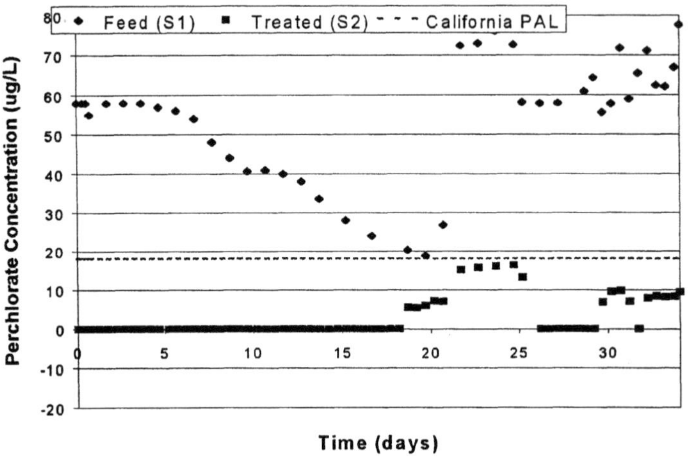


**Figure 2.** Removal of perchlorate from ground water using ISEP® system. Waste percent levels indicate ISEP® spent brine (7% w/w NaCl) generated as a percent of feed water treated, which was varied during the optimization process.

The steady state performance results obtained from the ISEP® system treating an actual contaminated ground water stream is shown in Figure 2. It is clear that the system continuously produced treated water with non-detectable level of perchlorate. This performance was maintained during the regenerant optimization phase and when the waste (spent brine effluent) from the system was reduced to 0.5% (v/v of feed water), perchlorate breakthrough was observed, although to a level below the California PAL. The feed perchlorate concentration was spiked to 60–80 μg $L^{-1}$ during the test as the well head concentration declined. At the optimal waste level of 0.75%, the system produced non-detectable perchlorate in treated water. This level of waste corresponds to a salt consumption of 160 kg $m^{-3}$ (9.8 lb $ft^{-3}$) of resin. In addition, the ISEP® system concurrently reduced the influent nitrate concentrations from ~22–28 mg $L^{-1}$ down to 5–14 mg $L^{-1}$ during the same period of operation. Moreover, influent sulfate concentrations of 45–60 mg $L^{-1}$ were also concurrently removed to < 2 mg $L^{-1}$ in treated water. Steady treatment of most low level perchlorate contaminated streams down to non-detectable level in addition to substantial removal of nitrate and sulfate was achieved by the ISEP® system at waste levels as low as 0.75%.

The second pilot test expanded on the results obtained from the earlier pilot study to evaluate the performance of the ISEP® system at higher perchlorate concentrations. Table 1 summarizes the perchlorate removal results and the optimized regenerant waste level generated at the different feed conditions during the pilot studies. Results indicated that for a modest increase in regenerant consumption, the system was able to produce non-detectable level of perchlorate in treated water even at influent perchlorate concentrations around 1200 $\mu g\ L^{-1}$. Near-complete removal of nitrate and sulfate was also observed in the treated water.

**Table 1.** Steady state ISEP® performance at different perchlorate feed concentrations.

| Influent perchlorate concentration $\mu g\ L^{-1}$ | Treated water perchlorate concentration $\mu g\ L^{-1}$ | ISEP Regeneration effluent (brine), as vol.% of feed water treated |
|---|---|---|
| 50–80 | < 4 | 0.75% |
| ~250 | < 4 | ~1.25% |
| ~1200 | < 4 | 1.75% |

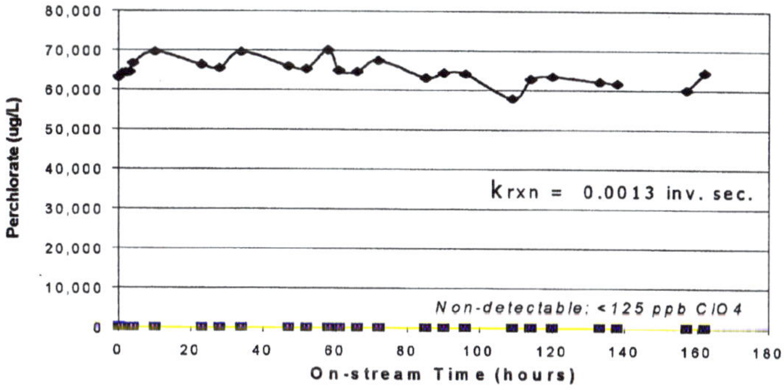


**Figure 3.** Destruction of perchlorate in ISEP® brine effluent by the catalytic reactor system in PNDM.

In addition to ISEP®, the PNDM tested at the second pilot site showed that the catalytic reactor system was effective in achieving destruction of perchlorate to a non-detectable level in the brine effluent from ISEP®, as shown in Figure 3. The detection limit of 125 μg/L for perchlorate indicated in Figure 3 is due to the interference effects of the chloride ion present in high concentrations (7% NaCl) in the brine stream. Destruction of nitrate to a non-detectable level was also achieved in the reactor system. It appears that the mechanism of perchlorate and nitrate reduction occurs through a series of steps of decreasing oxidation states of Cl and N, accompanied by oxidation of a suitable reductant such as ethanol or ammonia. A possible scheme of the oxidation-reduction (redox) reaction is shown below:

$$ClO_4^- \longrightarrow ClO_3^- \longrightarrow ClO_2^- \longrightarrow Cl^-$$

Oxid. state of Cl: $+7 \quad +5 \quad +3 \quad -1$

$$NO_3^- \longrightarrow NO_2^- \longrightarrow \ldots \longrightarrow N_2$$

Oxid. state of N: +5 +3 0

$$CH_3CH_2OH \longrightarrow \ldots\ldots \longrightarrow CO_2 + H_2O$$

or

$$NH_3 \longrightarrow \ldots\ldots \longrightarrow N_2$$

Reaction results indicated that both perchlorate and nitrate reduction reactions follow first order kinetics with nitrate reduction being much faster than perchlorate reduction. The reason for slower perchlorate destruction kinetics compared to nitrate destruction may be attributed to the relatively high activation energy for the reduction of perchlorate to chlorate, which is most likely the rate-determining step. This was determined indirectly from laboratory tests with chlorate as the starting reactant in a synthetic brine mixture, where the kinetics of chlorate reduction was found to be comparable to that of nitrate. These results are consistent with the observations of Urbansky[1] and Schock. Subsequent to the pilot test, several rate-enhancing and cost-effective catalyst compositions were developed leading to substantial improvement in the perchlorate reaction rate constant indicated in Figure 3. The steady state rate constant for perchlorate reduction was increased by almost four times to around 0.005 $s^{-1}$, during recent in-house pilot studies with an improved catalyst composition. Destruction of perchlorate and nitrate to undetectable levels are achievable within residence times of 15 min and 3 min, respectively, for many ion-exchange brine streams containing high concentrations of perchlorate and nitrate.

The NF system was able to recover 91% of its feed brine (coming from the catalytic reactor) at a 96% sulfate rejection rate, thus producing a 9% purge stream, which is now the only process waste stream. This indicates that the integrated ISEP+™ system is successful in reducing the already low ISEP® waste stream by an order of magnitude. In other words, for a 1200 $\mu g\ L^{-1}$ $ClO_4^-$ in feed water, at the ISEP® regeneration effluent level of 1.75%, the total process waste is around 0.17% of the feed water treated. In addition, this low waste stream (sulfate-laden brine) is free of perchlorate and nitrate and can be easily disposed. Alternatively, where water quality permits, the low waste stream generated may be blended either partially or completely with the treated water, thereby approaching a zero-waste process. For the feed water conditions observed in the second pilot study, this will result in an increase in chloride and sulfate concentrations in treated water by about 70 $mg\ L^{-1}$ and 50 $mg\ L^{-1}$, respectively. Considering that feed water already has sulfate in 45–60 $mg\ L^{-1}$ range, most of which is removed by ISEP®, the overall impact of blending the sulfate-laden brine waste with the treated water on discharged water quality will be a small increase in chloride concentration with nearly no change in sulfate concentration, in addition to near-complete removal of perchlorate and nitrate. Disposal of sulfate-laden brine waste might also be a viable option for many cases due to the low waste flow rate.

## CONCLUSIONS

The results obtained from laboratory and pilot studies indicate that perchlorate and other anions in ground water can be effectively and efficiently removed by the ISEP® system to produce a high quality treated water that far exceeds the current specifications. The superior performance is maintained while consuming low amounts of regenerant. In addition, the PNDM can achieve almost complete elimination of perchlorate and nitrate as well as substantial removal of sulfate from the ISEP® brine effluent. The purified brine stream from the PNDM can be recycled as regeneration feed back into the ISEP® system. The entire process (*ISEP+™*) produces a small waste stream that contains no perchlorate and nitrate and can be easily disposed. For feed streams containing very high concentrations of perchlorate (> 10 mg $L^{-1}$) and/or nitrate, use of the PNDM directly (without the ISEP®) to treat the feed water may be a viable option.

## ACKNOWLEDGMENTS

The authors would like to acknowledge the Main San Gabriel Watermaster and the National Aeronautics and Space Administration (NASA) for financial support of the pilot tests. We sincerely acknowledge Robert Maniet and David Engel for their commendable operational support as well as Chris Dixon and Tammy Veloski for their outstanding analytical support. Members of Calgon Carbon Corporation's senior management are also acknowledged for review of the manuscript. The editor's constructive comments regarding the manuscript are sincerely appreciated.

## REFERENCES

1. Urbansky, E.T.; Schock, M.R. "Issues in managing the risks associated with perchlorate in drinking water." *Journal of Environmental Management* **1999**, *56*, 79–95.

2. *Final Report of the Perchlorate Research Issue Group Workshop*. American Water Works Association Research Foundation, Ontario, CA, September 30–October 2, 1997.

3. *Ion Exchange Technology: Advances in Pollution Control*. A.K. Sengupta, Ed., Technomic: Lancaster, 1995; pp. xiii–xx.

4. Jackson, P.E.; Laikhtman, M.; Rohrer, J.S. "Determination of trace level perchlorate in drinking water and ground water by ion chromatography." *Journal of Chromatography A* **1999**, *850*, 131–135.

# Chapter 15

# THE DESIGN OF SELECTIVE RESINS FOR THE REMOVAL OF PERTECHNETATE AND PERCHLORATE FROM GROUNDWATER[⊗]

Gilbert M. Brown,*[①] Peter V. Bonnesen,[①] Bruce A. Moyer,[①] Baohua Gu,[②] Spiro D. Alexandratos,[③] Vijay Patel,[③] and Robert Ober[③]

① Chemical and Analytical Sciences Division, Oak Ridge National Laboratory, Oak Ridge, Tennessee 37831
② Environmental Sciences Division, Oak Ridge National Laboratory, Oak Ridge, Tennessee 37831
③ Department of Chemistry, University of Tennessee, Knoxville, Tennessee 37996

## INTRODUCTION

The radionuclide $^{99}Tc$ is a fission product of $^{235}U$ and $^{239}Pu$, and it has been found as a contaminant in groundwater at DOE sites in which these species have been processed. DOE's Paducah Gaseous Diffusion Plant (PGDP) site is typical of the contamination problem, where technetium is found in the groundwater at concentrations as high as 8 nmol $L^{-1}$.[1] In oxygenated groundwater, the predominant Tc species is the pertechnetate anion,[2] $TcO_4^-$, and pertechnetate salts are highly water-soluble and mobile in underground aquifers. When coupled with the long half-life of $^{99}Tc$ (213,000 years), the resulting probable transport into the biosphere makes the presence of this radioisotope in groundwater a great concern. In view of the low (typically nanomolar) concentrations of Tc involved, the development of resins with enhanced selectivity for the pertechnetate anion over other anions commonly found in groundwater such as chloride, sulfate, and nitrate will have a favorable impact on the economics of groundwater treatment. The objective of our DOE sponsored program was to develop an anion exchange resin which will selectively remove $TcO_4^-$ from groundwater while leaving behind other interfering anions.

[⊗]This work was presented at the 218th national meeting of the American Chemical Society as part of the Environmental Division symposium Perchlorate in the Environment, held August 22–24, 1999, in New Orleans, Louisiana.

*Author to whom correspondence should be directed. Phone: 423-576-2756. Fax: 423-574-4939. Electronic mail: BrownGM1@ORNL.gov.

*Perchlorate in the Environment*, edited by Urbansky.
Kluwer Academic/Plenum Publishers, New York, 2000. 

In the course of this work we prepared and evaluated new anion-exchange resins which were designed to be highly selective for pertechnetate, while at the same time maintaining favorable exchange kinetics and good capacity.

The perchlorate anion ($ClO_4^-$) has been discovered as a contaminant in ground and surface waters at sites where perchlorate salts were manufactured or used.[3] Perchlorate salts are highly soluble in water, and the physical and chemical properties of the perchlorate anion make it difficult to remove by conventional environmental treatment methods.[3] In view of the similarity of the chemical and physical properties between the perchlorate and pertechnetate species, the same resins that are selective for pertechnetate were expected to have good selectivity for perchlorate sorption.

In resin ion exchange, Moyer and Bonnesen pointed out that the factors that increase the affinity of an anion for the resin include large size, small charge-to-size ratio, and low hydration energy.[4] Since pertechnetate is larger and has a lower hydration energy than most other anions encountered in groundwater (such as bicarbonate, nitrate, chloride, and sulfate), there is a natural bias toward exchanging pertechnetate preferentially over the other anions in the solution. Table 1 shows the calculated hydration energy as a function of size for perchlorate, pertechnetate, and some other anions commonly found in groundwater. For ions with a single negative charge, the hydration energy systematically decreases as the size of the anion increases. Divalent anions such as sulfate have a higher charge density and consequently a higher hydration energy.

**Table 1.** Hydration energies of perchlorate, pertechnetate, and anions commonly found in groundwater

| Anion | Thermochemical radius, nm | $\Delta G°$(hydration),[a] kJ $mol^{-1}$ |
|---|---|---|
| $HCO_3^-$ | 0.126 | −388 |
| $Cl^-$ | 0.127 | −338 |
| $NO_3^-$ | 0.196 | −314 |
| $ClO_4^-$ | 0.240 | −259 |
| $TcO_4^-$ | 0.255 | −244 |
| $SO_4^{2-}$ | 0.230 | −1103 |

[a]Calculated as described by Moyer, B.A.; Bonnesen, P.V. "Physical Factors in Anion Separations." In *Supramolecular Chemistry of Anions*, A. Bianchi, K. Bowman-James, E. Garcia-Espana, Eds., Wiley: New York, 1997, p. 1.

Measurement of the distribution ratio for $TcO_4^-$ sorption to a series of resins has led to a scientific basis for pertechnetate selectivity based on systematic variations of the microenvironment of the exchange sites.[5] These measurements support an electrostatic model which indicates that selectivity for $TcO_4^-$ can be achieved by increasing the size of the cation and maintaining an environment of low polarity at the exchange site. The electrostatic model indicates that transfer of a large cation and a large anion from an aqueous to a non-aqueous phase is favored over the transfer of smaller cations and anions.

The distribution ratio for sorption of pertechnetate ion to a series of commercially available resins including Reillex™ HPQ, Amberlite® IRA-900 and IRA-904, Purolite® A-520E and A-850, and Sybron Ionac® SR-6 and SR-7 has been measured.[5] In that survey it was apparent that pertechnetate sorbed much more strongly to resins with a polystyrene backbone (Amberlite® IRA-900 and IRA-904, Purolite® A-520E, and Sybron Ionac® SR-6 and SR-7) than to resins with an acrylic backbone (Purolite® A-850) or a polyvinylpyridine backbone (Reillex™ HPQ). In general, the relative distribution ratios can be understood on the basis of the electrostatic model. However a number of factors such as the size of the groups responsible for the quaternary ammonium exchange sites and the nature and extent of cross-linking can also influence the selectivity for pertechnetate over the other anions in the solution. Our work on the development of perchlorate- and pertechnetate-selective resins has focused on resins prepared from a poly(vinylbenzylchloride) backbone in which the crosslinking has been carefully controlled by the addition of known amounts of divinylbenzene.

## EXPERIMENTAL

Resins were prepared in the laboratory of Professor Spiro Alexandratos at the University of Tennessee as previously described.[5] The total anion exchange capacity (TAEC) for the resins prepared at the University of Tennessee was determined by performing a Mohr titration on the chloride ion displaced when the resin was treated with excess sodium nitrate. These resins were tested along with several commercial resins for sorption of pertechnetate from a *groundwater test solution* consisting of pertechnetate at a concentration of 6.0 μM in a matrix of sodium chloride, sodium nitrate, and sodium sulfate (each at 60 mM). Although the salt content of this solution is much higher than that of actual groundwater, distribution ratios are within a measurable range. The bicarbonate anion is a major constituent of groundwater, but it was omitted because the affinity of this anion for type I resins[4] is much lower than the anions of this test solution. The affinity of pertechnetate for a resin was determined by measuring the batch distribution ratio $K_d$ for $TcO_4^-$ sorption on the resin following a 24-hr equilibration period. For some resins longer time periods were required for equilibrium to be approached. All resins were first equilibrated with the ionic matrix of the test solution and then shaken with the pertechnetate-containing test solution for the equilibration period. A quantity of moist resin equivalent to ~ 200 mg of *dry* resin (calculated on the basis of the percent solids of the particular moist sample) was shaken with 100 mL of the pertechnetate-containing test solution for the equilibration period on a reciprocating shaker. Aliquots of the test solution were withdrawn at the end of the equilibration period, and the $^{99}$Tc activity was determined using liquid-scintillation counting techniques. The amount of Tc sorbed onto the resin was determined by subtracting the Tc activity in the equilibrium solution from the total Tc activity in the starting solution. The distribution ratio $K_d$ in mL g$^{-1}$ was determined from the ratio of the number of moles of Tc sorbed to the resin to the number of moles remaining in solution at equilibrium, referenced per unit g of dry resin and mL of solution.

$$K_d = \frac{\dfrac{A(^{99}\mathrm{Tc})_{\mathrm{total}} - A(^{99}\mathrm{Tc})_{\mathrm{soln}}}{m_{\mathrm{resin}}}}{\dfrac{A(^{99}\mathrm{Tc})_{\mathrm{soln}}}{V_{\mathrm{soln}}}} = \frac{\dfrac{n^{\mathrm{resin}}_{\mathrm{TcO_4^-}}}{m_{\mathrm{resin}}}}{\dfrac{n^{\mathrm{soln}}_{\mathrm{TcO_4^-}}}{V_{\mathrm{soln}}}} \quad (1)$$

Sorption of $ClO_4^-$ on the resins was determined by bringing 0.1 g resin (dry weight equivalent) in contact with a solution containing initially 10 mg L$^{-1}$ of $ClO_4^-$ in 100 mL volume using methods described by Gu et al.[6] The test solution was a simulant of a typical contaminated groundwater and consisted of 3 mM $NaHCO_3$, 1 mM $CaCl_2$, 0.5 mM $MgCl_2$, 0.5 mM $Na_2SO_4$, and 0.5 mM $KNO_3$. Filtered samples of the solution following equilibration with the resin were analyzed for the $ClO_4^-$ concentration by means of an ion chromatograph (Dionex DX-500, Sunnyvale, California) equipped with a Dionex IonPac AS11 analytical column and an AG11 guard column. The detection limit was ~ 3 mg $L^{-1}$ $ClO_4^-$. The distribution ratio $K_d$ (mL $g^{-1}$) was calculated as the ratio of $ClO_4^-$ sorbed on the resin to the concentration of $ClO_4^-$ remaining in solution:

$$K_d = \frac{\text{perchlorate sorbed on resin, mg g}^{-1}}{\text{perchlorate in solution, mg mL}^{-1}} = \frac{(C_0 - C)/m}{C} \quad (2)$$

where $C$ is the observed concentration in solution and $m$ is the mass of resin. For both pertechnetate and perchlorate sorption, the value of $K_d$(Eq) was also computed in which the distribution ratio was corrected for the total anion exchange capacity (TEAC) of the resin, $K_d$/TEAC. $K_d$(Eq) is a better measure of selectivity for the sorption of a particular anion.

## RESULTS AND DISCUSSION

### Distribution Ratios for Pertechnetate Sorption

In an ion exchange resin, an environment favorable to $TcO_4^-$ was created within a divinylbenzene-crosslinked poly(vinylbenzylchloride)-based resin by reacting trialkylamines having long hydrocarbon chains with the unfunctionalized benzylchloride groups on the resin to create the quaternary ammonium ion exchange sites. The level of crosslinking and the size(s) of the alkyl substituents of the amine were varied for the macroreticular resin so that the distribution ratio for pertechnetate ion sorption over the other anions commonly found in groundwater was greatly enhanced.[5] The selectivity for large anions like pertechnetate over smaller anions generally increases with increasing size of the cationic site. The data in Table 2 shows that the value of $K_d$(Eq) for pertechnetate sorption at 24 h increases as the size of the R group increases in the series methyl, ethyl, *n*-propyl, *n*-butyl, but decreases for *n*-hexyl, and *iso*-octyl. Time dependent measurements of $K_d$ show that when equilibrium (or near equilibrium) is reached that there is a monotonic increase in $K_d$ with increasing radius of the alkyl group.[5] The decrease in apparent $K_d$ at 24 h for the hexyl and isooctyl derivatives is due to kinetic limitations. The electrostatic model suggests that the selectivity decreases as the dielectric constant of the medium increases, and the decrease in selectivity as triethylamine is replaced with triethanolamine is consistent with this trend. Also, it is known that an environment rich in hydrogen-bond donors will exhibit weaker bias toward large anions. It is not sufficient to have one long-chain alkyl group to achieve superior selectivity for pertechnetate sorption. The electrostatic model suggests that separation of the cationic and anion charges in a medium of low dielectric leads to superior selectivity. The presence of the two methyl groups in the resin based on dimethyl(dodecyl)amine exchange sites allows the pertechnetate anion to approach the center of positive charge too closely for superior selectivity.

Concomitant with this increased selectivity afforded by large alkyl groups, however, is a decrease in the rate of anion exchange (exchange kinetics), to the extent that it requires a few weeks for equilibrium to be reached for the resins where R is *n*-hexyl. As the length of the

**Table 2.** Effect of quaternary ammonium group size and type on $^{99}Tc$ distribution ratios

| Amine functional group[b] | Total anion exchange capacity (meq $g^{-1}$) | $K_d$ (mL $g^{-1}$)[a] (24 h) | $K_d$(Eq) (mL $meq^{-1}$)[a] (24 h) |
|---|---|---|---|
| $Me_3N$ | 3.77 | 6350 | 1690 |
| $Et_3N$ | 2.84 | 16,150 | 5690 |
| $Pr_3N$ | 2.33 | 22,340 | 9590 |
| $Bu_3N$ | 1.66 | 31,750 | 19,130 |
| $Hex_3N$ | 0.98 | 1540 | 1570 |
| *iso*-$Oct_3N$ | 0.70 | 942 | 1350 |
| $(HOCH_2CH_2)_3N$ | 2.71 | 1590 | 586 |
| $Me_2C_{12}H_{25}N$ | 2.08 | 20,310 | 9760 |

[a]All $K_d$ values were determined at 25 °C and have a ± 5% uncertainty.
[b]All resins are crosslinked with 10% divinyl

alkyl groups increases, the total anion exchange capacity (TEAC) for the functionalization of a common backbone decreased, reflecting the fact that steric effects limit the number of these large groups that can be substituted on a given mass of unsubstituted beads. For trihexylamine roughly half of the available chloromethyl groups on the polymer backbone remained unreacted, and it was apparent that a smaller trialkylamine could be reacted with these remaining chloromethyl sites to form a bifunctional resin. A consideration of the rate of approach to equilibrium in time-dependent distribution ratio measurements thus led us to prepare bifunctional resins that have exchange sites comprised of quaternary ammonium groups with large alkyl groups for high selectivity and small alkyl groups for improved kinetics.[5]

The observation that resins with superior pertechnetate selectivity and breakthrough performance in column flow-through experiments could be obtained by incorporating two different types of strong-base anion exchange sites into the resin is an important innovation. The first type of exchange site is derived from trialkyl amines in which the three identical alkyl groups each contain 5–8 carbon atoms; and the second type is the same except the alkyl groups each contain 2–4 carbon atoms (a methyl group is insufficient). The large alkyl groups enhance pertechnetate selectivity, whereas the smaller alkyl groups enhance the kinetics of the sorption process. In batch equilibrium measurements, the resin with the highest 24-h $K_d$ contained exchange sites derived from trihexyl- and tripropylamines. For column flow-through operations, the trihexylamine/triethylamine resins (for a given resin mesh size) processed more bed volumes for a given percent breakthrough than the trihexylamine/tripropylamine resins.[7–9]

These novel bifunctional resins are designed to incorporate together in one resin exchange sites that are highly selective for pertechnetate (but kinetically slow), with sites that are less selective (but kinetically fast). The resulting bifunctional resins have superior selectivity and the same or better exchange capacity compared to resins prepared from single tri-alkyl amines (such as tripropyl- or tributylamine alone), but that possess equivalent or faster exchange kinetics.

## Distribution Ratios for Perchlorate Sorption

The size and hydration energies of perchlorate and pertechnetate are very close (see Table 1), and it was anticipated that the same resins that were selective for pertechnetate would also be selective for perchlorate. Table 3 lists the $K_d$ values of a group of synthetic monofunctional and bifunctional resins at an initial $ClO_4^-$ concentration of ~ 10 mg $L^{-1}$. Included are data for the commercial-resin Purolite® A-520E that is among the best nitrate selective resins available. As observed for pertechnetate sorption, the bifunctional trihexylamine/triethylamine resin designed for high pertechnetate selectivity has superior selectivity for $ClO_4^-$ sorption compared to the commercial Purolite® A-520E resin, removing $ClO_4^-$ to below its detection limit (3 μg $L^{-1}$) from an initial concentration of 10 mg $L^{-1}$.

A systematic study of the $K_d$ values for $ClO_4^-$ sorption was conducted as a function of the length of the hydrocarbon chain (size of the alkyl groups). The results (Table 3) indicated that the $K_d$ values (24-h measurement) increased in the series of methyl < ethyl < propyl in the order expected from measurements on the sorption of pertechnetate. However the value for the resin with tributylamine functional groups shows a decrease. We did not observe this decrease in $K_d$ with pertechnetate sorption until the size of the alkyl group reached six carbons (trihexylamine).

**Table 3**. Distribution ratios for sorption of perchlorate to monofunctional and bifunctional resins

| Amine[a] | DVB fraction (%) | TAEC[c] (meq $g^{-1}$) | 24-h $K_d$[b] (mL $g^{-1}$) | 96-h $K_d$[b] (mL $g^{-1}$) |
|---|---|---|---|---|
| $Me_3N$ | 10 | 3.77 | 86,810 | 73,510 |
| $Et_3N$ | 10 | 2.84 | 166,020 | 162,160 |
| $Pr_3N$ | 10 | 2.33 | 587,580 | ND[d] |
| $Bu_3N$ | 10 | 1.66 | 319,030 | 350,710 |
| $Pr_3N$ | 10 | 2.33 | 587,580 | ND[d] |
| $Pr_3N$ | 15 | 1.58 | 168,790 | 191,420 |
| $Pr_3N$ | 25 | 1.32 | 154,500 | 153,620 |
| $Me_3N/Hex_3N$ | 5 | 3.20 | 137,990 | 136,840 |
| $Et_3N/Hex_3N$ | 5 | 2.53 | >3,300,000[e] | ND[d] |
| $Pr_3N/Hex_3N$ | 5 | 2.15 | 523,190 | >3,300,000[e] |
| Purolite® A-520E | – | 2.80 | 203,180 | 216,800 |

[a]All resins are with 100% chloromethylstyrene polymer backbone, crosslinked with divinylbenzene (DVB). [b]Initial $ClO_4^-$ concentration was 10 mg $L^{-1}$; all $K_d$ values have a ± 5% uncertainty. [c]TAEC = total anion exchange capacity. [d] ND = Not determined. [e] $K_d$ values were estimated as the equilibrium concentration approached zero (with a detection limit of ~ 3 μg $L^{-1}$).

Comparison of the 24-h and 96-h distribution ratios suggests that these values are at or near equilibrium, and the decrease in $K_d$ cannot be attributed to a failure to reach equilibrium in the 24-h time period. The values of $K_d$(Eq), in which the distribution ratios were corrected for the differences in TEAC ($K_d$/TEAC), are a better measure of selectivity for the sorption of perchlorate. These values also show a decrease in proceeding from the tripropylamine resin to the tributylamine resin. In contrast, the ratio of $K_d$(Eq) values for sorption of pertechnetate to the tripropylamine and tributylamine resins is nearly 2 (see Table 2). For the sorption of pertechnetate, a large jump in selectivity was observed as the size of the alkyl group increased from 3 to 4 carbons.

The influence of additional divinylbenzene to increase the level of crosslinking is to make the resin more rigid. The TEAC was observed to decrease as the %DVB increased, and the observed trend was that the distribution ratio for perchlorate sorption decreased as well. The decrease in TEAC is consistent with a blocking of the chloromethyl reaction sites such that a tripropylamine is too large to react. Increasing the rigidity of the polymer matrix was expected to reduce the affinity of the resin for the more hydrated anions, and it was anticipated that the selectivity for less hydrated anions such as perchlorate would be enhanced. In the case of pertechnetate sorption, the observed values of $K_d$ showed no increase beyond 5% DVB although $K_d$(Eq) was observed to increase slightly.[5] For the sorption of perchlorate, a decrease in both $K_d$ and $K_d$(Eq) is observed in going from 10% DVB to 15% and 25% DVB. For the sorption of pertechnetate and particularly for perchlorate, it is apparent that high levels of DVB are undesirable in creating an environment at the exchange site that favors sorption of these ions.

For perchlorate sorption to the bifunctional resins, the $K_d$ values in Table 3 indicate that the triethylamine-trihexylamine resin is the most selective of the bifunctional resins. This was not the case for pertechnetate sorption. The bifunctional resins based on the combinations of tripropylamine-trihexylamine and tributylamine-trihexylamine were more selective for 24-h measurements of $K_d$. As observed for the sorption of pertechnetate, the properties of the trimethylamine-trihexylamine combination are dominated by the smaller trialkylamine, resulting in a resin with reduced selectivity for perchlorate. The tripropylamine-trihexylamine combination shows excellent sorption characteristics after 96 h, but the triethylamine-trihexylamine combination is the preferred resin for shorter contact times. The selectivity of this resin for perchlorate sorption is observed to be at least an order of magnitude better than the commercially available nitrate selective resin (A-520E) which has triethylamine functional groups.

**Distribution Ratios for Pertechnetate Sorption as a Function of Sodium Chloride Concentration**

The perchlorate and pertechnetate ions are strongly sorbed to polystyrene-based anion exchange resins that have been evaluated. The conventional wisdom is that strong brine solutions are the only cost effective means of regenerating an anion exchange resin. We have observed that there is a dramatic difference in the selectivity of various general types of resin for pertechnetate sorption. It is presumed that these general trends will extend to perchlorate sorption as well. Figure 1 shows the distribution ratios for pertechnetate sorption to the strong base commercial resins Amberlite® IRA-900, Purolite® A-520E, Purolite® A-850, and the ORNL-designed triethylamine-trihexylamine bifunctional resin as a function of the concentration of NaCl in solution. The latter sample of bifunctional resin was prepared to our specifications by Purolite International (Purolite® D-3696). With the exception of the Purolite® A-850 resin, these macroporous resins are based on a divinylbenzene-crosslinked polystyrene backbone. The Purolite® A-850 resin has an acrylic backbone with trimethylammonium exchange sites. The Amberlite® IRA-900 and Purolite® A-520E resins have trimethylammonium and triethylammonium exchange sites, respectively. The measurements were made by equilibrating a sample of resin with a solution containing 6 μM pertechnetate and the indicated NaCl concentration for 24 h. The order of selectivity for these resins are as anticipated from previous work, and it is evident that these resins span over three orders of magnitude difference in selectivity. It may also be noted that the ~ −1-power dependence (i.e., slope ≈ −1) of the curves in Figure 1 is consistent with a simple anion exchange mechanism for univalent anions.

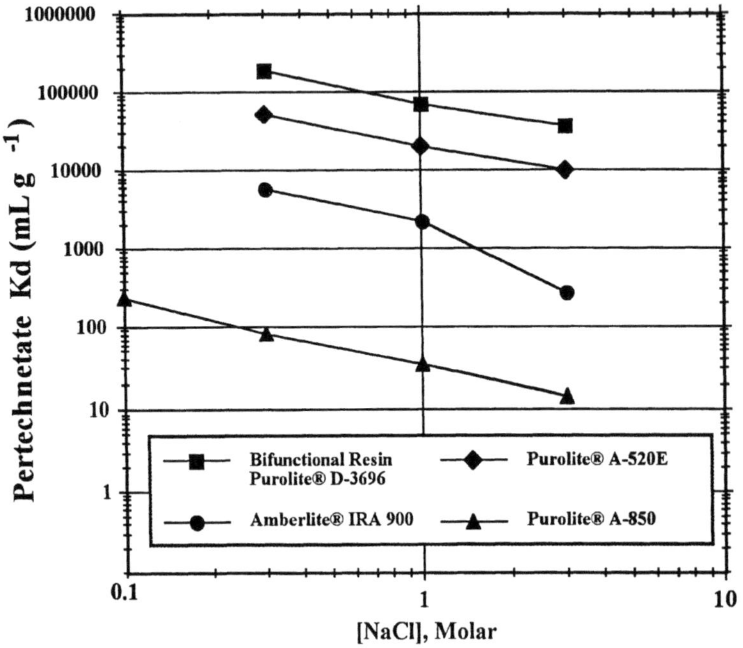


**Figure 1.** Distribution ratios for sorption of pertechnetate to resins after a 24-h equilibration period as a function of NaCl concentration.

The least selective resin, Purolite® A-850, is the only resin that has a small enough distribution ratio in 3 M NaCl to be effectively regenerated with a strong NaCl solution. The observed $K_d$ in 3 M NaCl is 17 mL $g^{-1}$, and the implications are that a given bed volume of resin will have to be contacted with approximately 10 bed volumes of 3 M NaCl to remove the pertechnetate. We did not measure a $K_d$ for this resin in the presence of a concentration of NaCl that is a close approximation to groundwater, but a reasonable extrapolation is a range of 1000–2000 mL $g^{-1}$. Sorption of pertechnetate on this resin followed by stripping with 3 M NaCl is the equivalent of concentration by a factor of ~ 100. The bifunctional resin is clearly the most selective resin for pertechnetate sorption, but it cannot be regenerated with a brine solution. A new method for resin regeneration is under development, and the preliminary results show that the bifunctional resin based on triethylammonium and trihexylammonium exchange sites can be economically regenerated with less than 20 bed volumes of solution.[6]

## CONCLUSIONS

Systematic variation of the chemical and physical properties of the exchange site of anion exchange resins has lead to experimental support of an electrostatic model of anion exchange. This model suggests that selectivity for large poorly-hydrated anions can be achieved by increasing the size of the organic resin-bound cation and by maintaining an environment of low polarity at the exchange site. Consideration of anion exchange kinetics led to the development of bifunctional anion exchange resins having two types of quarternary ammonium exchange sites, large alkyl groups which enhance pertechnetate selectivity and smaller alkyl groups which enhance the kinetics of the sorption process. These resins were developed for $TcO_4^-$ sorption, and there are strong parallels between the sorption characteristics of the $TcO_4^-$ and $ClO_4^-$ species. There are also differences between the sorption characteristics of the two species that merit further study. Strong base type I anion exchange resins with a polystyrene backbone cannot be regenerated with a simple NaCl brine, and work is in progress to develop alternative methodologies.

## ACKNOWLEDGMENTS

Development of a $TcO_4^-$-selective resin was sponsored by the Efficient Separations and Processing Crosscutting Program, Office of Science and Technology, Office of Environmental Management, U.S. Department of Energy, under contract DE-AC05-96OR22464 with Oak Ridge National Laboratory, which is managed by Lockheed Martin Energy Research Corporation. Funding for perchlorate sorption research was provided by Lockheed Martin Corporation, project number ERD-98-1644. We are grateful to Tom Blackman, David Jensen, and James H. O'Brien at Lockheed Martin Corporation for their continued support. We acknowledge the cooperation of J.A. Dale and S. Plant of Purolite International in the preparation of a commercial version of the bifunctional resin.

## REFERENCES

1. Clausen, J. L.; Zutman, J. L.; Pickering, D. A.; Farrow, N. D. Report No. KY/ER-66. Lockheed Martin Energy Systems, Kevil, Kentucky, 1995.

2. Pourbaix, M. *Atlas of Electrochemical Equilibria*. Pergamon: Oxford, England (UK) 1966, passim.

3. Urbansky, E.T. "Perchlorate chemistry: implications for analysis and remediation." *Bioremediation Journal* **1998**, *2*, 81–95.

4. Moyer, B.A.; Bonnesen, P.V. "Physical Factors in Anion Separations." In *Supramolecular Chemistry of Anions*, A. Bianchi, K. Bowman-James, and E. Garcia Espana, Eds. Wiley: New York, NY, 1997; p. 1.

5. Bonnesen, P.V.; Brown, G.M.; Alexandratos, S.D., Bavoux, L.B.; Presley, D.J.;.Patel, V.; and Moyer, B.A *Environmental Science and Technology* **1999**, submitted.

6. Gu, B.; Brown, G.M.; Alexandratos, S.D.; Ober, R.; Dale, J.A.; Plant, S. "Efficient Treatment of Perchlorate ($ClO_4^-$)-Contaminated Groundwater with Bifunctional Anion Exchange Resins." In THIS VOLUME: *Perchlorate in the Environment.*, E.T. Urbansky, Ed. Kluwer/Plenum: New York, NY, 2000, Ch. 16.

7. Gu, B.; Liang, L.; Brown, G.M.; Bonnesen, P.V.; Moyer, B.A.; Alexandratos, S.D.; Ober, R. *A Field Trial Of Novel Bifunctional Resins For Removing Pertechnetate From Contaminated Groundwater*. Oak Ridge National Laboratory: Oak Ridge, TN, 1998. ORNL Doc. No. ORNL/TM-13593.

8. Brown, G.M.; Bonnesen, P.V.; Presley, D.J.; Bates, L.M.; Moyer, B.A.; Alexandratos, S.D.; Patel, V.; Ober, R.; Gu, B.; Liang, L. Paper presented at the American Chemical Society National Meeting. San Francisco, April 13–17, 1997.

9. Brown, G.M.; Alexandratos, S.D.; Bates, L.M.; Bonnesen, P.V.; Hussain, L.A.; Moyer, B.A.; Patel, V. *Program and Abstracts, Ninth Symposium on Separation Science and Technology for Energy Applications*. October 22–26, 1995, p. 50.

# Chapter 16

# EFFICIENT TREATMENT OF PERCHLORATE ($ClO_4^-$)–CONTAMINATED GROUNDWATER WITH BIFUNCTIONAL ANION EXCHANGE RESINS⊗

Baohua Gu,[1,*] Gilbert M. Brown,[2] Spiro D. Alexandratos,[3] Robert Ober,[3] James A. Dale,[4] and Steven Plant[4]

[1]Environmental Sciences Division, Oak Ridge National Laboratory, Oak Ridge, Tennessee 37831
[2]Chemical and Analytical Sciences Division, Oak Ridge National Laboratory, Oak Ridge, Tennessee 37831
[3]Department of Chemistry, University of Tennessee, Knoxville, Tennessee 37996
[4]Purolite International, Cowbridge Road, Pontyclun, Wales CF72 8YL (UK)

## INTRODUCTION

The perchlorate ($ClO_4^-$) anion originates as a contaminant in the environment primarily from the disposal of solid salts of ammonium or sodium perchlorate, which are very soluble in water.[1,2] Although thermodynamically a strong oxidizing agent, the perchlorate anion is known to be kinetically inert in many redox reactions and noncomplexing in its interactions with typical metal ions found in the environment. These properties make the perchlorate ion exceedingly mobile in the subsurface soil environment. It can persist for many decades under typical groundwater and surface-water conditions because of kinetic barriers in its reactivity with other organic or inorganic constituents. Large volumes of perchlorate containing compounds have been disposed of in the environment since the 1950s.[1] However, the extent of the problem was not fully realized until 1997, shortly after the development of a sensitive ion chromatographic method for detecting $ClO_4^-$ in water.[3] A national survey indicates that 44 states have former perchlorate manufacturers or users; $ClO_4^-$ has now been detected in groundwater or surface

⊗This work was presented at the 218th national meeting of the American Chemical Society as part of the Environmental Division symposium Perchlorate in the Environment, held August 22–24, 1999, in New Orleans, Louisiana.

*Author to whom correspondence should be addressed. Phone: 423-574-7286. Fax: 423-576-8543. Electronic mail: gub1@ornl.gov.

*Perchlorate in the Environment*, edited by Urbansky.
Kluwer Academic/Plenum Publishers, New York, 2000. 

water in 14 states.[2] For example, water suppliers in California have detected $ClO_4^-$ in 144 public water-supply wells; 38 of these are above California's advisory action level of 18 μg $L^{-1}$ $ClO_4^-$.

Treatment technologies capable of removing $ClO_4^-$ from groundwater or surface water are therefore urgently needed. Water utilities, in particular, need treatment methods that can effectively reduce $ClO_4^-$ concentrations to low or non-detectable levels.[1] However, a drinking-water treatment method must be cost-effective and acceptable to regulatory agencies and the public, and must generate a minimum amount of secondary waste. In addition, a treatment method must not be adversely affected by other components in the water and must not be responsible for other water quality or distribution system problems. Some candidate treatment technologies have been proposed, and each of them has advantages and disadvantages.[1] Conventional ion-exchange techniques are capable of removing $ClO_4^-$ to low levels. However, a nonselective anion-exchange resin requires frequent regeneration for reuse, and a selective resin requires a large excess of brine for regeneration.[4,5] High operating costs for regenerant chemicals and regenerant disposal render this technology unattractive. Membrane-based technologies, such as reverse osmosis, are thought impractical, as are conventional anion-exchange resins. The cationic chemical nitron precipitates perchlorate efficiently and may be suitable for remediation of high concentrations of perchlorate. However, cost and waste-disposal issues also render this technology unacceptable. Given the chemical inertness of $ClO_4^-$ to reduction reactions and the low concentrations involved, it seems clear that chemical reduction is not feasible. Bioremediation methods appear to be relatively economical and technically feasible remediation methods.[6–8] However, a method based on living microorgan-isms and added nutrient gradients is also unlikely to be accepted by the drinking-water industry or the public.[1,9]

A new class of highly selective anion-exchange resins has recently been developed to effectively remove large, poorly hydrated anions [such as $ClO_4^-$ and pertechnetate ($TcO_4^-$)] to a non-detectable level.[10–14] These resins are called *bifunctional* anion-exchange resins because they have two quaternary ammonium groups: the first has long alkyl chains for higher selectivity and the second has shorter alkyl chains for improved reaction kinetics. Their compositions are based on a systematic study of the relationship between resin selectivity and alkyl chain length of the quaternary ammonium groups. The resin was initially developed to remove low levels of $TcO_4^-$ (~ 1 μg $L^{-1}$ range) from contaminated groundwater at the U.S. Department of Energy's (DOE's) Paducah Gaseous Diffusion Plant site in Kentucky.[10–12] A pilot-scale field trial successfully demonstrated that the new bifunctional resin was able to treat >700,000 bed volumes (BV) of the contaminated groundwater (running at a flow rate of ~ 6 BV $min^{-1}$) before a 3% $TcO_4^-$ breakthrough occurred, despite the fact that the concentration of $TcO_4^-$ ion was ~ 6 orders of magnitude lower than that of the other competing anions found in groundwater (such as $Cl^-$, $HCO_3^-$, $SO_4^{2-}$, and $NO_3^-$).[11] Because of the chemical similarities between $ClO_4^-$ and $TcO_4^-$ (both are large anions with a low hydration energy), we anticipated that the same bifunctional resins could be effectively used for removing $ClO_4^-$ from the contaminated groundwater or surface water. Previous studies also indicated that $ClO_4^-$ strongly compete with $TcO_4^-$ for adsorption on anion exchange resins.[15] The present study was therefore undertaken to determine the selectivity and long-term performance of the new bifunctional resins to remove $ClO_4^-$ in both laboratory and field flow-through experiments. Additionally, regeneration of the bifunctional resins was investigated.

## EXPERIMENTAL

All anion-exchange resins investigated contained chloromethyl reaction sites that were functionalized by reactions with various trialkylamine groups to create quaternary ammonium strong-base exchange sites (Table 1). The bifunctional resin, RO-02-119, was synthesized in the Chemistry Department at the University of Tennessee, Knoxville.[10,11] A commercial scale-up version of the bifunctional resin (Purolite D-3696) was prepared by Purolite International, Inc. Details of the preparation of the bifunctional resins and their characteristics are published elsewhere.[10,11,13] Three commercially available monofunctional anion-exchange resins were also selected for investigation, and the performance of these resins forms the basis for comparison with the bifunctional resins (Table 1). In particular, our initial laboratory screening studies indicated that Purolite® A-520E was one of the best monofunctional anion-exchange resins with respect to $ClO_4^-$ removal.[11] It is currently being used for removing $TcO_4^-$ from the contaminated groundwater at the Paducah Gaseous Diffusion Plant site.

Sorption of $ClO_4^-$ on resins was determined by bringing 0.1 g resin (dry weight equivalent) in contact with a solution containing varying amounts of $ClO_4^-$ in 100 mL of test solution. The test solution, a simulation of typical contaminated groundwater, consisted of 3 mM $NaHCO_3$, 1 mM $CaCl_2$, 0.5 mM $MgCl_2$, 0.5 mM $Na_2SO_4$, and 0.5 mM $KNO_3$. The initial $ClO_4^-$ concentration varied from 0.1 to 100 mg $L^{-1}$ and was therefore about 1 to 4 orders of magnitude lower than that of the background anions, which included $Cl^-$, $HCO_3^-$, $SO_4^{2-}$, and $NO_3^-$. Samples were equilibrated on a reciprocal shaker, and at different time intervals an aliquot of sample was taken and filtered through a PTFE syringe filter (0.45 μm). The filtered samples were then analyzed for $ClO_4^-$ concentration by means of an ion chromatograph

**Table 1.** General properties of synthetic resins[a]

| Resin | First amine functional group | Second amine functional group | Mesh size | TAEC[b] (meq $g^{-1}$) |
|---|---|---|---|---|
| *Monofunctional anion-exchange resins* | | | | |
| Amberlite® IRA-900 | Trimethylamine | None | 16–50 | 3.56 |
| Purolite® A-520E | Triethylamine | None | 16–50 | 2.80 |
| Sybron® SR-6 | Tributylamine | None | 16–50 | 1.80 |
| *Bifunctional anion-exchange resins* | | | | |
| VP-02-217 | Trihexylamine | Tripropylamine | 60–200 | 2.06 |
| RO-02-119 | Trihexylamine | Triethylamine | 40–60 | 2.53 |
| Purolite D-3696 | Trihexylamine | Triethylamine | 40–60 | 2.40 |

[a]All bifunctional resins used 100% Chloromethylstyrene backbone with 5% divinylbenzene (DVB) crosslinking. The DVB crosslinking density of the monofunctional resins ranged from 3 to 5 %.
[b]TAEC = Total anion-exchange capacity in meq $g^{-1}$ dry resin.

(Dionex DX-500, Sunnyvale, California) equipped with a Dionex IonPac AS11 analytical column and an AG11 guard column. The detection limit was approximately 3 μg $L^{-1}$ $ClO_4^-$ (or ~ 3 × $10^{-8}$ M).

Laboratory column flow-through experiments were performed with small glass chromatographic columns (10 × 40 mm). Resins were wet-packed in columns as slurries, and a high precision high performance liquid chromatograph (HPLC) pump (Alltech 426, Deerfield, Illinois) was used for feeding the $ClO_4^-$ test solution at a constant flow rate of 30 mL $min^{-1}$ or ~ 9.5 BV $min^{-1}$. Two types of the influent solutions were used: (1) the simulated groundwater test solution, and (2) actual groundwater from Redlands, California (with 16 μg $L^{-1}$ $ClO_4^-$, 0.6 mM $Cl^-$, 1 mM $NO_3^-$, 0.6 mM $SO_4^{2-}$, and 3 mM $HCO_3^-$).[14] Both the $ClO_4^-$ test solution and the groundwater were spiked with ~ 1100 μg $L^{-1}$ $ClO_4^-$ in order to accelerate the breakthrough of $ClO_4^-$ in the laboratory flow-through experiments. Columns were typically run continuously for ~ 8–10 h, and the effluent samples were collected at a given time interval and analyzed for $ClO_4^-$ concentration.

On the basis of both laboratory batch and column studies, a commercial version of the bifunctional resin (Purolite D-3696) was prepared and was tested for its effectiveness in removing $ClO_4^-$ under realistic groundwater conditions. Two glass columns (25 × 115 mm) were configured in a lead-and-lag series with groundwater passing through one column before it passed through the second. The lead-and-lag configuration allowed for sampling between the two columns so that the breakthrough of $ClO_4^-$ in the lead column could be monitored independently. The contaminated groundwater was pumped directly through the resin columns without pre-treatment at an initial flow rate of ~ 200 mL $min^{-1}$ (or ~ 1.8 BV $min^{-1}$). The flow rate was kept relatively constant (between 150 and 210 mL $min^{-1}$) by adjusting the inlet pressure (or flow rate) periodically. Concentrations of $ClO_4^-$ in the influent and the effluents of the lead and lag columns were monitored three times a week for ~ 12 weeks. Additionally, two columns of the Purolite® A-520E resin (also in a lead-and-lag configuration) were tested in parallel to provide a basis for comparison on the relative effectiveness of the bifunctional resin in removing $ClO_4^-$ from the contaminated groundwater.

## RESULTS AND DISCUSSION

### $ClO_4^-$ Adsorption Selectivity and Kinetics

Perchlorate is a large, poorly hydrated anion in comparison with most of the other anions encountered in groundwater (such as $Cl^-$, $HCO_3^-$, $SO_4^{2-}$, $NO_3^-$) and has a chemical bias toward adsorption and will be preferentially exchanged over the other anions in the aqueous solution. The distribution coefficient ($K_d$, mL $g^{-1}$) is an indication of such chemical bias or adsorption selectivity of a given anion and is defined here as the ratio of the amount of $ClO_4^-$ sorbed (as mg $g^{-1}$ dry resin) divided by the concentration of $ClO_4^-$ remaining in the equilibrium solution (as mg $mL^{-1}$):[15]

$$K_d = \frac{ClO_4^- \text{ sorbed on resin (mg g}^{-1})}{ClO_4^- \text{ in solution (mg mL}^{-1})}.$$

Obviously, the $K_d$ value depends on a number of factors in addition to the chemical nature of the given anion. Table 2 lists the $K_d$ values determined at different equilibrium time periods for both the bifunctional and monofunctional anion-exchange resins. In general, $K_d$ values increased with equilibrium time because most of these resins have large internal micropores, as are expected for the macroreticular resins. More importantly, however, the time-dependent adsorption of $ClO_4^-$ by these anion-exchange resins may be related to the size of the trialkylamine functional groups on the resin. The Sybron® SR-6 has the largest trialkyl functional group (i.e., tributylamine) and exhibited the slowest exchange kinetics among the three monofunctional anion-exchange resins (Table 2). The 168-h $K_d$ was more than 4 times higher than the 1-h $K_d$ value of the Sybron® SR-6. On the other hand, the $K_d$ values for the Amberlite® IRA-900 resin (with small trimethylamine groups) increased less than ~ 24% within the same equilibrium time period.

An increase in the size of the trialkylamine functional group however greatly enhanced the adsorption selectivity ($K_d$) of the resin for $ClO_4^-$ adsorption (Table 2). The Amberlite® IRA-900 resin (with trimethylamine) exhibited the least selectivity, and its $K_d$ values were 2–6 times lower than that of the Sybron® SR-6 resin, depending on the equilibration time. These observations were consistent with our previous results with pertechnetate uptake by anion-exchanges resins[11,16] and formed the scientific basis for the formulation of the bifunctional resins as discussed in detail in Chapter 15.[13] Superior performance of these bifunctional resins was thus achieved through the use of two trialkylamine functional groups, one having long chains (trihexylamine) for higher selectivity for poorly hydrated large anions and one having shorter chains (e.g., triethylamine) for enhanced exchange kinetics.

**Table 2.** Distribution coefficients ($K_d$) of $ClO_4^-$ on monofunctional and bifunctional anion-exchange resins[a]

| **Resin** | **Amine functional groups** | **1-h $K_d$** (mL $g^{-1}$) | **24-h $K_d$** (mL $g^{-1}$) | **168-h $K_d$** (mL $g^{-1}$) |
|---|---|---|---|---|
| | | *Monofunctional anion-exchange resins* | | |
| Amberlite® IRA-900 | Trimethyl | 34,960 | 40,100 | 43,030 |
| Purolite® A-520E | Triethyl | 97,000 | 203,000 | 217,000 |
| Sybron® SR-6 | Tributyl | 65,400 | 250,000 | 282,000 |
| | | *Bifunctional anion-exchange resins* | | |
| VP-02-217 | Trihexyl/tripropyl | 408,500 | 714,000 | >3,330,000[b] |
| RO-02-119 | Trihexyl/triethyl | 285,000 | >3,330,000[b] | >3,330,000[b] |
| Purolite D-3696 | Trihexyl/triethyl | 164,800 | 1,877,000[b] | 1,842,000[b] |

[a]The initial $ClO_4^-$ concentration was 10 mg $L^{-1}$, and the equilibrium concentration ranged from 0 to 0.4 mg $L^{-1}$. All $K_d$ values have a ± 5% uncertainty.
[b]The equilibrium concentration was between 0 and 6 µg $L^{-1}$ and was below or near the detection limit, ~ 3 µg $L^{-1}$. The $K_d$ value should be considered as an estimate.

As illustrated in Table 2, all bifunctional resins exhibited a higher selectivity (~3–10 times higher $K_d$ values) than did the three commercial monofunctional resins. The $K_d$ values of these bifunctional resins ranged from ~ 165,000 to > 3,300,000 mL $g^{-1}$, depending on the equilibration time. On the other hand, the three commercial monofunctional resins showed a lower selectivity; $K_d$ values ranged from ~ 35,000 to 280,000 mL $g^{-1}$. At an initial concentration of 10 mg $L^{-1}$, the bifunctional resins were able to remove more than 99.8% of $ClO_4^-$ from the solution after 24 h of equilibration. At a lower initial concentration (i.e., ~ 1 mg $L^{-1}$ $ClO_4^-$), they removed all $ClO_4^-$ to below the detection limit within 1 h (data not shown). The Purolite® A520E and the Sybron® SR-6 resins took ~ 1 week to remove $ClO_4^-$ to less than the detection limit. The 24-h $K_d$ for the bifunctional resin (VP-02-217) was slightly lower than that of the RO-02-119 or Purolite D-3696 bifunctional resins, although the resin had a smaller bead size (60 – 200 mesh) (Table 1). This result may also be attributed to the fact that the VP-02-217 resin contained a relatively large tripropylamine group in comparison with the other two bifunctional resins, which contained triethylamine groups.

Perchlorate adsorption on both the bifunctional and monofunctional resins also appears to be concentration dependent (Figure 1). Nonlinear adsorption isotherms were observed, particularly at relatively high $ClO_4^-$ concentration ranges (up to 100 mg $L^{-1}$ $ClO_4^-$ in the initial solution) or with a relatively high loading of $ClO_4^-$. The adsorption isotherms exhibited a linear partitioning only within a limited low-concentration range, with a higher $K_d$ at a lower $ClO_4^-$ concentration or a lower $K_d$ at a higher $ClO_4^-$ concentration (Figure 1). The nonlinearity of $ClO_4^-$ adsorption may be due to the limited anion-exchange capacity of the resins and because of the presence of other competing anions in the background solution (e.g., $Cl^-$, $HCO_3^-$, $SO_4^{2-}$, $NO_3^-$). These competing anions were present in concentrations ~ 1–4 orders of magnitude higher than that of the $ClO_4^-$ concentration in the test solution.

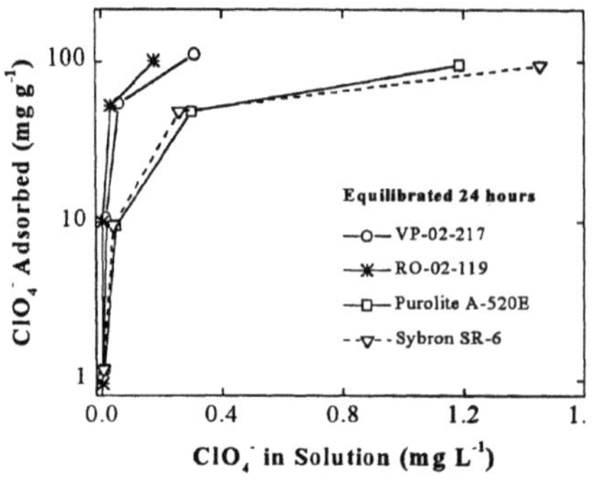


**Figure 1.** Adsorption of $ClO_4^-$ on synthetic resins in a background solution of 3 mM $NaHCO_3$, 1 mM $CaCl_2$, 0.5 mM $MgCl_2$, 0.5 mM $Na_2SO_4$, and 0.5 mM $KNO_3$.

Therefore, observations from the batch equilibrium studies suggest that an overall removal rate and efficiency of $ClO_4^-$ by anion-exchange resins will depend on a balance of the following factors, including the adsorption selectivity (which is related to the chemical nature of anions and the size of trialkylamine functional groups), the equilibrium time (or exchange kinetics), the initial $ClO_4^-$ concentration, and the background competing-anion concentrations. The concept may be further illustrated by determining $ClO_4^-$ removal rates at varying initial concentrations (Figure 2) and by performing column flow-through experiments (discussed in the following section). Figure 2a indicates that, at a low initial $ClO_4^-$ concentration but with a relatively high concentration of competing anions (molar ratio of the competing anions to $ClO_4^-$ = 70), the bifunctional resins exhibited faster overall reaction kinetics and a high removal efficiency for $ClO_4^-$. However, at a high $ClO_4^-$ concentration but a relatively low competing anion concentration (molar ratio of the competing anions to $ClO_4^-$ ~ 3.5), the monofunctional resin with small triethylamine functional groups (i.e., Purolite® A520E) showed an overall fast removal rate and the best performance (Figure 2b). On the other hand, the RO-02-119 bifunctional resin showed the slowest reaction kinetics and poor performance to remove $ClO_4^-$ from the test solution initially (within first 5 h). The monofunctional Sybron® SR-6 resin has the longest trialkyl functional groups and therefore exhibited the slowest reaction kinetics and poor performance after 5 h of equilibration. These observations suggest that the bifunctional resins

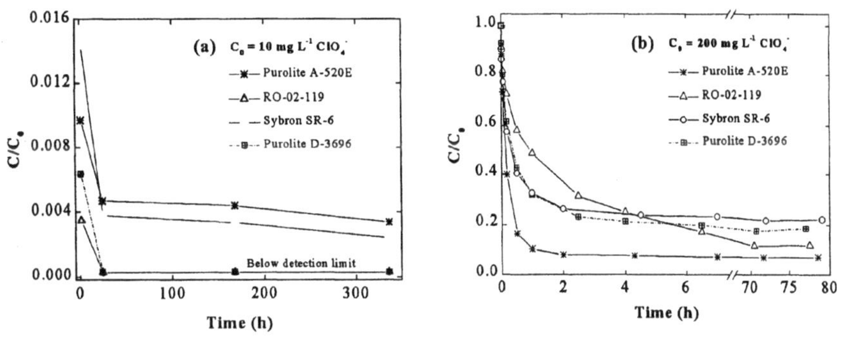


**Figure 2.** $ClO_4^-$ exchange kinetics on anion-exchange resins in a background solution of 3 mM $NaHCO_3$, 1 mM $CaCl_2$, 0.5 mM $MgCl_2$, 0.5 mM $Na_2SO_4$, and 0.5 mM $KNO_3$. (a) Initial $ClO_4^-$ concentration = 10 mg $L^{-1}$. (b) Initial $ClO_4^-$ concentration = 200 mg $L^{-1}$.

are particularly effective in removing the low levels of $ClO_4^-$ in aqueous solution commonly encountered in natural groundwater or surface water.

**Column Flow-Through Studies and Field Performance Issues**

On the basis of adsorption equilibrium studies, the trihexyl/triethyl bifunctional resin (RO-02-119) was further evaluated for $ClO_4^-$ removal in laboratory column flow-through experiments along with two commercial monofunctional resins (Purolite® A520E and Sybron® SR-6) (Figure 3). Results indicated that all of these anion-exchange resins performed well in removing $ClO_4^-$ at an initial concentration of 1100 μg $L^{-1}$ from both the laboratory-synthesized test solutions and actual groundwater. Less than 8% breakthrough of $ClO_4^-$ was observed after ~ 5000 BV of the test solution were passed through these columns at a rapid flow rate (~ 9.5 BV $min^{-1}$). The bifunctional resin, RO-02-119, performed the best, as predicted from the previous batch-adsorption studies. No $ClO_4^-$ breakthrough was observed within the instrument detection limit (~3 μg $L^{-1}$) until ~2000 BV of test solution had been treated with this resin column; less than ~ 0.5% breakthrough of $ClO_4$ was observed after ~ 5000 BV. The two commercial resins, however, showed a 2% breakthrough of $ClO_4^-$ following treatment of < 100 BV of the test solution, and about 8% breakthrough of $ClO_4^-$ after ~ 5000 BV of test solution had passed through the two columns. These results are consistent with previous observations when these anion-exchange resins were used to remove $TcO_4^-$ from contaminated groundwater.[11,16] Results also confirmed our previous conclusion that the new bifunctional resins have a superior selectivity and removal efficiency for removing large anions such as $TcO_4^-$ and $ClO_4^-$.[11,16]

It is noted that $C/C_0$ (Figure 3) increased with time (or with the number of bed volumes) but to a much lesser extent with the bifunctional resin than with the two commercial monofunctional resins. Early breakthroughs (~ 2%) of $ClO_4^-$ were observed in the Purolite and Sybron columns in comparison with breakthroughs of the bifunctional resins. Similarly, when a higher initial $ClO_4^-$ concentration (10 mg $L^{-1}$) was used as the influent solution, a 9% breakthrough of $ClO_4^-$ (< 50 BV) was immediately observed on the Purolite® A520E column. On the other hand, the bifunctional resin (RO-02-119) treated ~ 2,200 BV of the test solution before a 9% breakthrough of $ClO_4^-$ occurred (data not shown). This property of the bifunctional resins is of particular importance and is again attributed to their fast reaction kinetics and high selectivity for $ClO_4^-$. Because a relatively fast flow rate was applied (with a residence time under 6 s), the high selectivity and fast reaction kinetics of resins for sorption of large, poorly hydrated anions are probably the most significant factors in removing trace quantities of $ClO_4^-$ from contaminated groundwater.

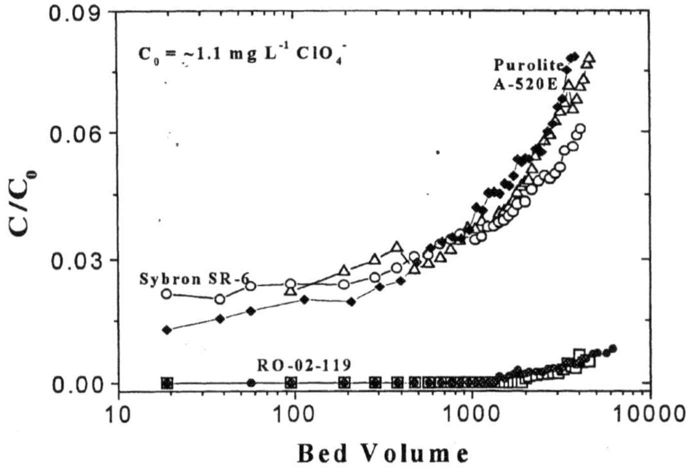


**Figure 3.** Breakthrough of $ClO_4^-$ on anion-exchange resins in a background solution of 3 mM $NaHCO_3$, 1 mM $CaCl_2$, 0.5 mM $MgCl_2$, 0.5 mM $Na_2SO_4$, and 0.5 mM $KNO_3$ (open symbols). The Purolite A-520E and RO-02-119 resins were also tested in a groundwater spiked with ~1.1 mg $L^{-1}$ $ClO_4^-$ (solid symbols).

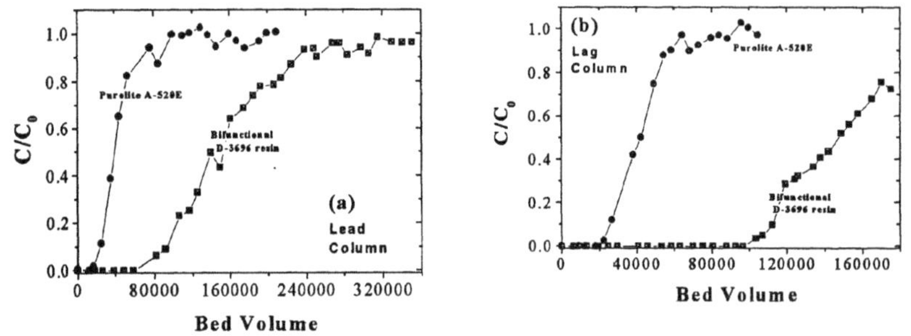


**Figure 4.** Breakthrough of $ClO_4^-$ on the lead and lag columns of the bifunctional resin (D-3696) and the mono-functional resin (Purolite A-520E). **(a)** Flow rate (~3.6 BV $min^{-1}$) and bed volumes were calculated based on one column (i.e., the lead column with a total bed volume = 57.4 mL); **(b)** Flow rate (~1.8 BV $min^{-1}$) and bed volumes were calculated based on two columns connected in a series (total bed volume = ~115 mL).

To determine the effectiveness of the bifunctional resins subjected to field groundwater conditions, a commercial scale-up version of the bifunctional resin (Purolite D-3696) was tested in parallel with the Purolite® A520E resin at a field site in northern California. Figures 4a,b illustrate the breakthroughs of $ClO_4^-$ in both the lead and lag columns of the Purolite D-3696 and A-520E columns from the field experiment. As indicated previously, both the lead and lag columns were connected in a series, although the breakthrough of $ClO_4^-$ on the lead column was monitored independently. As a result, the volumetric flow rate in the lead column (~3.6 BV $min^{-1}$, based on a single column of 57.4 mL BV) should be considered to be twice as much in the lag column (~ 1.8 BV $min^{-1}$; calculations based on two columns). Results for the lag column (Figure 4b) indicated that a breakthrough of $ClO_4^-$ on the Purolite® A-520E columns occurred after ~ 20,000 BV of groundwater had passed through the column. On the other hand, $ClO_4^-$ breakthrough occurred at ~100,000 BV in the D-3696 resin column. At a 10% breakthrough, the bifunctional D-3696 resin treated up to ~ 110,000 BV of groundwater containing ~ 50 μg $L^{-1}$ $ClO_4^-$ and running at 1.8 BV $min^{-1}$. The Purolite® A-520E treated up to ~ 24,000 BV of the groundwater under the same experimental conditions. Interestingly, on the lead columns of the Purolite® A-520E and the D-3696 resins (Figure 4a), a 10% breakthrough of $ClO_4^-$ also occurred when ~24,000 and ~100,000 BV of groundwater had passed through these columns, respectively. These results suggest that the volumetric flow rates (within 2–4 BV/min) did not appear to significantly affect the capability of the resins to remove $ClO_4^-$ from the contaminated groundwater. The same amount of groundwater may be treated per bed volume of resin or the breakthrough of $ClO_4^-$ occurs whenever the resins reach their adsorption capacity for $ClO_4^-$. Similarly, a greater volume of groundwater may be treated with the same resin bed if the groundwater contains a lower concentration of $ClO_4^-$, or vice versa.

It is pointed out that the apparent adsorption capacity of resins is not equivalent to their inherent anion-exchange capacities listed in Table 1. Under field flow-through conditions and in the presence of high concentrations of competing anions, the ability or capacity of these anion-exchange resins to remove $ClO_4^-$ depends on the selectivity ($K_d$) and $ClO_4^-$ concentration ($C$) in the groundwater (~ 50 μg $L^{-1}$); i.e., adsorption capacity = $K_d \cdot C$ (by assuming a linear partitioning of $ClO_4^-$ on resins). A mass balance analysis of the field experimental results indicated that, at the end of the field experiment (with > 70% breakthrough of $ClO_4^-$, Figure 4), the Purolite® A520E and the bifunctional D-3696 resins retained an average of ~ 5.8 and 23 mg $g^{-1}$ $ClO_4^-$ (or ~ 0.058 and 0.23 meq $g^{-1}$), respectively, on a dry-weight basis. This was equivalent to ~ 2% and ~ 10% of the anion-exchange capacity of the Purolite® A520E and the bifunctional D-3696 resins (Table 1). In other words, more than 98% of the anion-exchange sites on the monofunctional resins and ~ 90% of the sites on the bifunctional resins were adsorbed with competing anions such as $HCO_3^-$, $NO_3^-$, $Cl^-$, and $SO_4^{2-}$ in the groundwater. The $K_d$ values for the Purolite® A520E and the bifunctional D-3696 resin columns were thus estimated to be ~ 117,000 and ~ 460,000 mL $g^{-1}$, respectively, which fall between the 1-h and 24-h $K_d$ values determined in laboratory batch studies (Table 2). However, it must be pointed out that, unlike laboratory batch equilibrium studies, a non-equilibrium condition exists under field flow-through conditions because the residence time of groundwater passing through these treatment columns was only ~ 10 s, assuming an effective pore volume of 0.31 in the resin bed.[16] It can also be anticipated that the amounts of $ClO_4^-$ retained by these resin columns were not uniformly distributed across the resin bed, as had been observed for $TcO_4^-$ treatment by these anion-exchange resins.[12] Nevertheless, results of this study indicate that the inherent anion-exchange capacities of these resins (Table 1) do not determine and are not proportional to the ability and performance of these resins to remove $ClO_4^-$ under specific environmental conditions. However, these observations confirm that high selectivity and exchange kinetics are of paramount importance in determining the ability of anion-exchange resins to remove $ClO_4^-$

from only slightly contaminated groundwater, e.g., < 0.1 mg $L^{-1}$ $ClO_4^-$. The bifunctional resins are about ~ 5 times more selective than one of the best commercial nitrate-selective resins (Purolite® A-520E).

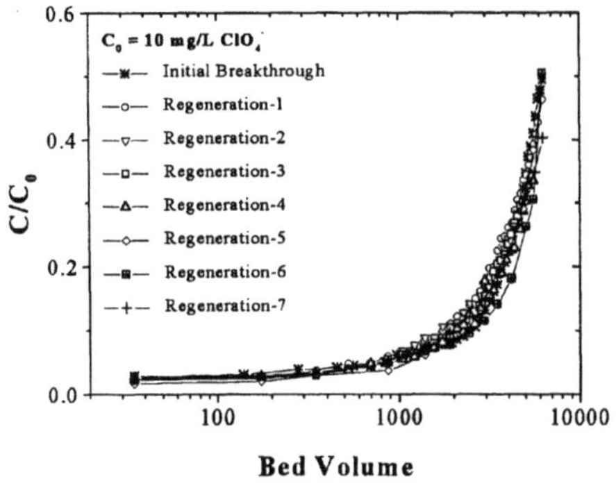


**Figure 5.** Comparison of breakthrough curves of $ClO_4^-$ between the untreated Purolite D-3696 resin column and the same column after repeated regenerations (column dimensions: 10 × 22 mm; flow rate of 30 mL $min^{-1}$).

## Resin Regeneration and Application for Water Treatment

As stated previously, high operating costs for regenerant chemicals and their disposal render conventional ion-exchange techniques unattractive for water treatment to remove $ClO_4^-$ contamination. Moreover, the conventional regeneration technique by washing with a brine solution (e.g., 2 M NaCl) is obviously ineffective for regenerating highly selective anion-exchange resins such as the bifunctional resins.[4,5,13] Therefore, a new paradigm for resin regeneration was investigated and developed in our laboratory. At the end of the field experiment, the Purolite D-3696 bifunctional resin was recovered and was subjected to regeneration with a proprietary regeneration scheme. The experiment was again performed in a small glass chromatographic column (10 × 22 mm) but with a high initial $ClO_4^-$ concentration (10 mg $L^{-1}$ $ClO_4^-$ in the background test solution) in order to accelerate the laboratory breakthrough experiments. Results of $ClO_4^-$ breakthroughs on the regenerated columns were compared with that of the untreated Purolite D-3696 resin column (or the initial breakthrough) so that the effectiveness of the regeneration process could be directly assessed (Figure 5). Seven regeneration cycles were applied on the same resin column, and we achieved a complete or nearly complete regeneration of the resin without an indication of significant deterioration or loss of performance for $ClO_4^-$ removal. On the basis that the bifunctional resin (Purolite D-3696) is able to treat ~ 100,000 BV of groundwater containing ~ 50 µg $L^{-1}$ $ClO_4^-$ (running at ~ 2 BV $min^{-1}$) (Figure 4), we estimate that < 5 BV of regenerant waste (equivalent to < 0.005% of treated groundwater) may be produced by applying our innovative regeneration process. Additionally, if the contaminated water contains less $ClO_4^-$, more water may be treated with the same resin bed, and the process thus requires even fewer regeneration cycles.

In summary, a new class of bifunctional anion-exchange resins with improved selectivity and sorption kinetics has been evaluated in both laboratory and field experiments for sorption of $ClO_4^-$. Results indicate that the bifunctional resins performed ~5 times better than one of the best commercial monofunctional resins (Purolite® A-520E) and were particularly effective in removing trace quantities of $ClO_4^-$ in groundwater. The present study also revealed that the overall performance of these anion-exchange resins to remove $ClO_4^-$ from water does not rely on the inherent anion-exchange capacity but on the exchange kinetics and the adsorption selectivity of resins. A high selectivity of the resin will depend on a number of factors, including the chemical nature of anions, the size of trialkylamine functional groups on the resin, the initial $ClO_4^-$ concentration, and background competing anion concentrations. This study also demonstrates that the bifunctional anion-exchange resins may provide an efficient and cost-effective solution to treat $ClO_4^-$-contaminated groundwater or surface water because they can process ~ 5 times or more water than monofunctional anion-exchange resins at a high flow rate

(~ 2–4 BV $min^{-1}$). Other advantages are that highly selective bifunctional resins require no pretreatment and no addition or removal of unwanted organic or inorganic components in the water.

## ACKNOWLEDGMENTS

We are grateful to T. Blackman, D. Jensen, and J. H. O'Brien at Lockheed Martin Corporation for their continued support and K. McCracken at the Oak Ridge National Laboratory (ORNL) for technical assistance. We also thank K. Baltz, M. DeLaCruz, and R. Pedlar at Radian International for site support and assistance for conducting the field experiment. The synthesis and development of bifunctional resins were sponsored by the Efficient Separations and Processing Cross-Cutting Program, Office of Science and Technology, Office of Environmental Management, U.S. DOE. ORNL is managed by Lockheed Martin Energy Research Corp. for the U.S. DOE under contract DE-AC05-96OR22424.

## REFERENCES

1. Urbansky, E.T. "Perchlorate chemistry: implications for analysis and remediation." *Bioremediation Journal* **1998**, *2*, 81–95.

2. Damian, P.; Pontius, F.W. "From rockets to remediation: the perchlorate problem." *Environmental Protection.* **1999**, 24–31.

3. Jackson, P.E.; Laikhtman, M.; Rohrer, J.S. "Determination of trace level perchlorate in drinking water and ground water by ion chromatography." *Journal of Chromatorgraphy A* **1999**, *850*, 131–135.

4. Batista, J.R.; McGarvey, F.X.; Vieira, A.R. "The removal of perchlorate from waters using ion exchange resins." In THIS VOLUME: *Perchlorate in the Environment*, E.T. Urbansky, Ed. Kluwer/Plenum: New York, NY, 2000; Ch. 13.

5. Tripp, A.R.; Clifford, D.A. "The treatability of perchlorate in groundwater using ion exchange technology." In THIS VOLUME: *Perchlorate in the Environment*, E.T. Urbansky, Ed. Kluwer/Plenum: New York, NY, 2000; Ch. 12.

6. Rikken, G.B.; Kroon, A.G.M.; Van Ginkel, C.G. "Transformation of (per)chlorate into chloride by a newly isolated bacterium: reduction and dismutation." *Applied Microbiology and Biotechnology* **1996**, *45*, 420–426.

7. Herman, D.C.; Frankenberger, W.T. Jr. "Microbial-mediated reduction of perchlorate in groundwater." *Journal of Environmental Quality* **1998**, *27*, 750–754.

8. Wallace, W. "Perchlorate reduction by a mixed culture in an up-flow anaerobic fixed-bed reactor." *Journal of Industrial Microbiology and Biotechnology* **1998**, *20*, 0126–0131.

9. Urbansky, E.T.; Schock, M.R. "Issues in managing the risks associated with perchlorate in drinking water." *Journal of Environmental Management* **1999**, *56*, 79–95.

10. Alexandratos, S.D.; Hussain, L.A. "Bifunctionality as a means of enhancing complexation kinetics in selective ion exchange resins." *Industrial Engineering Chemistry Research* **1995**, *34*, 251–258.

11. Bonnesen, P.V.; Brown, G.M.; Bavoux, L.B.; Presley, D.J.; Moyer, B.A.; Alexandratos, S.D.; Patel, V.; Ober, R. "Development of bifunctional anion exchange resins with improved selectivity and sorptive kinetics for pertechnetate. 1. Batch-equilibrium experiments." *Environmental Science and Technology*, submitted.

12. Gu, B.; Liang, L.; Brown, G. M.; Bonnesen, P. V.; Moyer, B. A.; Alexandratos, S. D.; Ober, R. *A Field Trial Novel Bifunctional Resins For Removing Pertechnetate ($TcO_4^-$) From Contaminated Groundwater*. Oak Ridge National Laboratory, Oak Ridge, TN, 1998; ORNL Doc. No. ORNL/TM-13593.

13. Brown, G.M.; Bonnesen, P.V.; Moyer, B.A.; Gu, B.; Alexandratos, S.D.; Patel, V.; Ober, R. "The design of selective resins for the removal of pertechnetate and perchlorate from groundwater." In THIS VOLUME: *Perchlorate in the Environment,* E.T. Urbansky, Ed. Kluwer/Plenum: New York, NY, 2000; Ch. 15.

14. Gu, B.; Brown, G. M.; Alexandratos, S. D.; Ober, R.; Patel, V. *Selective Anion Exchange Resins for the Removal of Perchlorate ($ClO_4^-$) from Groundwater*. Oak Ridge National Laboratory, Oak Ridge, TN, 1999; ORNL Doc. No. ORNL/TM-13753.

15. Kawasaki, M.; Omori, T.; Hasegawa, K. "Adsorption of pertechnetate on an anion exchange resin." *Radiochimica Acta* **1993**, *63*, 53–56.

16. Gu, B.; Brown, G.M.; Bonnesen, P.V.; Liang, L.; Moyer, B.A.; Ober, R.; Alexandratos, S.D. "Development of novel bifunctional anion-exchange resins with improved selectivity for pertechnetate. 2. Column breakthrough and field studies." *Environmental Science and Technology*, submitted.

# Chapter 17

# LONG-TERM RELEASE OF PERCHLORATE AS A POTENTIAL SOURCE OF GROUNDWATER CONTAMINATION[⊗]

Tracey C. Flowers and James R. Hunt*

Department of Civil and Environmental Engineering, University of California–Berkeley, Berkeley, California 94720-1710

## INTRODUCTION

Recently, perchlorate has been found at appreciable levels in the drinking water supplies of California and other southwestern states.[1] Water suppliers have detected concentrations at part per billion levels, but some have found concentrations as high as 280 $\mu g\ L^{-1}$. Perchlorate has also been detected in Lake Mead and the Colorado River at ppb levels. The source of the perchlorate ion is the salt, ammonium perchlorate, which is used predominately as an oxidizer in solid rocket fuel and fireworks. Perchlorate has been released to the environment from manufacturing facilities and solid rocket booster testing and maintenance sites. Perchlorate has also been reported in certain fertilizers at concentrations up to 0.84 % w/w.[2]

Perchlorate from ammonium perchlorate is expected to be highly mobile in surface and groundwaters. The water solubility of ammonium perchlorate is estimated from Schumacher[3] to be 200 $g\ L^{-1}$. The sodium, calcium, and magnesium salts are even more soluble. An aqueous solution in equilibrium with solid ammonium perchlorate has an estimated density of 1.11 $g\ cm^{-3}$, substantially more dense than freshwater and seawater. Perchlorate is not expected to sorb to mineral surfaces under near neutral pH conditions and there is little evidence of biological transformations under natural conditions. However, researchers have had moderate success with transforming perchlorate using bacteria[4, 5] and plants.[6]

[⊗] This work was presented at the 218th national meeting of the American Chemical Society as part of the Environmental Division Symposium Perchlorate in the Environment, held August 22–24, 1999, in New Orleans, Louisiana.

*Author to whom correspondence should be directed. Phone: 510-642-0948. Fax: 510-642-7483. Electronic mail: hunt@ce.berkeley.edu

*Perchlorate in the Environment*, edited by Urbansky.
Kluwer Academic/Plenum Publishers, New York, 2000.

Research directions on perchlorate are following a similar path as most newly recognized contaminants. Risk assessments for perchlorate in water and food have been dependent upon rather limited data from the therapeutic application of perchlorate to decrease thyroid hormone production.[7] Perchlorate and iodine have similar ionic size, and thus, perchlorate blocks iodine uptake and reduces hyperthyroidism. Also, perchlorate can potentially cause congenital hypothyroidism, which is a preventable cause of mental retardation in developing fetuses.[8] Extrapolation of effects at high doses on adults to low doses for other subpopulations leads to uncertainties. As a consequence, the California Department of Health Services has established a provisional drinking water action level of 0.018 mg $L^{-1}$.[1]

Assessing environmental exposure to perchlorate requires analytical methods for detection and quantification. Improvements in ion chromatography over the past few years have reduced the detection limit from 1 mg $L^{-1}$ to 0.004 mg $L^{-1}$. The improved analytical capabilities and low provisional action level have led to widespread detection in surface and groundwater supplies of drinking water. In fact, levels in southern California have been traced to former manufacturing facilities in southern Nevada. Discharges into Las Vegas wash were transported through Lake Mead, down the Colorado River, and diverted to southern California. Besides being a health concern, perchlorate has become a tracer to remind us how connected our water systems have become.

Given the evidence for occurrence of perchlorate and the possible health impacts, exposure reduction is warranted. The focus of most research on exposure reduction has been the investigation of technologies for treating perchlorate-contaminated drinking water. The general inertness of perchlorate has required considerable creativity and new technology development. An alternative approach that has received less attention is to consider the source of perchlorate contamination at manufacturing and rocket testing facilities. The properties of a high aqueous solubility and resulting dense brine can result in a long-term source of groundwater contamination. The seven order of magnitude difference between the solubility and the provisional action level suggests source control is desirable.

In this chapter, transport models are used to determine the subsurface distribution of perchlorate following the release of a concentrated solution. Analytical solutions to the diffusion equation are combined to estimate the time scale for recovery of perchlorate by groundwater extraction assuming a relatively simple representation of the subsurface. The analysis demonstrates that recovery by groundwater extraction will be mass-transfer limited with a time scale of approximately 100 years. These results suggest that more attention should be given to source identification and control.

## PERCHLORATE TRANSPORT PROCESSES

The model predictions to follow will consider the initial flow of a concentrated brine in the subsurface following its discharge. Experimental observations and theories are utilized to predict brine emplacement. The long-term release of perchlorate from the brine is modeled by considering mass transfer processes expected in idealized heterogeneous porous media.

### Density Driven Flow

Brine transport has been considered in a number of groundwater systems. The most studied example is the interface of saltwater and freshwater in coastal aquifers where lighter freshwater overlies denser seawater.[9, 10] Dense contaminant plumes generated by landfills have also been recognized as significant sources of groundwater contamination both at the field scale[11] and the laboratory scale.[12] There have also been extensive modeling studies related to brine formation over salt domes, which are being considered as repositories of radioactive waste.[13]

In Russia, Samsonova and Drozhko[14] found liquid wastes from nuclear material processing contained a high amount of dissolved salts. These wastes originated from the now defunct nuclear chemical processing facilities of Chelyabinsk–70. They found that the density of these waste streams varied from 1.001 to 1.1 g $cm^{-3}$. They studied the peculiarities of the migration of these industrial waste solutions both in the subsurface environment and in laboratory experiments. The density difference causes gravity to dominate the subsurface flow rather than normal groundwater flow driven by regional gradients in water pressure. The waste plume of concentrated brine emplaced in fractured media under Lake Karachai, Russia, was largely undiluted. Researchers have also recognized this as being an issue with the Hanford Tanks.[15]

Schincariol and Schwartz[16] performed an experimental investigation of variable density groundwater flow in homogeneous and heterogeneous porous media. They found that a density difference between the brine and the ambient groundwater as little as 0.8 mg $cm^{-3}$ would lead to the creation and propagation of gravitational instabilities at realistic groundwater velocities. They studied two configurations of heterogeneity: lenticular and layered. They found that, in the case of layered heterogeneity, the dense plume would accumulate along the bedding interface. In the case of lenticular heterogeneity, the presence of the heterogeneity led to increased mixing and dilution of the contaminant plume.

Transport models can be used to understand the subsurface transport behavior of a brine once it has reached the water table compared to the regional groundwater flow. Horizontal flow in saturated porous media is quantified by Darcy's Law.[17] The horizontal groundwater velocity is directly proportional to the local hydraulic gradient:

$$U_h = -\frac{kg}{\nu}\frac{\mathrm{d}h}{\mathrm{d}x} \tag{1}$$

Where $U_h$ is the horizontal velocity, $k$ is the permeability of the medium, $g$ is gravitational acceleration, $\nu$ is the kinematic viscosity, and $\mathrm{d}h/\mathrm{d}x$ is the local hydraulic gradient.

The vertical velocity of a dense fluid in porous media is directly proportional to the density difference between the brine and the ambient groundwater:[18]

$$U_v = \frac{kg}{\nu}\frac{\Delta\rho}{\rho_w} \tag{2}$$

Where $U_v$ is the vertical velocity, $\Delta\rho$ is the density difference between the brine and the ambient groundwater, and $\rho_w$ is the density of the ambient groundwater. As the brine density increases, the vertical velocity will increase.

The mixing of the dense brine with ambient groundwater depends on the ratio of the vertical, eq 1, to the horizontal, eq 2, velocity given by

$$\frac{U_v}{U_h} = \frac{\Delta\rho/\rho}{\left|\mathrm{d}h/\mathrm{d}x\right|} \tag{3}$$

For the case of ammonium perchlorate, the normalized density difference is 0.11 and under environmentally realistic conditions, the hydraulic gradient would be at most 0.01. Therefore, the ratio of the vertical to the horizontal velocity is 10 or greater.

Based on the work of Schincariol and Schwartz,[16] a brine with a density contrast with respect to ambient groundwater of 0.11 g $cm^{-3}$ released into an aquifer with layered heterogeneity, would be expected to form an undiluted pool on the bedding interface. Evidence of this lack of mixing is further given by the research conducted by List.[18] List's work determined the potential stability of a flow system in which a more dense fluid was introduced above a less dense fluid flowing through a porous medium. List found that these flows were potentially stable. The factors governing the stability of this flow system were the density difference between the brine and ambient groundwater and the ratio of the longitudinal to the transverse dispersivity. As density difference is increased, the flow is less likely to be stable.

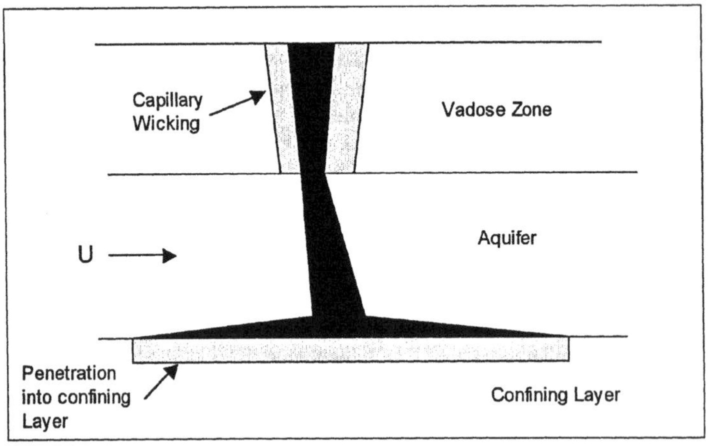


**Figure 1.** Conceptual model of subsurface brine transport. The cross-hatching indicates long-term sources.

Figure 1 outlines the important physical transport mechanisms of a perchlorate brine in the subsurface. The figure assumes that a dense brine has been released from a surface disposal site. Gravity drives the vertical migration of the brine in the vadose zone and capillarity causes horizontal migration. Once the brine reaches the water table, it will rapidly sink through the aquifer. This will lead to the creation of a pool of undiluted brine on a lower permeability formation. Density differences will cause the brine to sink into the lower-permeability confining layer as well, but at a slower velocity.

After the discharge has been stopped at the surface, the brine pool at the aquifer bottom will cease growing and will either sink into the less permeable region or be flushed out of the aquifer. The brine emplaced in the less permeable region can diffuse back into the aquifer and represent a long-term source of perchlorate contamination.

**Mass-Transfer Limited Release**

Once brine has been flushed from the aquifer, additional brine is expected to be retained in the vadose zone and within the less-permeable region along the bottom of the aquifer.

These regions are illustrated in Figure 2 and represent long-term sources of brine for the flowing groundwater. Releases will occur from the vadose zone by infiltration events and by the rise of the water table into the capillary bound brine. While these vadose zone processes are likely important, they will not be considered further in this chapter. The emphasis here is quantifying the release from the less permeable region into the aquifer through diffusion. Figure 2 defines a coordinate system with $x = 0$ representing the upstream start of the brine pool that is of length $L$. The vertical coordinate is $z$ positive upwards from the bottom of the aquifer, the aquifer thickness is $H$ and the brine pool thickness is $b$. The goal is to predict the downstream concentration as would be detected in a well that sampled over the complete depth of the aquifer.

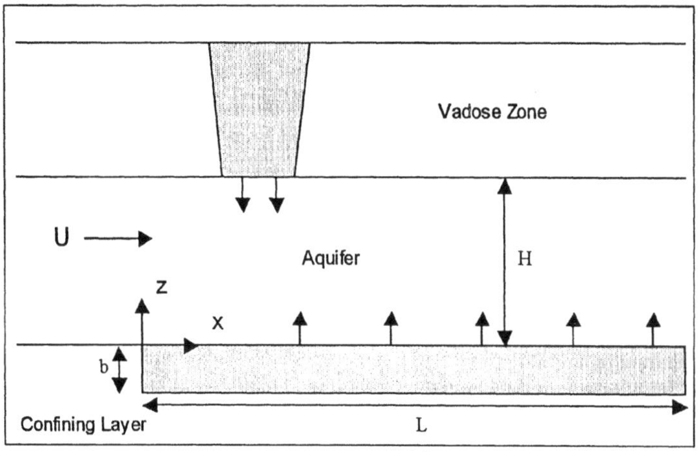


**Figure 2.** Schematic model of long-term release.

In order to utilize analytical solutions for this complex problem, mass transfer from the less-permeable layer is modeled in three phases. The first phase assumes steady state release of brine into the aquifer with saturated brine at the interface of the aquifer and the confining layer. The second phase accounts for one-dimensional diffusion from an infinitely deep confining layer into the aquifer. The third phase accounts for the finite depth of the confining layer. For phase 1, where the aquifer limits the diffusive loss of the brine, the steady-state advection-diffusion equation in the aquifer can be written as:

$$U\frac{\partial C}{\partial x} = D_{\mathrm{PM}}\frac{\partial^2 C}{\partial z^2} \qquad z \geq 0 \tag{4}$$

Subject to the boundary conditions:

$$\begin{aligned} &C(x=0,z)=0 && z>0 \\ &C(x,z=0)=C_{sol} && 0 \leq x \leq L \\ &C(x,z\rightarrow\infty)=0 && x \geq 0 \end{aligned} \tag{5}$$

Where $D_{PM}$ is the molecular diffusivity ($D_{mol}$) corrected for porosity ($n$) and tortuosity ($\tau$) and $C_{sol}$ is the aqueous solubility of ammonium perchlorate. The solution to this system will give the concentration in the aquifer as a function of length along the spill and vertical position in the aquifer. While the actual geometry for the aquifer is of finite thickness, the solution to the semi-infinite diffusion equation provides a reasonable approximation:[19]

$$C(x,z) = C_{sol}\,\mathrm{erfc}\left(\frac{z}{2\sqrt{\frac{D_{PM}x}{U}}}\right) \tag{6}$$

Where erfc(η) represents the complimentary error function. The complimentary error function is defined as[20]

$$\mathrm{erfc}(\eta) = 1 - \mathrm{erf}(\eta) = 1 - \frac{2}{\sqrt{\pi}}\int_0^{\eta} e^{-\xi^2}\,\mathrm{d}\xi \tag{7}$$

The average concentration leaving the spill zone is calculated by integrating the concentration profile in eq 6 at $x = L$ over the entire height of the aquifer:

$$\overline{C} = C_{sol}\left[\mathrm{erfc}(\omega) + \frac{1-e^{-\omega^2}}{\omega\sqrt{\pi}}\right] \qquad \omega = \frac{H}{2\sqrt{\frac{D_{PM}L}{U}}} \tag{8}$$

The aquifer-controlled solution is valid for times less than the time for the first flush.

Once there is removal of brine from the interface between the aquifer and the confining layer, further releases are limited by diffusion within the confining layer. These later time solutions are phases 2 and 3. The confining layer is assumed to be homogenous along the length of the pool and brine concentrations are described by the one-dimensional diffusion equation:

$$\frac{\partial C}{\partial t} = D_{PM}\frac{\partial^2 C}{\partial z^2} \tag{9}$$

Subject to the initial and boundary conditions:

$$\begin{aligned} &C(z,t=0) = C_{sol} && -b \le z \le 0 \\ &C(z=0,t) = 0 && \\ &\left.\frac{\partial C}{\partial z}\right|_{z=-b} = 0 && t > 0 \end{aligned} \tag{10}$$

The solution under these conditions is:

$$C(z,t)=\sum_{n=1}^{\infty}\frac{2C_{sol}}{\pi\left(n-\frac{1}{2}\right)}\sin\left(\lambda_n z\right)e^{-\lambda_n^2 D_{PM}t}$$
$$\text{where: } \lambda_n=-\left(n-\frac{1}{2}\right)\frac{\pi}{b} \qquad (11)$$

The mass flux into the aquifer is a function of time and can be determined by Fick's Law evaluated at the interface of the aquifer and confining layer:

$$F(t)=-D_{PM}\left.\frac{\partial C}{\partial z}\right|_{z=0^-}=\frac{2D_{PM}C_{sol}}{b}\sum_{n=1}^{\infty}e^{-\lambda_n^2 D_{PM}t} \qquad (12)$$

Finally, the average concentration leaving the spill zone as a function of time can be calculated by mass balance:

$$\overline{C(t)}=\frac{F(t)\cdot L}{U\cdot H}=\frac{2\cdot L\cdot D_{PM}\cdot C_{sol}}{U\cdot H\cdot b}\sum_{n=1}^{\infty}e^{-\lambda_n^2 D_{PM}t} \qquad (13)$$

The infinite series in eq 13 does not converge well for times less than the time to diffuse from the bottom of the confining layer ($b^2\tau/2D_{mol}$). Phase 2 of the analytical model solves this problem by assuming that the confining layer is infinite. Under these conditions, the concentration profile in the confining layer can be described by an error function solution:

$$C(z,t)=C_{sol}erf\left(\frac{-z}{2\sqrt{D_{PM}t}}\right) \qquad -b\le z\le 0,\ \frac{L}{4U}\le t\le\frac{b^2\tau}{2D_{mol}} \qquad (14)$$

The mass flux into the aquifer is:

$$F(t)=C_{sol}\sqrt{\frac{D_{PM}}{\pi\cdot t}} \qquad (15)$$

The average concentration leaving the spill zone is:

$$\overline{C}=\frac{C_{sol}\cdot L}{U\cdot H}\sqrt{\frac{D_{PM}}{\pi\cdot t}} \qquad (16)$$

Phase 3 of the analytical model accounts for a finite confining layer and for times greater than the time to diffuse from the bottom of the confining layer. For these times, the infinite series in eq 13 can be truncated the first two terms with very little error. Therefore, the average concentration of perchlorate leaving the spill zone as a function of time can be summarized as:

$$\overline{C(t)} = \begin{cases} C_{sol}\left[\operatorname{erfc}(\omega) + \dfrac{1-e^{-\omega^2}}{\omega\sqrt{\pi}}\right] & 0 \le t \le \dfrac{L}{4U} \\ \dfrac{C_{sol}\cdot L}{U\cdot H}\sqrt{\dfrac{D_{PM}}{\pi t}} & \dfrac{L}{4U} \le t \le \dfrac{b^2\tau}{2D_{mol}} \\ \dfrac{2\cdot C_{sol}\cdot L\cdot D_{PM}}{U\cdot H\cdot b} e^{-\frac{\pi^2 D_{PM} t}{4b^2}}\left(1+e^{-2\frac{\pi^2 D_{PM} t}{b^2}}\right) & t \ge \dfrac{b^2\tau}{2D_{mol}} \end{cases} \quad (17)$$

## RESULTS AND DISCUSSION

The model represented in equation (17) was developed based on established transport processes thought to be applicable for perchlorate in water saturated porous media. In order to present a solution, typical values for the aquifer geometry and contaminant were chosen. A spill length ($L$) of 20 m, an aquifer height ($H$) of 5 m, a confining layer depth ($b$) of 1 m, a molecular diffusivity ($D_{mol}$) of $10^{-9}$ $m^2$ $s^{-1}$, and a groundwater velocity ($U$) of 1 m $day^{-1}$ were assumed. Transverse hydrodynamic dispersion in the aquifer was neglected in this idealized model. A porosity of 0.4 and a tortuosity of 1.4 were assumed giving a porous media dispersivity of approximately 0.009 $m^2$ $yr^{-1}$.

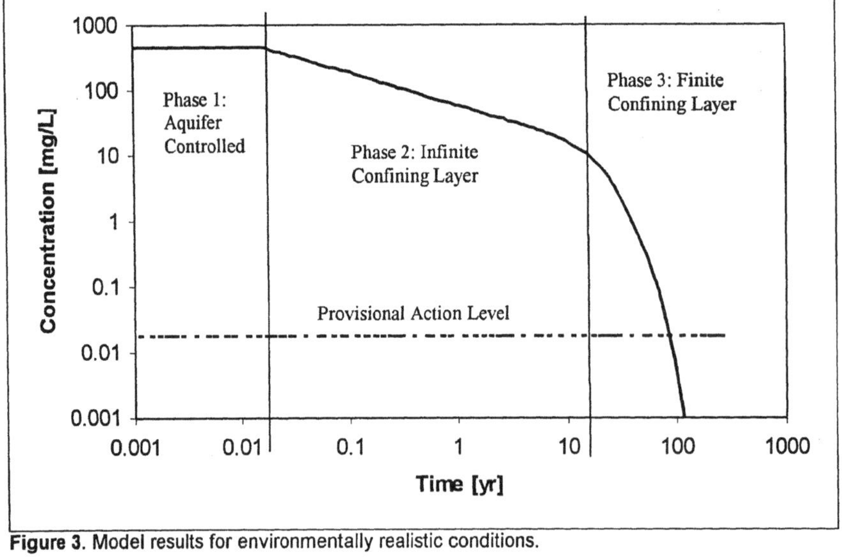


**Figure 3.** Model results for environmentally realistic conditions.

Figure 3 gives the results of the model in equation (17) for the assumed conditions. The average concentration in Phase 1, where release is controlled by the aquifer, is constant and about 450 mg $L^{-1}$. The average concentration remains constant for approximately 5 days, when the transition to Phase 2 occurs. The average concentration in Phase 2 decreases as the square root of time and appears linear on this log-log plot from about 5 days until approximately 16 years, where the transition to Phase 3 occurs. The average concentration in phase 3 decreases exponentially and the provisional action level of 0.018 mg $L^{-1}$ is reached after approximately 100 years.

A highly idealized representation of aquifer heterogeneity was assumed in the development of this model. It was assumed that the heterogeneity of the system could be described by a homogenous aquifer overlying a lower-conductivity confining layer. In order to speculate on the applicability of this model to field conditions, an assumption of the nature of the heterogeneity at the field scale would have to be made. Based on the laboratory experimentations program of Schincariol and Schwartz,[17] if layered heterogeneity is assumed, the brine is expected to collect at the bedding interface. If a lenticular heterogeneity is assumed, the brine would be expected to be distributed more evenly through the aquifer.

The sensitivity of the model to the various parameters can be used as a first approximation to understanding field-scale behavior of the contaminant plume. General trends are that if the length of the spill $L$ increases or the aquifer height $H$ decreases, the average concentration will increase. If a long plume is developed in a thin aquifer, the constant initial concentration described by phase 1 of the model is expected to be high. Additionally, if the groundwater velocity $U$ decreases the average concentration in phase 2 is expected to increase. If the depth of the confining layer $b$ increases, the transition from phase 2 to phase 3 is expected to occur at a later time and the concentrations in phase 3 are expected to increase.

These issues of the slow release of a contaminant from a concentrated source emplaced in less mobile regions have been recognized previously for dense nonaqueous phase liquids (DNAPLs). Consider the organic solvent trichloroethylene (TCE) *[79-01-6]*. TCE has a solubility in water of 1100 mg $L^{-1}$ [21] and a drinking water standard of 0.005 mg $L^{-1}$, [22] a difference of 5.5 orders of magnitude. In the beginning researchers were finding dilute downstream concentrations of TCE in water supply wells. The existence of a pure liquid phase was not appreciated even though such liquids were released to the subsurface from leaking tanks and pits. It was found that pure phase TCE was trapped at the source in subsurface pools and ganglia. Extensive research at laboratory and field scales has shown recovery from these pools is mass transfer limited and time scales for remediation by groundwater extraction is decades.

Table 1 further describes the analogy between DNAPLs and brines using TCE and perchlorate as representative examples. Both TCE and perchlorate have aqueous solubilities many orders of magnitude greater than drinking water standards. Both were released to the subsurface in a concentrated form. While DNAPLs are emplaced by capillary forces and brines by density, both have recoveries severely limited by mass transfer. As a consequence, groundwater extraction alone is inefficient for recovery and time scales are on the order of a hundred years.

**Table 1.** Analogy of perchlorate contamination to TCE contamination.

| | **TCE** | **Perchlorate** |
|---|---|---|
| Solubility | 1100 mg $L^{-1}$ | 200,000 mg $L^{-1}$ |
| DW Standard | 0.005 mg $L^{-1}$ | 0.018 mg $L^{-1}$ |
| Source | Pure Liquid | Brine |
| Emplacement | Lens/Pool | Confining Layer |
| Recovery | Mass Transfer | Mass Transfer |
| Time Scale | ~100 Years | ~100 Years |

## SUMMARY AND CONCLUSIONS

Ammonium perchlorate will form dense brines. A large density contrast with respect to ambient groundwater will dominate subsurface migration and cause the brine to sink through the aquifer until pooling on a lower-permeability confining layer. Density differences will cause the brine to emplace in this lower-permeability confining layer leading to the creation of a long-term source. Release from these sources will be mass transfer limited and will have a time scale of approximately 100 years. While these results are model predictions, they do suggest that recovery of perchlorate and other brines will be inefficient by groundwater extraction alone. Some consideration needs to be given to locating the source of the brine and either isolating or removing the source term. Long-term pump and treat approaches are not going to be economically efficient or protective of groundwater basins as a drinking water source.

## ACKNOWLEDGMENTS

This work was supported by the National Institute of Environmental Health Sciences Superfund Basic Research Program at University of California–Berkeley (NIH P42 ES04705) and by the American Geophysical Union Horton Research Grant.

## REFERENCES

1. California Department of Health Services. *Perchlorate in California Drinking Water.* September 1997.

2. Susarla, S.; Collette, T.W.; Garrison, A.W.; Wolfe, N.L.; McCutcheon, S.C. "Perchlorate identification in fertilizers." *Environmental Science and Technology* **1999**, *33*, 3469–3472.

3. Schumacher, J.G. *Perchlorates: Their Properties, Manufacture, and Uses.* Reinhold: New York, NY, 1960; passim.

4. Herman, D.C.; Frankenberger, W.T. Jr. "Microbial-mediated reduction of perchlorate in groundwater." *Journal of Environmental Quality* **1998**, *27*, 750–754.

5. Logan, B.E. "A review of chlorate- and perchlorate-respiring microorganisms." *Bioremediation Journal* **1998**, *2*, 69–79.

6. Nzengung, V.A.; Wang, C.; Harvey, G. "Plant–mediated transformation of perchlorate into chloride." *Environmental Science and Technology* **1999**, *33*, 1470–1478.

7. Lamm, S.H.; Braverman, L.E.; Li F.X.; Richman, K.; Pino, S.; Howearth, G. "Thyroid health status of ammonium perchlorate workers: A cross–sectional occupational health study." *Journal of Occupational and Environmental Medicine* **1999**, *41*, 248–260.

8. Lamm, S.H.; Doemland, M. "Has perchlorate in drinking water increased the rate of congenital hypothyroidism?" *Journal of Occupational and Environmental Medicine* **1999**, *41*, 409–411.

9. Freeze, R.A.; Cherry, J.A. *Groundwater*. Prentice-Hall: Englewood Cliffs, NJ, 1979; Ch. 8.

10. Galeati, G.; Gambolati, G.; Neuman, S.P. "Coupled and partially coupled Eulerian-Langrangian model of freshwater-seawater mixing." *Water Resources Research* **1992**, *28*, 149–165.

11. Kimmel, G.E.; Braids, O.C. "Preliminary findings of a leachate study on two landfills in Suffolk County, New York." *Journal of Research of the U.S. Geological Survey* **1975**, *3*, 273–280.

12. Oostrom, M.; Hayworth, J.S.; Dane, J.H.; Guven, O. "Behavior of dense aqueous phase leachate plumes in homogeneous porous media." *Water Resources Research* **1992**, *28*, 2123–2134.

13. Oldenberg, C.M.; Pruess, K. "Dispersive transport dynamics in a strongly coupled groundwater–brine flow system." *Water Resources Research* **1995**, *31*, 289–302.

14. Samsonova, L.M.; Drozhko, E. *Deep Injection Disposal of Hazardous and Industrial Waste*. Academic: San Diego, CA, 1996; Ch. 41.

15. Shepard, C.L.; Burghard, B.J.; Friesel, M.A.; Hildebrand, B.P.; Moua, X.; Diaz, A.A.; Enderlin C.W. "Measurement of density and viscosity of one- and two-phase fluids with torsional waveguides." *IEEE Transactions on Ultrasonics, Ferroelectrics, and Frequency Control* **1999**, *46*, 536–548.

16. Schincariol, R.A.; Schwartz, F.W. "An experimental investigation of variable density flow and mixing in homogeneous and heterogeneous media." *Water Resources Research* **1990**, *26*, 2317–2329.

17. Hubbert M.K. "The theory of groundwater motion." *Journal of Geology* **1940**, *8*, 785–944.

18. List E.J. *The Stability and Mixing of a Density–Stratified Horizontal Flow in a Saturated Porous Medium*, California Institute of Technology: Pasadena, CA, 1965; Report Number KH–R–11.

19. Hunt, J.R.; Sitar, N.; Udell, K.S. "Nonaqueous phase liquid transport and cleanup 1. Analysis of mechanisms." *Water Resources Research* **1988**, *24*, 1247–1258.

20. Abramowitz, M.; Stegun, I.A. *Handbook of Mathematical Functions*, 1st ed. Dover: New York, NY, 1970, p. 295.

21. Schwarzenbach, R.P.; Gschwend, P.M.; Imboden, D.M. *Environmental Organic Chemistry*. Wiley: New York, NY, 1993, p. 622.

22. Pontius, F.W. *Water Quality and Treatment*, 4th ed. McGraw-Hill, New York, NY, 1990, p. 35.

# Chapter 18

# EVALUATION OF BIOLOGICAL REACTORS TO DEGRADE PERCHLORATE TO LEVELS SUITABLE FOR DRINKING WATER[⊗]

Bruce E. Logan*

Department of Civil and Environmental Engineering, The Pennsylvania State University, University Park, Pennsylvania 16802

## INTRODUCTION

Perchlorate ($ClO_4^-$) contamination of groundwater has been estimated to potentially affect the drinking water supplies of at least 12 million people in the United States.[1,2] Perchlorate is used as an oxidizer with solid missile and rocket fuels (up to 70% w/w), in automobile air bag inflation systems, and is reported to be present in lawn fertilizers (< 0.84%).[3,4] Perchlorate is a human health concern due to its ability at high doses to interfere with iodine uptake and the ability of the thyroid to regulate hormones and metabolism. There is currently no federal drinking water standard for perchlorate, but many states have adopted an interim provisional drinking water standard of 18 ppb. Perchlorate is stable and extremely soluble in water, and is not efficiently removed by conventional activated carbon and ion exchange processes. However, perchlorate is used as an electron acceptor by a number of bacterial strains under anoxic conditions.[2]

The use of a biological water treatment system in the U.S. presents both societal and engineering challenges. The societal challenge results from the fact that biological treatment of drinking water is not generally accepted or practiced by the water industry. There is currently only one specific-contaminant biological treatment system in the U.S. that treats water for potable uses.[6] This system, located in Coyle, Oklahoma, is designed to biologically reduce

[⊗]A portion of this work was presented at the 218th national meeting of the American Chemical Society as part of the Environmental Division Symposium, Perchlorate in the Environment, held August 22–24, 1999, in New Orleans, Louisiana.

*Phone: 814-863-7908. Fax: 814-863-7304. Electronic mail: blogan@psu.edu.

nitrate in water from average concentrations of 12 mg $L^{-1}$ to <10 mg $L^{-1}$. The biological treatment system removes nitrate to lower levels (2 mg $L^{-1}$), but in order to minimize treatment costs part of the water is bypassed around the system and recombined with treated water to achieve system design specifications. In Europe, there are many biological systems have been used to pretreat drinking water and remove nitrate.[7, 8] All around the world unintentional biological treatment of water occurs indirectly as a result of water filtration in a sand or granular activated carbon (GAC) filter, unless deliberate steps are taken to periodically chlorinate the filter media.[8] Bacteria naturally occur at concentrations of about $10^5$ - $10^7$ cells per mL in streams, lakes and rivers. When this water is filtered, both the bacteria and particulate organic matter are removed, producing a bacterial biofilm on the support medium. These attached bacteria can grow and degrade filtered particles and dissolved organic matter in the water. Under such conditions, biological treatment of the water occurs resulting in lower effluent concentrations of organic matter. In order to limit the distribution of potentially pathogenic bacteria in the water distribution system, water is disinfected before leaving the treatment plant.

The biological treatment of water to remove perchlorate presents some unique engineering challenges relative to nitrate. Although biological systems have been designed to remove nitrate, firms in the U.S. have little experience in designing such systems. A more important factor in design than experience of a particular firm, however, is the engineering challenge that arises from designing a system to achieve the level of treatment required for perchlorate compared to that for nitrate. Nitrate must be removed to only <10 mg $L^{-1}$ for drinking water standards. Most systems are therefore designed to treat nitrate in water from concentrations less than one order-of-magnitude higher to a concentration that is still relatively large compared to a perchlorate concentration of <0.018 mg $L^{-1}$. Perchlorate concentration in contaminated sources vary from these levels up to on the order of $10^3$ mg $L^{-1}$, thus potentially requiring removal over five orders of magnitude (from $10^3$ to $10^{-2}$). The engineering challenge for the design of perchlorate bioreactors is therefore how to ensure this level of treatment efficiency over such a large concentration range.

There are few alternatives to biological treatment of perchlorate contaminated water, whether the water is used as a potable water source or simply treated and released back into the environment. Physical removal using ion exchange, for example, is possible, but removal must be coupled to a processe (chemical or biological) to treat the contaminated brine solution. Even if the perchlorate in the brine is destroyed, the concentrations of other anions in high-salinity brine solutions present residual treatment difficulties. There is recent work on suggesting that ammonia-based regeneration solutions could be used instead of salt to eliminate the production of salty brine solutions,[9] but this technology is not fully developed. Although biological treatment therefore appears to be the most cost effective method of perchlorate treatment, there remain engineering issues on how to design the most effective treatment system. The purpose of this paper is therefore to discuss the options for different types and configurations of reactors, and to comment on advantages and disadvantages of these methods to treat perchlorate contaminated water to drinking water levels.

## EXPERIMENTAL

Perchlorate is used by bacteria under anoxic conditions (in the absence of oxygen) as an electron acceptor. Typically, in situ organic substrate concentrations under natural conditions limit bacterial growth. To achieve perchlorate reduction in a reactor requires the absence of oxygen (or its consumption by the bacterial assemblage), a source of energy (either organic matter or inorganic sources such as hydrogen gas), and carbon (organic matter or dissolved

$CO_2$). Thus, the primary purpose of the reactor is to selectively stimulate the growth of perchlorate reducing microbes (PRMs) over the growth of other microbes capable of growing in the reactor.

**Reactor Kinetics**

Bacteria growth can be assumed to limited by the concentration of one substrate (electron and energy donor) and one electron acceptor (such as oxygen or perchlorate). The rate of cells growth, $r_X$, is

$$r_X = (\mu - b)X = \left[\left(\mu_{max} \frac{S}{K_S + S} \frac{E}{K_E + E}\right) - b\right] X \qquad (1)$$

where $\mu$ is the growth rate, $b$ the decay rate, $\mu_{max}$ the maximum growth rate, $X$ is the biomass concentration, $S$ the substrate and $E$ the electron acceptor concentrations, and $K_S$ and $K_E$ the half saturation constants of the electron donor and acceptor. Assuming bacterial growth is coupled to uptake (i.e cells do not store substrate without cell division) and that by definition no substrate is used during cell decay, the rate of substrate utilization is

$$r_s = -\frac{\mu X}{Y_{X/S}} = -\frac{\mu_{max} X}{Y_{X/S}} \frac{S}{K_S + S} \frac{E}{K_E + E} \qquad (2)$$

where $Y_{X/S}$ is the cell yield defined as the mass of cells produced per mass of substrate used. The utilization rates of substrate and electron donor ($r_E$) are assumed to be coupled and cell decay requires cell respiration, so that

$$r_E = Y_{S/E}\,(r_S + bX) \qquad (3)$$

where $Y_{S/E}$ is the mass of substrate used per mass of electron acceptor. This assumes that all biomass is degradable, although typically 20% of the cell mass is not.[10]

There is also a minimum concentration of substrate necessary to maintain cells due to energy requirements for cell maintenance, which can be calculated[10] as

$$S_{min} = \frac{b\,K_S}{\mu_{max} - b} \qquad (4)$$

With this relationship we can see that some oxidizable substrate will remain in the reactor.

The detention time necessary in a CSTR for a given substrate concentration is:

$$\theta = \frac{1}{\mu - b} \qquad (5)$$

Because a CSTR is completely mixed, $\mu$ is the cell growth rate at the effluent substrate concentration.

**Kinetic Constants**

There is little information available on perchlorate degradation kinetics, and therefore for the purposes of comparing potential reactor advantages and disadvantages, we will have to allow for assumption of typical values. A typical endogenous decay constant[10] is $b = 0.01\ d^{-1}$ and we can assume a typical cell yield of $Y = 0.5$ g cells (g acetate)$^{-1}$ and perchlorate yield[11] of 1.9 mg acetate (mg-perchlorate)$^{-1}$. Kinetics constants recently measured in our laboratory suggest values of $\mu_{max} = 0.2\ h^{-1}$, $K_P = 20$ mg perchlorate $L^{-1}$ and $K_A$=300 mg acetate $L^{-1}$.

**Table 1.** Possible bioreactor types for water treatment

| Suspended Cell Reactors | Fixed-film Reactors |
|---|---|
| - Completely mixed reactor (no cell recycle): CSTR | - Packed bed-saturated flow: PFR |
| - Activated sludge- mixed reactor with cell recycle (many configurations): CSTR and PFR | - Packed bed- unsaturated flow (trickling filter): PFR |
| - Upflow anaerobic sludge blanket: PFR | - Fluidized bed: CSTR |
| | - Rotating biological contactor: CSTR |

**Reactor Types**

The different types of reactors that could be used to treat perchlorate contaminated water are listed in Table 1 according to whether they are based on suspended cells or fixed-films attached to either a carrier particle or fixed medium. These reactor types are also indicated as a plug flow reactor (PFR) or completely mixed/stirred constant-flow reactor (CSTR). Of the different reactors listed in Table 1, only a few have been used for biological water treatment for nitrate or perchlorate removal (Table 2). These reactors are all fixed film systems, with both PFR (packed bed) and CSTR (fluidized bed) reactors having been used to date.

## RESULTS AND DISCUSSION

To compare the potential for different bioreactor systems to treat drinking water, it was assumed that perchlorate needed to be removed to essentially non-detectable levels and that the acetate feed had to be removed to very low levels. Low levels of acetate can also be achieved outside the main reactor in a downstream polishing unit (for example an aerobic filter) but for economic reasons (to minimize purchasing costs of acetate) we would like to have the lowest possible acetate concentrations entering and leaving the reactor. Using the above equations and estimates of kinetic constants, a suspended growth reactor does not appear to be feasible for perchlorate degradation. If we assume for the moment that only acetate limits the growth rate of the cells, then the detention time would be calculated as:

$$\mu = \frac{\mu_{max}\, S}{K_S + S} = \frac{(0.2\,h^{-1})\ (1\,mg\,L^{-1})}{(300\,mg\,L^{-1}) + (1\,mg\,L^{-1})}\ \frac{24\,h}{d} = 0.016\,d^{-1} \tag{6}$$

$$\theta = \frac{1}{\mu - b} = \frac{1}{(0.016\,d^{-1}) - (0.01\,d^{-1})} = 62\,d \tag{7}$$

Thus, we would have to have a cell detention time of 62 days to remove acetate and perchlorate to low levels. This detention time would not permit any type of suspended growth reactor unless the reactor had a cell recycle line such as an activated sludge system. However, this is an unusually long detention time (2 months) and therefore makes even an activated sludge system questionable. If effluent substrate concentrations are allowed to be higher, the required detention time could be dramatically reduced making this system feasible for

**Table 2.** Reactors used to remove nitrate and perchlorate

| Named Reactor | Type of system |
|---|---|
| Biological Activated Filter (BAF) | Packed bed reactor using an ethanol feed, located in Eragny, France.[7] |
| "Denipour" | Fluidized bed system fed ethanol using buoyant carriers in Germany.[7] |
| "Denitropour" | Packed bed reactor in Germany with a hydrogen gas feed.[7] |
| "BioDen" | Packed bed reactors in series (anoxic then aerobic) based on an acetate feed. First drinking water nitrate removal system in U.S..[7] |
| Fluidized bed reactor | GAC and sand supported fluidized bed studies.[12, 13] |
| PSU-O4 Systems | Packed bed reactor systems. One system is a saturated-flow packed bed reactor system fed acetate; the other is an unsaturated packed bed reactor fed hydrogen gas.[14] |

wastewater-type treatment. For example, a suspended growth reactor system was developed to degrade perchlorate in wastewaters where brewer's yeast was used as a feed source.[15] Perchlorate needed to be reduced only to less than 0.5 mg $L^{-1}$ level, and a high organic matter concentration in the effluent was not a concern since wastewater could be further treated to removed this material. Thus, while a CSTR system is possible for wastewater treatment, from this brief analysis it appears that its use would be limited for water treatment. Therefore, only fixed-film systems need to be further considered.

**Fixed Film Reactor Analysis**

Among the fixed-film systems, the main factor is whether the system is completely mixed or not. Of the four types of fixed film reactors listed in Table 1, RBCs need not be further considered as they have not been previously used for water treatment. If we neglect consider the unsaturated flow reactor as a variation on a packed bed reactor, that leaves two types of fixed film systems: a packed bed PFR reactor, and a fluidized bed CSTR system. A detailed modeling comparison of these two types of reactors is beyond the scope of this analysis. However, with some simplifications, the advantages and disadvantages of these two reactors can be considered.

First, let us consider the fluidized bed reactor. In a fluidized bed reactor bacteria grow as a thick biofilm on a packing material, and the contents of the reactor (water and biofilm) are recirculated fast enough to keep the system well mixed. As a result, all of the bulk water phase in the reactor is at the same concentration. The concentration of the acetate and perchlorate at the surface of the biofilm is therefore quite low because it is equal to the desired effluent concentration. For nitrate removal, low effluent concentrations allow reactions to proceed at appreciable rates in the reactor biofilm because nitrate must only be removed to < 10 mg $L^{-1}$. However, because extremely low effluent concentrations are required for perchlorate water treatment, there will be correspondingly low reaction rates at the surface of the biofilm. As the chemicals diffuse into the biofilm, these concentrations are further reduced resulting in even lower growth rates. The minimum substrate concentration that can be achieved under steady growth conditions for acetate removal based on the kinetics given above is

$$S_{min} = \frac{b\ K_S}{\mu_{max} - b} = \frac{(0.01\ d^{-1})\ (300\ mg\ L^{-1})}{(0.2\ h^{-1})\ (24\ h\ d^{-1}) - (0.01\ d^{-1})} = 0.6\ mg\ L^{-1} \tag{8}$$

This indicates the lowest acetate concentration that could be achieved is still only around 1 mg $L^{-1}$. At this concentration, the rate of cell growth in a CSTR will be very low since all cells must be growing in the CSTR at the bulk substrate concentration or lower.

The primary disadvantage of the fluidized bed reactor is therefore the slow growth rates of all cells in the reactor due to the desired low effluent substrate concentration. If the substrate concentration is raised to keep growth rates in the reactor higher, there can be other problems in the reactor. When substrate is present at large concentrations relative to cell growth rates, substrate is not oxidized but instead is processed by the bacteria rapidly into excreted exopolymers. This production of copious amounts of polymer can result in gelatinous particles which stick together and cause the process to fail. If a fermentable substrate is used, it is possible to gradually displace the perchlorate- or nitrate-reducing population with a biofilm that just ferments the substrate. This can cause a loss of reactor performance despite considerable biofilm growth and substrate transformation to other products.

The main advantage of the fluidized bed reactor is the ability to maintain very high biomass concentrations in the reactor creating the potential for small, highly efficient, reactors. However, biomass concentrations can get too high. In nitrate removal systems bed particles are recycled out of the reactor in order to shear off excess biofilm. In perchlorate systems operated at very low effluent perchlorate concentrations, the long detention times of the biomass for a suspended growth CSTR (calculated above) suggest that much of the biomass will decay in the reactor. A second advantage of a fluidized bed reactor is that they are well known to produce suspended cell concentrations in their effluent (because particles are cleaned in a side process) and therefore post treatment filtration is often not necessary. However, the need to ensure complete removal of unused substrate makes it likely that any such system would have a biologically aerated filter following the process.

The packed bed reactor also has advantages and disadvantages. The main advantages of the packed bed reactor configuration result from its operation as a PFR. Because concentrations of substrate and electron donor are large at the reactor entrance, the perchlorate degradation rates at the PFR entry are very large compared to those occurring in a CSTR like the fluidized bed reactor and theoretically a PFR reactor under these conditions can be smaller than a CSTR system. Using equation 2 and assuming that we have an identical biomass concentration and yield and maximum kinetic constants, and neglecting decreases in chemical concentrations in the biofilm, we can compare these rates in the fixed bed to fluidized bed as:

$$\frac{r_s\ \text{(packed bed)}}{r_s\ \text{(fluidized bed)}} = \frac{[S/(K_S + S)]\ [E/(K_E + E)]|_{\text{packed bed}}}{[S/(K_S + S)]\ [E/(K_E + E)]|_{\text{fluidized bed}}} \tag{9}$$

For comparison, we will assume that both reactor achieve a perchlorate reduction from 3 to 0.010 mg $L^{-1}$ and that acetate is reduced from 6.7 to 1 mg $L^{-1}$ based on our observed yield. For the fluidized bed reactor, the effluent concentrations are present everywhere in the reactor, but for the packed bed reactor the concentrations vary over the reactor length. Substituting in the values of the constants, we have at the beginning of the reactor:

$$\frac{r_s\ \text{(packed bed)}}{r_s\ \text{(fluidized bed)}} = \frac{[6.7/(300 + 6.7)]\ [3/(20 + 3)]|_{\text{packed bed}}}{[1/(300 + 1)]\ [0.010/(20 + 0.010)]|_{\text{fluidized bed}}} = 1720 \tag{10}$$

Thus, we can see that we potentially have rates 1720 times greater at the inlet of the packed bed reactor than the inlet of the fluidized bed reactor. At the reactor exit the rates in the reactors would be equal. These high rates help achieve a small reactor size and help perchlorate-reducing conditions to predominate over other less desirable rates such as fermentation.

The main disadvantage of the PFR is the potential for reactor plugging at the reactor entry, and insufficient biofilm at the reactor exit. Because substrate oxidation rates are very large at the entrance, the biofilm will grow faster and accumulate to a greater extent at the entrance than at the exit. In order to reduce clogging due to biofilm buildup, packed bed reactors are equipped with some method to remove excess biofilm. During the biofilm removal process, flow to the reactor is temporarily halted and the biofilm is removed. One approach to remove biofilm, if the media size is small, is to "backwash" the bed like a water treatment filter. During backwashing, the media is fluidized and the biofilm is knocked off the particles and removed from the reactor. Internal rakes can be placed in the bed to help break up the particles. When heavy packing material such as plastic rings or saddles are used as a support structure, the bed is usually cleaned by an air-scour stream.[6] Although this exposes the biofilm to air, most perchlorate reducing microbes are facultative and will rapidly remove the oxygen when the reactor is placed back into operation. The biofilm produced from either process can be settled out and disposed of as a non-hazardous sludge, and the wash water recycled back to the reactor feed.

**Operating Experience with Perchlorate-Degrading Reactor Systems**

The treatment of perchlorate in drinking water is a relatively new problem, and therefore there is little data available in the refereed literature at this point to base comments on larger scale system performance. Research at Penn State has centered on two different fixed-film biological treatment processes to determine their feasibility for drinking water treatment. These systems are: a packed bed (slow sand filter) amended with soluble microbial carbon sources (acetate); and a hydrogen gas fed four-phase (hydrogen gas, water, biofilm, and support media), unsaturated trickle-type packed column.[16–18] The development of the packed bed reactor is more advanced at this time and is being field tested. Our laboratory experiments have demonstrated that perchlorate can be completely removed, from 20 mg $L^{-1}$ to less than < 0.004 mg $L^{-1}$, at detention times of 13 to 48 min in a sand-packed reactor.

A fluidized bed reactor system using granular activated carbon (GAC) was tested at a site in California.[12] Perchlorate was reduced from 40 µg $L^{-1}$ to below detectable levels (< 4 µg $L^{-1}$) at the same time nitrate removals of 99% were achieved (40–50 mg $L^{-1}$ reduced to < 0.45 mg $L^{-1}$). Ethanol was used for the reactor feed and it was noted that a narrow range in the ethanol feed concentration (75 to 100 mg $L^{-1}$) was necessary to achieve ethanol concentrations below the analytical reporting limit (< 5 mg $L^{-1}$). Oxygen had to be reduced quickly in the reactor to < 0.1–0.2 mg $L^{-1}$ for optimum reactor performance. Clumping of the GAC was noted when acetate concentrations exceeded 200 mg $L^{-1}$, presumably to polymer production by the biomass as discussed above.

A series of laboratory tests using GAC and sand in a methanol- and ethanol-fed fluidized bed reactors.[13] GAC performed better than sand and ethanol was superior to methanol in laboratory tests. A lack of substrate (ethanol) in the reactor effluent limited perchlorate destruction. Field tests are planned to test this system at large scale.

In conclusion, it appears that fixed film systems, such as packed and fluidized bed reactors can be used to treat perchlorate contaminated water to the very low levels suitable for drinking water. It is still too soon to decide what the optimal reactor configuration is for stable, long-term treatment of perchlorate contaminated water. The preliminary comparison of the fluidized

# Chapter 19

# AN AUTOTROPHIC SYSTEM FOR THE BIOREMEDIATION OF PERCHLORATE FROM GROUNDWATER[⊗]

Tara L. Giblin, David C. Herman, and William T. Frankenberger, Jr.*

Department of Soil and Environmental Sciences, University of California–Riverside
Riverside, California 92521

## INTRODUCTION

In the past 5 years, advances in ion chromatography have allowed the detection of the perchlorate ion ($ClO_4^-$) at levels as low as 4 μg $L^{-1}$.[1] The California Department of Health Services (CDHS) has advised that wells containing more than 18 μg $L^{-1}$ $ClO_4^-$ not be used as a source of drinking water, even though there is no federal drinking water standard for $ClO_4^-$. According to the CDHS, 144 wells in California have $ClO_4^-$ at detectable levels and 38 wells exceed the CDHS action level of 18 μg $ClO_4^-$ $L^{-1}$.[2] In Riverside and San Bernardino, California, some drinking water wells contain up to 216 μg $L^{-1}$ $ClO_4^-$ and 9 wells have been closed in this area.[2] Because of the unknown health effects of $ClO_4^-$, which may include adverse effects on the thyroid gland,[3,4] there is an urgent need to devise methods to remove $ClO_4^-$ from groundwater and drinking water.[5,6]

Some $ClO_4^-$ remediation strategies implement physical methods for removal of $ClO_4^-$ from water.[7,8] These methods non–selectively remove many salts from the water, concentrating them into a smaller volume. As a result, the high salt, high $ClO_4^-$ containing concentrate must still be disposed. Bioremediation is the preferred strategy because it relies on microorganisms that can completely transform $ClO_4^-$ into chloride,

[⊗]This work was presented at the 218th national meeting of the American Chemical Society as part of the Environmental Division symposium Perchlorate in the Environment, held August 22–24, 1999, in New Orleans, Louisiana.

*Author to whom correspondence should be directed. Phone: 909-787-3405. Fax: 909-787-2954. Electronic mail: williamf@orange.ucr.edu

thereby eliminating the contaminant from the environment.[9–11] Several research programs are currently developing $ClO_4^-$ bioremediation strategies, with at least two pilot scale reactors under construction.[12–15] These reactors make use of heterotrophic bacteria isolated from anaerobic digesters to degrade $ClO_4^-$. Carbon sources such as acetate, ethanol or yeast extract must be included in the reactor to encourage heterotrophic bacterial growth for optimal reactor performance.[11,16]

An important factor in bioreactor design for groundwater treatment is the choice of oxidizable substrate used as reactor feed. The choice should be economically feasible as well as lend itself to rapid removal of excess substrate from the waste stream to eliminate excessive microbial growth in post–reactor effluent.[11] European bioreactors designed for denitrification of drinking water often employ hydrogen gas as the electron source because it meets these criteria.[17, 18] With a solubility of only 1.6 mg $L^{-1}$, hydrogen would be unlikely to persist in the water after passing through the bioreactor. Additionally, hydrogen could be more cost effective than acetate, ethanol or methanol.[11] Hydrogen will support the growth of autotrophic bacteria, but relatively few organisms can grow with hydrogen and bicarbonate as substrates, so the growth conditions created within the bioreactor are highly selective.[19] Also the biomass yields within the reactor are often lower than yields achieved by heterotrophic systems, a factor which prevents clogging of the system.[20]

The objective of this work was to compare two bacterial based systems that would remediate very low levels of $ClO_4^-$, such as those found in Southern California drinking water, to less than the CDHS action level, and preferably to less than detectable levels. This work is a comparison of a $ClO_4^-$ bioremediation system designed to use a heterotrophic (acetate–utilizing) organism with a system that uses an autotrophic (hydrogen oxidizing) consortium. For the heterotrophic system, the bacterium of interest, perclace, has been described in previous studies.[9, 21] For the autotrophic system (AS), a consortium of 5 organisms was used. Both systems are capable of removing nitrate and $ClO_4^-$ simultaneously.

## EXPERIMENTAL

### Cultivation of Microorganisms

The heterotroph, perclace, was maintained in culture by monthly transfers into a mineral salts medium [FTW[10] or BMS ($KH_2PO_4$ [20 mg $L^{-1}$], $Na_2HPO_4$ [75 mg $L^{-1}$], $NH_4Cl$ [60 mg $L^{-1}$], $MgSO_4{\bullet}7H_2O$ [10 mg $L^{-1}$], $CaCl_2{\bullet}2H_2O$ [10 mg $L^{-1}$], $FeCl_2{\bullet}4H_2O$ [2 mg $L^{-1}$])] plus a mixture of trace metals under nitrogen. Perchlorate ion (500 mg $L^{-1}$) and acetate (as sodium acetate, 1000 mg $L^{-1}$) were added. The autotrophic consortium was maintained in the same medium but the acetate was replaced with 500 mg $L^{-1}$ bicarbonate (as sodium bicarbonate) and the headspace gas was 80% v/v $H_2$ + 20% v/v $CO_2$. Although the redox potential necessary for $ClO_4^-$ reduction was not measured, it was estimated to be below −110 mV. This is known to be the $E_h$ at which the redox indicator resazurin is reduced from pink to clear, a visible change that was always necessary for $ClO_4^-$ reduction to occur.[22]

**Perchlorate, Acetate, and Nitrate Analysis**

A $ClO_4^-$ specific electrode (model 93–81, Orion Research, Boston, MA) was used to detect $ClO_4^-$ concentrations between 1 and 1000 mg $L^{-1}$. Detection of $ClO_4^-$ to 5 µg $L^{-1}$ was performed by ion chromatography using an IonPac AS11 column (Dionex, Sunnyvale, CA)[23]. Nitrate and acetate were also determined using the IonPac AS11 column with 10mM NaOH as the eluent and a sampling loop of 10 µL. The suppressor (ASRS–II) was run in the self–regenerating mode operating at 50 mA.

**Groundwater**

Two separate samples of groundwater were pumped from $ClO_4^-$ contaminated wells in the San Gabriel Valley, CA. The initial sample of water contained 200 µg $L^{-1}$ $ClO_4^-$ and the second 550 µg $L^{-1}$ $ClO_4^-$. Both samples had a pH of 7.5 and contained 26 mg $L^{-1}$ $NO_3^-$. Two mg $L^{-1}$ phosphorus (as $KH_2PO_4$) and 20 mg $L^{-1}$ nitrogen (as $NH_4Cl$) were added as well as a suitable carbon source (either bicarbonate or acetate). The groundwater was filtered through a 0.45 µm filter (Gelman Sciences, Ann Arbor, MI), rather than autoclaving, for use in column studies.

**Column studies**

Column studies used Celite R–635 (Celite Corporation, Lompoc, CA) as the solid support packed into PVC pipe to produce a column with a void volume of 300 mL for the heterotrophic system or 120 mL for the AS reactor. Side-ports were evenly spaced along the column for sampling purposes. Following saturation with water, the column was inoculated with a washed cell suspension of either the AS or heterotrophic culture (perc1ace). $ClO_4^-$ in the column eluent was monitored with the Orion electrode. Samples were taken daily from each column eluent (2–5 samples per day), and filtered prior to ion chromatographic analysis. Tables 1–4 describe each column in detail. In every case, day 1 is the beginning of column operation while the time used for loading and conditioning the column is indicated as a negative.

**Heterotrophic system 1**

The purpose of this system, detailed in Figure 1a, was two-fold: to determine the capacity of perc1ace to remove $ClO_4^-$ from groundwater in a flow-through system and to determine the residence time necessary for $ClO_4^-$ removal to below CDHS action levels.

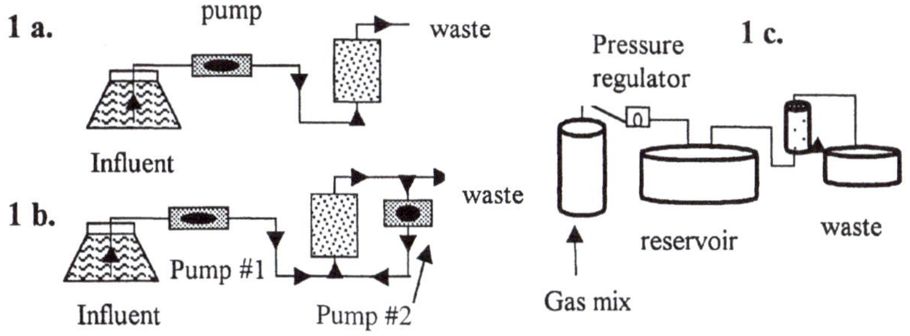


**Figure 1.** Design of bioreactor systems for heterotrophic (a and b) and autotrophic (c) bacteria.

The concentration of $ClO_4^-$ in groundwater was raised from 0.200 mg $L^{-1}$, the level of natural contamination, to 0.738 mg $L^{-1}$ using $NaClO_4^-$. When mineral salts medium was used, nitrogen gas was continually bubbled into the medium to maintain anaerobic conditions. When groundwater was the influent, nitrogen gas was used to sparge only the headspace of the reservoir to limit undesirable pH changes in the water.

**Table 1.** Parameters for operation of heterotrophic column 1

| Day | Flow rate mL $min^{-1}$ | $[ClO_4^-]$ mg $L^{-1}$ | $[NO_3^-]$ mg $L^{-1}$ | $[CH_3CO_2^-]$ mg $L^{-1}$ | Medium | Comments |
|---|---|---|---|---|---|---|
| -11 to -9 | 0.25 | n/a | n/a | n/a | $H_2O$ | Column saturation |
| -8 to -7 | 0.25 | n/a | n/a | n/a | FTW | Load perclace |
| -6 to 0 | 0.25–3.0 | 100 | n/a | 1000 | FTW | Establish biomass |
| 1–6 | 0.5 | 0.738 | 26 | 250 | GW | 20 mg $L^{-1}$ P and 2 mg $L^{-1}$ N added |
| 7–12 | 1 | 0.738 | 26 | 250 | GW | none |
| 13–18 | 2 | 0.738 | 26 | 250 | GW | none |
| 19–21 | 2 | 0.040 | n/a | n/a | GW | Pass GW eluent through 2nd time |
| 23–25 | 2 | 0.674 | n/a | 250 | BMS | Sideport samples taken |

n/a: none added; BMS: basal mineral salts medium; GW: groundwater; $CH_3CO_2^-$: acetate

**Heterotrophic system 2**

The objective of this experiment was to determine whether rapid recirculation of the treated water would affect the removal of $ClO_4^-$. As shown in Figure. 1b, a second pump that produced flow rates of 100 mL $min^{-1}$ was added to the system to rapidly cycle the water through the column while maintaining an overall flow rate through the column of 1 to 3 mL $min^{-1}$. For this experiment, groundwater was obtained with a higher level of contamination, 0.550 mg $L^{-1}$ $ClO_4^-$, so it did not require perchlorate amendment. Table 2 details the running parameters of this column.

**Table 2.** Parameters for operation of heterotrophic column 2 with recycling

| Day | Flow rate mL $min^{-1}$ | $[ClO_4^-]$ mg $L^{-1}$ | $[NO_3^-]$ mg $L^{-1}$ | $[CH_3CO_2^-]$ mg $L^{-1}$ | Medium | Comments |
|---|---|---|---|---|---|---|
| -6 to -4 | 0.25 | n/a | n/a | n/a | $H_2O$ | Saturate column |
| -4 to -2 | 0.25 | n/a | n/a | n/a | BMS | Load perclace |
| -1 to 0 | 0.25–3.0 | 100 | n/a | 1000 | BMS | Recycling pump set at 100mL $min^{-1}$ |
| 1–7 | 3 | 0.672 | n/a | 150 | BMS | Day 4 + 7 sideport samples taken |
| 8–10 | 3 | 0.672 | n/a | 150 | BMS | No recycling |
| 11–12 | 3 | 0.672 | n/a | 150 | BMS | Recycling resumes day 11 |
| 13–15 | 3 | 0.550 | 26 | 150 | GW | none |
| 16–18 | 2 | 0.550 | 26 | 150 | GW | none |
| 19 | 1 | 0.550 | 26 | 150 | GW | none |

n/a: none added; BMS: basal mineral salts medium; GW: groundwater; $CH_3CO_2^-$: acetate

**AS system 1**

This system required a different design, as Figure. 1c shows, so that hydrogen could be delivered to the bacteria in the column. The water was placed in a pressurizable carboy with a gas line connected to a supply of 95/5 (v/v) $H_2/CO_2$. When pressurized, the hydrogen would dissolve slightly in the medium or groundwater and in this manner would be delivered to the bacteria in the column. The pressure on the liquid in the reservoir also determined the flow rate, as the pressure increased the flow rate into the column also increased. The details of the operation of this column are shown in Table 3.

**Table 3.** Parameters for operation of AS system 1

| Day | $[ClO_4^-]$ mg $L^{-1}$ | Flow rate mL $min^{-1}$ | Medium | Comments |
|---|---|---|---|---|
| -20 to -18 | n/a | 0.25 | FTW | load cells into column |
| -17 to -8 | 10 mg | 0.15 | FTW | condition with FTW |
| -8 – 0 | 10 mg | 0.15 | BMS | condition with BMS |
| 1–7 | 0.700 | 0.15–1 | BMS | increase flow rate to 1 mL $min^{-1}$ |
| 8–15 | 0.700 | 1 | BMS | check $ClO_4$ removal |
| 16–23 | 0.700 | 2 | BMS | check $ClO_4$ removal |
| 24–30 | 0.700 | 3 | BMS | check $ClO_4$ removal |
| 31–40 | 0.700 | 3 | GW | N, P and $HCO_3$ added |

n/a: none added; BMS: basal mineral salts medium; GW: groundwater

**AS System 2**

On days 31–40 of the operation of reactor 1, the influent was switched from BMS to groundwater amended with nitrogen, phosphorus and bicarbonate. This resulted in loss of viability of the column, possibly due to a drastic change in pH when groundwater was pressurized with hydrogen.

The column was repacked with Celite and re–inoculated to create reactor 2. In this case, 0.1 g $L^{-1}$ yeast extract (Difco, Sparks, MD) was added to the conditioning medium to promote growth of the facultative anaerobes so that these organisms might promote anaerobic conditions in the column more rapidly. On day 37 (reactor 2), the groundwater was amended only with nitrogen and phosphorus, to test the capacity of the organisms to remove $ClO_4^-$ from the water with no added carbon source. The groundwater itself contains 50 mg $L^{-1}$ bicarbonate, occurring naturally. Operation of this reinoculated column is described in Table 4.

**Table 4.** Parameters for operation of AS system 2

| Day | $[ClO_4^-]$ mg $L^{-1}$ | Flow rate mL $min^{-1}$ | Medium | Comments |
|---|---|---|---|---|
| -10 to -8 | n/a | 0.25 | FTW | load cells into column |
| -8 – 0 | 10 | 0.25 | FTW | 0.1 g $L^{-1}$ yeast extract added |
| 1–11 | 0.700 | 0.15–1 | BMS + GW | mineral salts medium with groundwater |
| 12– 15 | 0.700 | 1 | BMS + GW | mineral salts medium with groundwater |
| 16 – 22 | 0.700 | 1 | GW | groundwater alone |
| 23–29 | 0.700 | 2 | GW | groundwater alone |
| 30–36 | 0.700 | 3 | GW | groundwater alone, samples taken for plating |
| 37–45 | 0.700 | 1 | GW | no added bicarbonate |

n/a: none added; BMS: basal mineral salts medium; FTW: mineral salt medium; GW: groundwater

**Biomass distribution**

Following completion of the flow–through experiments, the distribution of biomass within the columns was determined. The celite was removed and samples of celite were placed in test tubes containing 1 mL of 0.1 M NaOH and heated to 95°C for 10 minutes to lyse the bacteria attached to the celite. Tubes were vortexed and an aliquot of the supernatant, which contained the bacterial lysate, was analyzed for protein content using the Lowry method.[24] The Celite in each test tube was then oven–dried at 90°C and the final protein content expressed as μg protein per g Celite (dry weight).

## RESULTS

Figure 2 shows that perchlorate was completely removed from 500 mg $L^{-1}$ to less than 5 μg $L^{-1}$ by the AS in 96 hours when grown at 30°C. The removal of $ClO_4^-$ by perc1ace is slightly faster, requiring only 48 hours. However, perc1ace grows to a higher density than the AS, which had a tendency to clump. Batch studies also indicated that 62 mg $L^{-1}$ nitrate was removed simultaneously with $ClO_4^-$ by both systems.

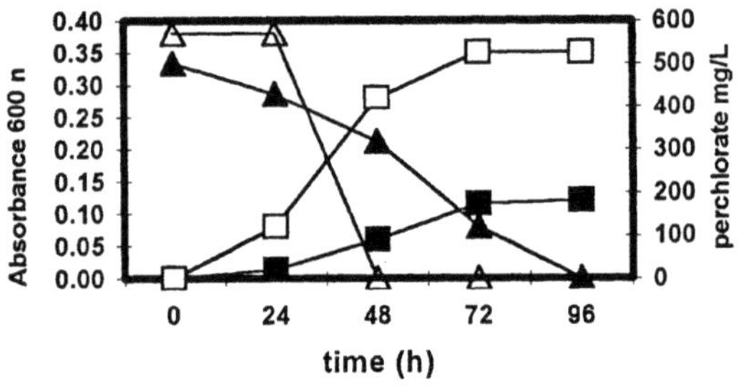


**Figure 2.** Removal of 500 mg $L^{-1}$ $ClO_4^-$ by perc1ace (ρ) AS (π) and growth rate at 30 °C of perc1ace (□) and AS (■).

Figure 3 demonstrates the removal of $ClO_4^-$ from contaminated groundwater by the first heterotrophic column. At flow rates of 0.5 and 1.0 mL $min^{-1}$, representing residence times of 10 and 5 hours respectively, $ClO_4^-$ was removed from 0.738 mg $L^{-1}$ to below detectable levels by perc1ace. When the flow rate was increased to 2 mL $min^{-1}$, which is a residence time of 2.5 hours, 92–95 % of the perchlorate was removed. At this flow rate, perchlorate was detectable at levels in the eluent of 0.04–0.06 mg $L^{-1}$, levels which are just slightly above the CDHS action level of 0.018 mg $L^{-1}$. A portion of the column eluent from days 13–18, containing low levels of $ClO_4^-$ breakthrough, was filter sterilized and again passed through the reactor resulting in $ClO_4^-$ removal to below detectable levels (not shown). This indicates that potential mass transfer limitations are prohibiting complete removal of $ClO_4^-$ from groundwater at fast flow rates.

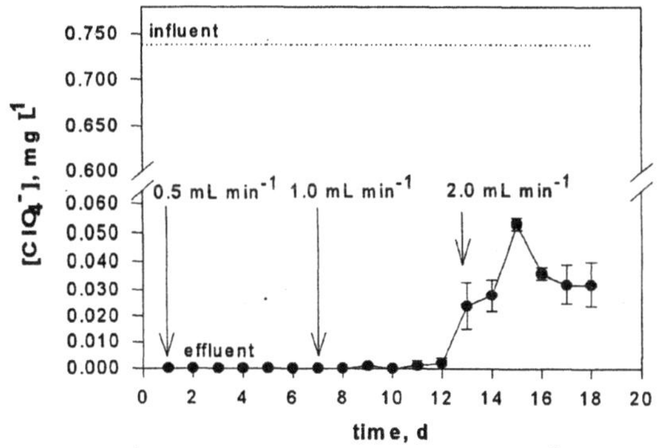


**Figure 3.** Removal of perchlorate from San Gabriel Valley groundwater by the first heterotrophic system. The flow rate was increased form 0.5 to 2.0 mL min$^{-1}$ over a period of 18 days.

A second heterotrophic reactor was constructed with a primary goal of passing the contaminated water multiple times through the reactor in an attempt to counteract mass transfer limitations. A second pump, which had a flow rate of 100 mL min$^{-1}$ was added. The pump maintaining flow into and out of the column (pump #1) had a flow rate never exceeding 3 mL min$^{-1}$. For the first eight days, 0.672 mg L$^{-1}$ perchlorate in BMS medium was degraded to less than detectable levels at flow rates of 3 mL min$^{-1}$ (not shown). When groundwater, contaminated with 0.550 mg L$^{-1}$ $ClO_4^-$, was loaded into the column at 3 mL min$^{-1}$ (a residence time of 1.6 hours) 95% of the $ClO_4^-$ was removed (Figure 3). When the flow was reduced to 2 mL min$^{-1}$, $ClO_4^-$ was removed to below the CDHS action level. Only at 1 mL min$^{-1}$ was $ClO_4^-$ removed to less than detectable levels of 0.005 mgL$^{-1}$.

The AS bioreactor was also capable of removing 0.700 mg L$^{-1}$ perchlorate to less than detectable levels from mineral salts medium at fast flow rates (Figure 4.). At a retention time in the column of 0.6 hours, a flow rate of 3 mL min$^{-1}$, $ClO_4^-$ was removed to nondetectable levels from BMS medium.

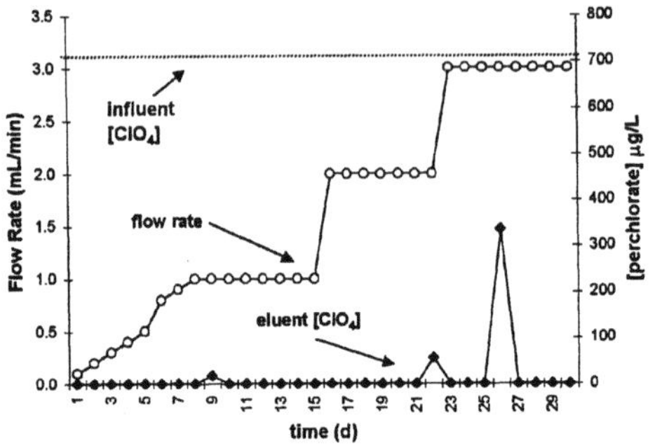


**Figure 4.** Removal of perchlorate from mineral salt medium by the AS at increasing flow rate by the AS. Eluent [$ClO_4$], ♦. Flow rate, O.

When the influent was switched to groundwater and the reactor was operated for 7 days, the microbes were inhibited to the extent that the column had to be repacked and re-inoculated (data not shown). The pH of the groundwater when pressurized with the $H_2$:$CO_2$ mix was as low as 5.9, whereas the mineral salts medium was pH 7.0.

As shown in Figure 5, the repacked reactor was fed groundwater amended with BMS medium for days 1–15, and the flow rate was again increased from 0.25 to 1.00 mL $min^{-1}$.

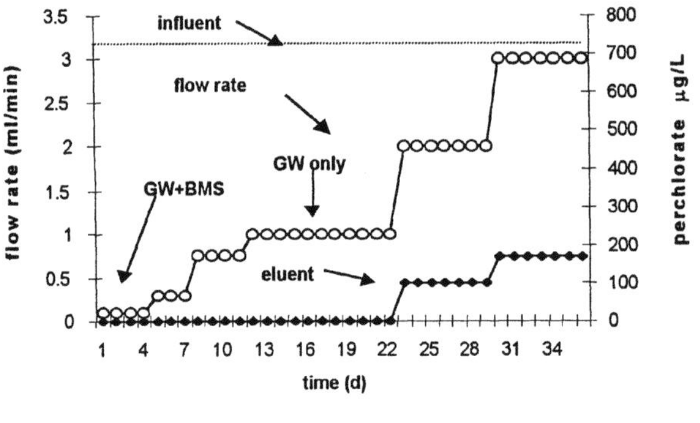


**Figure 5.** Removal of perchlorate from groundwater by the second AS reactor. Groundwater was amended with BMS for the first 15 days then only with N, P, and bicarbonate. Eluent [$ClO_4$], ♦. Flow rate, O.

Once the AS had become conditioned to groundwater, only N, P and $HCO_3$ were added. As the flow rate was increased from 1 to 3 mL $min^{-1}$, the amount of $ClO_4^-$ removed from the water decreased. Only at flow rates of 1 mL $min^{-1}$ (representing a residence time of 2 hours) was $ClO_4^-$ removed to below detectable levels. When the flow rate was increased to 2 and 3 mL $min^{-1}$ (residence times of 1 and 0.6 hours) $ClO_4^-$ breakthrough was 100 and 200 μg $L^{-1}$, respectively. Analysis of nitrate in the column effluent revealed that at all flow rates, nitrate was removed from 25 mg $L^{-1}$ to below detectable levels, indicating that the consortium was capable of nitrate reduction in the presence of $ClO_4^-$ (not shown).

Figure. 6 demonstrates the capacity of the organisms in the AS bioreactor to remove $ClO_4^-$ from groundwater to which no extra carbon source was added when the flow rate was 1 mL $min^{-1}$. On days 37–42, the first 6 days in which no bicarbonate was added, $ClO_4^-$ breakthrough ranged from 0 to 120 μg $L^{-1}$. For days 43–45, perchlorate was removed to nondetectable levels.

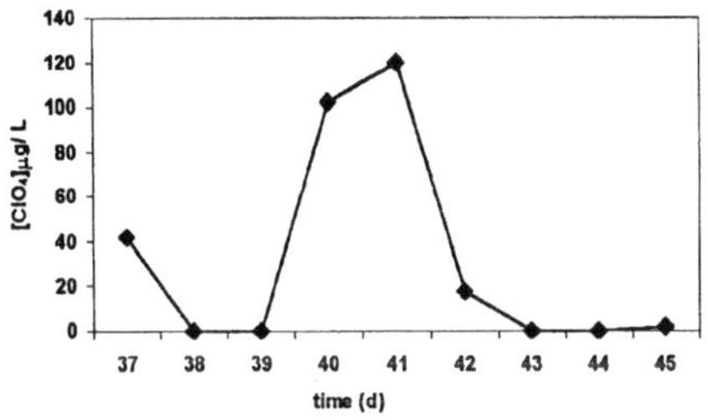


**Figure 6.** Perchlorate breakthrough from the AS reactor at flow rate of 1 mL $min^{-1}$ when the reactor was fed groundwater without added carbon source.

Inefficient $ClO_4^-$ removal at faster flow rates in all of the reactors tested could be the result of inefficient biomass distribution within the column. Figure 7 demonstrates that most of the biomass was present near the inlet of the AS reactor. The data for the first heterotrophic system also indicated that the majority of the bacteria were colonized in the first quarter of the column, near the inlet. This region contained 60 μg of protein per g celite as opposed to the upper 75% of the column, which averaged about 2 μg protein per g celite (not shown). Results of side–port eluent sampling from the heterotrophic systems supported the protein analysis data. Most of the $ClO_4^-$ was removed from the eluent before it reached the first sampling port, indicating that the bacterially active zone of the column was near the inlet.

## DISCUSSION

Successful remediation of $ClO_4^-$ in groundwater using immobilized bacteria in a bioreactor was first demonstrated 25 years ago.[25,26] Recent work has described groundwater treatment systems that treat lower levels of $ClO_4^-$ contamination and commercial operations are already being tested.[14,15,27,28] These systems utilize autotrophic (hydrogen oxidizing) and heterotrophic (acetate oxidizing) processes to reduce 20 mg $L^{-1}$ $ClO_4^-$ to below detectable levels.

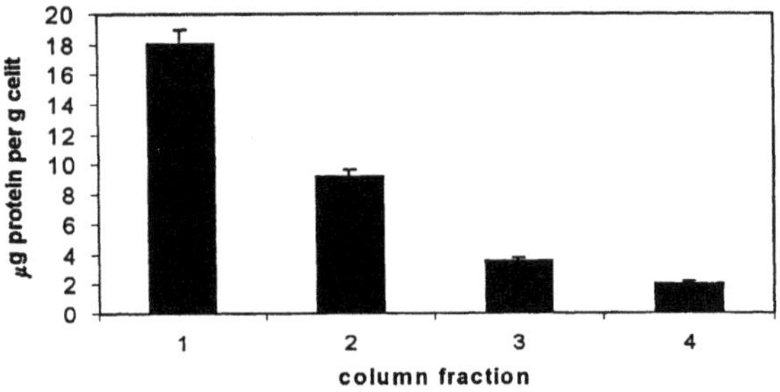


**Figure 7.** Distribution of biomass based on protein analysis of the biomass attached to celite pellets using a modified Lowry method. The autotrophic reactor 2 was divided into four sections. 1= inlet, 2 + 3 = middle portions, 4 = outlet.

The current work compares two systems for $ClO_4^-$ removal which differ in the type of bacterial metabolism used by the organisms within the reactor. The results indicate that celite-packed columns inoculated with the AS or perclace have the potential to remove $ClO_4^-$ to below CDHS action levels (0.018 mg $L^{-1}$). Removal of $ClO_4^-$ to less than 0.005 mg $L^{-1}$ from groundwater only occurred at the slowest flow rates for both systems. Another study also reported an increase in $ClO_4^-$ breakthrough with an increase in loading rates using a sand-packed column inoculated with mixed consortium of bacteria.[29]

The heterotrophic systems were capable of removing $ClO_4^-$ to nondetectable levels at a residence time of 5 hours. The AS could achieve nondetectable levels in two hours. The heterotrophic system removed $ClO_4^-$ to below the CDHS action limit in 2.5 hours (a breakthrough of 10 µg $L^{-1}$). As the flow rate was increased in both systems, the level of $ClO_4^-$ breakthrough also was increased.

The inefficient $ClO_4^-$ removal at fast flow rates could be caused by the non-uniform distribution of biomass within the column. Biomass distribution indicated that most of the bacteria were concentrated near the inlet of the reactor. This is a phenomenon that has been described previously [21] and results from a lack of nutrients reaching the farthest points in the column. Poor nutrient availability may be exacerbated in the AS reactors by the poor solubility of hydrogen, preventing this energy source from reaching the biomass near the outlet of the column. Another factor contributing to inefficiency is the dramatic pH change when mineral salts medium was replaced with groundwater as the influent. The pressurization of the liquid with hydrogen gas had little effect on the pH of the mineral salts medium, because it was buffered, but the water alone often had a pH decrease from 7.2 to less than 6 when under hydrogen pressure.

This is the first report of a hydrogen utilizing-microbial consortium that is capable of completely removing low levels of $ClO_4^-$ from groundwater. Other reports have described the partial removal of $ClO_4^-$ from simulated groundwater by hydrogen utilizing organisms using a slightly different hydrogen delivery system[15] as well as a hydrogen based, chlorate-reducing reactor that was capable of partial removal of $ClO_4^-$[28].

Hydrogen-utilizing organisms have been used for denitrification of drinking water in Europe.[19,20,29] Hydrogen is a preferred electron source for several reasons. As Vanbrabant describes, heterotrophic conditions promote high biomass which can clog a system and overdosing of the carbon source can result in need for additional treatment of reactor effluent. These same problems could plague a reactor built for $ClO_4^-$ removal from water. The development of a $ClO_4^-$ removal strategy that can use hydrogen as the electron donor could be a cost–effective advance in perchlorate remediation technology.

## CONCLUSION

Other reports have described the removal of $ClO_4^-$ from simulated[15,30] and naturally contaminated groundwater.[21] The autotrophic system described here may have an advantage over these other systems because of the ability to use bicarbonate, already present in the groundwater, as the carbon source. As the final experiment showed, the consortium was capable of removing perchlorate from groundwater with no added carbon source at a residence time of 2 hours.

## REFERENCES


1. California Department of Health Services (CDHS). "Determination of Perchlorate by Ion Chromatography." Sanitation and Radiation Laboratories Branch, Berkeley, CA. Rev. 0, June 3, 1997.

2. California Department of Health Services (CDHS). "Perchlorate in California drinking water." 1998. URL: www.dhs.cahwnet.gov/ps/ddwem/chemicals/perchl/perchlindex.htm.

3. Capen, C.C. "Mechanisms of chemical injury of thyroid gland." In *Progress in Clinical and Biological Research. Receptor Mediated Biological Processes: Implications for Evaluation Carcinogenesis.* Wiley–Liss: New York, 1994; pp. 173–191.

4. Von Burg, R. "Toxicology update–perchlorates." *Journal of Applied Toxicology* **1995**, *15*, 237–241.

5. Urbansky, E.T. "Perchlorate chemistry: implications for analysis and remediation." *Bioremediation Journal* **1998**, *2*, 81–95.

6. Urbansky, E.T.; Schock, M.R. "Issues in managing the risks associated with perchlorate in drinking water." *Journal of Environmental Management* **1999**, *56*, 79–95.

7. Betts, K.S. "Technology update: rotating ion–exchange system removes perchlorate." *Environmental Science and Technology* **1998**, *32*, 454A–455A.

8. Gu, B.; Brown, G.M.; Alexandratos, S.D.; Ober, R.; Patel, V. *Selective Anion Exchange Resins for the Removal of Perchlorate ($ClO_4^-$) from Groundwater*. U.S. Department of Energy, Oak Ridge National Laboratory: Oak Ridge, TN, 1999. ORNL Doc. No. ORNL/TM–13752; Environmental Sciences Division Publication No. 4863.

9. Herman, D.C.; Frankenberger, W.T. Jr. "Microbial–mediated reduction of perchlorate in groundwater." *Journal of Environmental Quality* **1998**, *27*, 750–754.

10. Herman, D.C.; Frankenberger, W.T. Jr. "Bacterial reduction of perchlorate and nitrate in water." *Journal of Environmental Quality* **1999**, *28*, 1018–1024.

11. Logan, B.E. "A review of chlorate– and perchlorate–respiring microorganisms." *Bioremediation Journal* **1998**, *2*, 69–79.

12. Attaway, H.; Smith, M. "Propellant wastewater treatment process." U.S. Patent No. 5,302,285; 1994.

13. Green, M.; Pitre, M.P. "Treatment of groundwater containing perchlorate using biological fluidized bed reactors with GAC or sand media." *Preprints of Extended Abstracts, Division of Environmental Chemistry.* 218th American Chemical Society Meeting, August 22–26, 1999, New Orleans, LA, *39* (2), pp 105–107.

14. Catts, J. G. "The biochemical reduction of perchlorate at low concentrations in water–technology application for groundwater in San Gabriel Basin, California." In *Proceedings of the Southwest Focused Groundwater Conference: Discussing the Issue of MTBE and Perchlorate in the Ground Water*, 1998, pp. 144–147.

15. Logan, B.E.; Kim, K., Kijung, K.; Miller, J; Mulvaney, P; Wu, J.; Zhang, H. "Factors affecting biodegradation of perchlorate contaminated waters." *Preprints of Extended Abstracts, Division of Environmental Chemistry.* 218th American Chemical Society Meeting, August 22–26, 1999, New Orleans, LA, *39* (2), pp. 112–114.

16. Wallace, W.; Beschear, S.; Williams, D.; Hospadar, S.; Owens, M. "Perchlorate reduction by a mixed culture in an up-flow anaerobic fixed bed reactor." *Journal of Industrial Microbiology and Biotechnology* **1998**, *20*, 126–131.

17. Selenka, F.; Dressler, R. "Microbiological and chemical investigations on a biological autotrophic denitrification plant using hydrogen as an energy source." *Journal of Water Services Research and Technology—Aqua* **1990**, *39*, 107–116.

18. Kurt, M.; Dunn, I.J.; Bourne, J.R. "Biological denitrification of drinking water using autotrophic organisms with $H_2$ in a fluidized bed biofilm reactor." *Biotechnology and Bioengineering* **1987**, *24*, 493–501.

19. Vanbrabant, J.; De Vos, P.; Vancanneyt, M.; Leissens, J.; Verstraete, W.; Dersters, K. "Isolation and identification of autotrophic and heterotrophic bacteria from an autohydrogenotrophic pilot plant for denitrification of drinking water." *Systematic and Applied Microbiology*, **1993**, *16*, 471–482.

20. Gros, H.; Schnoor, G.; Rutten, P. "Biological denitrification process with hydrogen-oxidizing bacteria for drinking water treatment." *Water Supply* **1988**, 6, 193–198.

21. Giblin, T.; Herman, D.C.; Deshusses, M.A.; Frankenberger, W.T., Jr. "Removal of perchlorate in water with a flow through bioreactor." *Journal of Environmental Quality* **1999**, in press.

22. Jacob, H.–E. "Redox potential." In *Methods in Microbiology.* J.R. Norris and D.W. Ribbons, Eds. Academic: New York, NY, 1970; Vol. 2, pp. 84–86.

23. Wirt, K.; Laikhtman, M.; Rohrer, J.; Jackson, P.E. "Low–level perchlorate analysis in drinking water and ground water by ion chromatography." *American Environmental Laboratory*, **1998**, *10*, pp.1,5.

24. Daniels, L, R.; Hanson, S.; Phillips, J.A. "Chemical analysis." In *Methods for General and Molecular Bacteriology*. Gerhardt, P., R.G.E. Murray, W. A. Wood, and N. R. Krieg, Eds. American Society for Microbiology:Washington, D.C., 1994.

25. Yakovlev, S.V.; Voronov, J.V.; Korenkov, V.N.; Nevsky, A.B.; Bobrikova, V.A.; Karjurkhina, T.A.; Churbanova, I.N.; Laskov, J.M. " Method for biochemical treatment of industrial waste water." US patent 3,755,156, 1973.

26. Korenkov, V. N.; Romanenko, V.I.; Kuznetsov, S.I.; and Voronov. J.J.V. "Process for purification of industrial waste waters from perchlorates and chlorates." US patent 3,943,055, 1976.

27. Giblin, T.; Frankenberger, W.T. Jr. Unpublished data.

28. Van Ginkel, C.G.; Kroon, A.G.M.; Rikken, G.B.; Kengen, S.W.M. "Microbial conversion of perchlorate, chlorate and chlorite." In *Proceedings of the Southwest Focused Groundwater Conference: Discussing the Issue of MTBE and Perchlorate in the Ground Water.* 1998, pp. 92–95.

29. Kurt, M.; Dunn, I.J.; Bourne, J.R. "Biological denitrification of drinking water using autotrophic organisms with $H_2$ in a fluidized bed biofilm reactor." *Biotechnology and Bioengineering*, **1987**, *24*, 493–501.

30. Logan, B.E.; Kim, K. "Microbiological treatment of perchlorate contaminated groundwater". In *Proceedings of the Southwest Focused Groundwater Conference: Discussing the issue of MTBE and perchlorate in the ground water*, 1998, pp. 87–90

# Chapter 20

# RISK ASSESSMENT OF PERCHLORATE IN BIOTA, SOIL, AND GROUNDWATER AT AGRICULTURAL SITE IN SOUTHERN CALIFORNIA[⊗]

Heriberto Robles*

LFR Levine Fricke Inc., 1920 Main Street, Suite 750, Irvine, California 92614-7211

## INTRODUCTION

An agricultural company in Southern California was informed that one of their irriga-tion wells had been contaminated with perchlorate. Concerned about the health and legal implications of using water from the contaminated well to irrigate edible vegetables, the farming company commissioned LFR Levine Fricke to assess the potential health risks posed by the presence of perchlorate in water, soil, and vegetables. This paper provides a brief description of the sequence of events that led to the contamination of an agricultural field, the toxicological properties of perchlorate, the results of the investigation, and recommendations made.

The source of the perchlorate in groundwater was traced to an aerospace company located six miles upgradient from the site. The perchlorate concentration in well water ($110 \pm 14$ $\mu g\ L^{-1}$) was found to exceed the California Department of Health Services Provisional Action Level (in 1997) of 18 $\mu g\ L^{-1}$.

## EXPERIMENTAL

Soil and vegetable samples were collected at random from the agricultural field and submitted to a State-certified analytical laboratory for perchlorate analysis (EPA Method 300.0 Modified). Results of the analysis plus the groundwater data that had been previously collected were used to conduct a risk assessment. The risk assessment was conducted following

[⊗]This work was presented at the 218th national meeting of the American Chemical Society as part of the Environmental Division symposium Perchlorate in the Environment held August 22–24, 1999, in New Orleans, Louisiana.

*Phone: 949-955-1390. Fax: 949-955-0683. Electronic mail: heriberto.robles@lfr.com.

*Perchlorate in the Environment*, edited by Urbansky.
Kluwer Academic/Plenum Publishers, New York, 2000.

Environmental Protection Agency risk assessment guidance.[2] In addition, a bench-scale laboratory study was conducted to estimate perchlorate's half-life in native soil.

**Risk Assessment**

Perchlorate, when taken at high enough doses, has the ability to block the uptake of iodine by the thyroid. This ability has been exploited in medicine to treat patients who suffer from Graves' disease.[3] Nonetheless, perchlorate's effect on the thyroid is of concern when the chemical is accidentally introduced into the environment because thyroid hormones are essential to regulate animal metabolism, growth, and development. Accidental consumption of large doses of perchlorate may interfere with human and animal growth and development by interfering with the normal production of thyroid hormones.[4]

The EPA's National Center for Environmental Assessment has classified perchlorate as "an indirect anti-thyroid chemical." Based on current toxicological data available, the EPA proposed an oral reference dose (RfD) of 0.9 $\mu g\ kg^{-1}.day^{-1}$.[5] This proposed RfD was used in this risk assessment.

The potential receptors in this assessment were assumed to be farm workers at the site and off-site consumers of vegetables grown at the parcel. In this analysis it was assumed that farm workers were exposed to perchlorate in soil through the dermal, oral, and particle inhalation routes as a result of their normal farming activities. Farm workers at the parcel studied do not consume vegetables grown there. Therefore, the on-site consumption of vegetables was not considered to be a realistic exposure pathway for on-site workers. The potential dose received by off-site consumers was calculated by assuming that all the kale, lettuce, and spinach consumed by a person was produced at the farm under study. Data on the average vegetable consumption patterns of U.S. consumers was obtained from the EPA.[6] The assumption that a person obtains their daily vegetable ration from the parcel is likely to result in an overestimation of potential risks. All the farm production is sold to a specialized distributor, who in turn sells it to Chinese restaurants in the U.S. and Canada. The distributor receives produce grown throughout Southern California, Arizona, and Mexico.

In accordance with EPA risk assessment guidelines, both an average and a reasonable maximum exposure (RME) scenario were considered in this evaluation. The average exposure scenario was evaluated using the arithmetic mean chemical concentrations in soil and vegetables combined with average intake values describing the extent, frequency, and duration of exposure. To determine potential exposures associated with the RME scenario, the 95th upper confidence limits of the arithmetic mean concentration values in soil were used to represent the exposure point concentration, combined with reasonable maximum intake values describing the extent, frequency, and duration of exposure.

**Perchlorate Half-Life in Native Soil Study**

A 12-kg soil sample was obtained at a random location from the agricultural field. The soil sample was allowed to air-dry at room temperature for three days and then sieved using a 7.5-mm sieve. In addition to the soil, 2 L of a 400 mg $L^{-1}$ sodium perchlorate solution were prepared using reagent grade sodium perchlorate and distilled water.

On day zero of the study, an 8.16-kg soil portion was obtained from the dried and sieved soil. The soil aliquot was placed in a clean 5-gallon plastic bucket. Then, the 2 L of 400 mg $L^{-1}$ perchlorate solution were added to the soil. The soil and perchlorate solution were thoroughly mixed by hand for about 30 min. After mixing, the soil containing the perchlorate solution was transferred to a clay planting pot. The planting pot was kept indoors at room temperature for the duration of the study. After transfer of the soil to the planting pot, two soil samples were collected in 5.0-cm (2″) stainless steel tubes. The tubes were sealed using PTFE tape and capped with plastic caps. The soil samples were immediately labeled and delivered to an analytical laboratory to determine their perchlorate concentration. Subsequent soil samples were collected in pairs on days 1, 2, 4, 7, 10, 15, and 21 as described above.

Soil in the planting pot remained undisturbed for the duration of the study. Soil samples were collected from the center of the pot during the sampling events. However, sampling results obtained on day 15 indicated that perchlorate concentrations in soil were not uniform. Perchlorate concentrations increased toward the bottom of the planting pot. Therefore, before sampling on day 21, the soil was removed from the planting pot, sieved, and mixed by hand for about 30 minutes. The sieving and remixing of day 21 was conducted to eliminate bias in the soil sampling and to obtain soil samples representative of all soil in the study.

## RESULTS AND DISCUSSION

Results of the risk assessment indicated that perchlorate in soil and vegetable material at the concentrations measured at the site did not represent a health risk to agricultural workers at the site or to consumers of vegetables grown there. Table 1 presents the average and RME perchlorate concentrations in soil, the estimated average daily doses received through the inhalation, oral, and dermal routes, and the hazard index (HI) estimated for this receptor. The average HI was estimated to be $3.07 \times 10^{-5}$ under the average exposure scenario and $3.68 \times 10^{-5}$ under the RME scenario. Table 2 presents the average and RME perchlorate concentrations in vegetables, the mean per capita kale, lettuce, and spinach consumption, the average and RME perchlorate dose received by the hypothetical consumers, and the HI estimated for off-site consumers. The average and RME HI estimated for off-site consumers were 0.113 and 0.189, respectively. HI values below 1.0 are considered acceptable by the EPA.

Table 1. Risk estimates for on-site workers

| Parameter | Units | Average scenario | RME scenario |
|---|---|---|---|
| perchlorate concentration in soil | mg kg$^{-1}$ | $3.40 \times 10^{-2}$ | $3.80 \times 10^{-2}$ |
| inhalation daily dose[a] | mg kg d$^{-1}$ | $5.27 \times 10^{-11}$ | $2.35 \times 10^{-10}$ |
| oral daily dose[a] | mg kg d$^{-1}$ | $1.62 \times 10^{-8}$ | $1.81 \times 10^{-8}$ |
| dermal daily dose[a] | mg kg d$^{-1}$ | $1.14 \times 10^{-8}$ | $1.47 \times 10^{-8}$ |
| total dose (all routes) | mg kg d$^{-1}$ | $2.76 \times 10^{-8}$ | $3.31 \times 10^{-8}$ |
| reference dose (RfD)[5] | mg kg d$^{-1}$ | $9.00 \times 10^{-4}$ | $9.00 \times 10^{-4}$ |
| hazard index[b] | unitless | $3.07 \times 10^{-5}$ | $3.68 \times 10^{-5}$ |

[a]Average daily dose = [$ClO_4^-$] × (intake factor). [b] Hazard index = total dose/RfD.

**Table 2.** Risk estimates for off-site consumers

| Parameter | Units | Average scenario | RME scenario |
|---|---|---|---|
| perchlorate concentration in vegetables [a] | mg $kg^{-1}$ | 1.87 | 3.14 |
| mean per capita kale consumption [b] | kg $kg^{-1}$ $day^{-1}$ | $1.50 \times 10^{-6}$ | $1.50 \times 10^{-6}$ |
| mean per capita lettuce consumption [b] | kg $kg^{-1}$ $day^{-1}$ | $9.20 \times 10^{-6}$ | $9.20 \times 10^{-6}$ |
| mean per capita spinach consumption [b] | kg $kg^{-1}$ $day^{-1}$ | $4.35 \times 10^{-5}$ | $4.35 \times 10^{-5}$ |
| average daily dose (ADD) [c] | mg $kg^{-1}$ $day^{-1}$ | $1.10 \times 10^{-4}$ | $1.70 \times 10^{-4}$ |
| reference dose (RfD) [5] | mg $kg^{-1}$ $day^{-1}$ | $9.00 \times 10^{-4}$ | $9.00 \times 10^{-4}$ |
| hazard index [e] | unitless | $1.13 \times 10^{-1}$ | $1.89 \times 10^{-1}$ |

[a] Average or 95% UCL of mean perchlorate concentration. [b] Values taken from EPA 1995. [c] Average daily dose (ADD) = perchlorate concentration times total vegetable intake per capita. [d] Hazard index = ADD/RfD.

Results of the laboratory study indicated that perchlorate degradation in soil takes place at a relatively fast pace. When the perchlorate concentration in soil was plotted against time after application (Figure 1), it became evident that perchlorate degradation in soil occurred in two distinct phases. In Phase I, there is a rapid breakdown of perchlorate in soil. In Phase II, the perchlorate degradation rate is much slower to negligible. These results indicate that active perchlorate degradation in soil occurred mostly during Phase I and that optimal conditions for perchlorate degradation were not present during Phase II, and therefore minimal degradation occurred during this second phase. Under the conditions of this study, the perchlorate half-life in soil during Phase I was estimated to be 52 h.

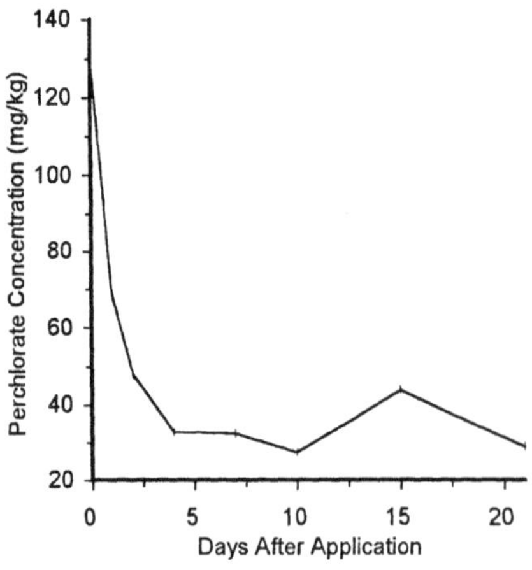


**Figure 1.** Perchlorate concentration in soil over time after application.

Perchlorate is known to degrade in soil in the presence of organic matter, reducing agents, and/or suitable microbial flora. The organic carbon content in the tested soil was 0.7 percent.

The presence and concentration of reducing agents and suitable bacteria were not determined. Therefore, the actual mechanism(s) responsible for the perchlorate degradation in soil was (were) not determined.

## CONCLUSION

The lack of an adequate analytical detection method for perchlorate has been noted elsewhere.[5] This problem was also noted in this study. Analytical interference by natural organic acids, soil salts, as well as agricultural chemicals could not be ruled out. Commercial analytical laboratories consulted did not have an analytical method (other than EPA 300.0 modified) to corroborate their analytical results. Thus, the presence of false positives in the analytical results could not be ruled out. This problem was not critical in this study, as the estimated health risks were below regulatory levels of concern. However, the possibility of false perchlorate detection in soil and biological samples should be given due consideration when risk reduction or regulatory action is considered based solely on uncorroborated analytical data.

LFR recommended that the contaminated well be monitored at regular intervals to make sure that its perchlorate concentrations do not increase. At the time the risk assess-ment was conducted, it was not clear whether the concentration of perchlorate in ground-water was stable, increasing or decreasing. Follow-up soil and water analyses have revealed that the perchlorate concentration in groundwater is stable and that, as predicted by the bench-scale degradation study, perchlorate is not accumulating in soil.

## REFERENCES

1. "Perchlorate in California Drinking Water." California Department of Health Services, Division of Drinking Water and Environmental Management, Drinking Water Program., 1997.

2. "Risk Assessment Guidance for Superfund." In *Human Health Evaluation Manual,* Vol. 1, Part A, Interim Final. Environmental Protection Agency: Washington, D.C., December 1989. EPA Doc. No. EPA/540/1-89/002.

3. Von Burg, R. "Toxicology update: derchlorates." *Journal of Applied Toxicology* **1995**, *15*, 237–241.

4. Wolff, J. "Perchlorate and the thyroid gland." *Pharmacological Reviews* **1998**, *50*, 89–105.

5. *Perchlorate Environmental Contamination: Toxicological Review and Risk Characterization Based on Emerging Information* (External Review Draft). Environmental Protection Agency. Office of Research and Development. 1999. URL: http://www.epa.gov/ncea/perch.htm.

6. *Exposure Factors Handbook.* Environmental Protection Agency: Washington, D.C., June 1995. EPA Doc. No. EPA/600/P-95/002A.

# Chapter 21
# INFLUENCES ON PHYTOREMEDIATION OF PERCHLORATE-CONTAMINATED WATER®

Valentine A. Nzengung①* and Chuhua Wang②

①Department of Geology, University of Georgia, Athens, Georgia 30602
②Technic, Inc., 1 Spectacle Street, Cranston, Rhode Island 02910

## INTRODUCTION

Ammonium perchlorate is an oxygen-adding component in propellants for rockets, missiles and fireworks. To ensure that rockets and missiles operate effectively and safely, old propellant mixes must be continuously replaced in rockets and missiles with fresh supplies. High-pressure washout of solid propellant generates large volumes of perchlorate contaminated wastewater. Perchloric acid and perchlorate salts are also used extensively in many commercial and industrial processes, such as wet digestions, organic syntheses, and electropolishing of metals; animal feed additives, explosives, pyrotechnics, and herbicides. As a result, perchlorate manufacturers and users have disposed of large amounts of this chemical, since the 1950s. In addition, perchlorate has been found as a contaminant in certain fertilizers and bulk water treatment chemicals.[1–2]

Being a salt, ammonium perchlorate is readily soluble in water and remarkably stable. Perchlorate has been found as a contaminant in many water systems throughout the United States. The current extent of perchlorate contamination of groundwater and soils in the United States is still to be determined. EPA estimates that perchlorate has been either manufactured or used in 44 states, and it has been measured in surface or groundwater in 13 states. However, the widest outbreak of perchlorate contamination has been recorded in California. Twelve facilities in this area manufacturing or testing solid rocket fuels for the Department of Defense and/or the National Aeronautics and Space Administration have been accredited with the environmental pollution.

®This work was presented at the 218th national meeting of the American Chemical Society as part of the Environmental Division symposium Perchlorate in the Environment, held August 22–24, 1999, in New Orleans.

*Author to whom correspondence should be directed. Phone: 706-542-2699. Fax: 706-542-2425. Electronic mail: vnzengun@uga.edu.

*Perchlorate in the Environment*, edited by Urbansky.
Kluwer Academic/Plenum Publishers, New York, 2000.

The anion perchlorate poses potential health concerns because its ionic radius and charge are similar to that of iodine, which allows perchlorate to competitively block thyroid iodine uptake. At relatively high doses, perchlorate is known to interfere with the thyroid's ability to produce hormones and regulate metabolism.[3] Potential health concerns related to this hormone disruption include carcinogenic, neurodevelopmental, developmental, reproductive, and immunotoxic effects. In 1998, the United States government formed the Interagency Perchlorate Steering Committee (IPSC), bringing together government representatives from the EPA, DoD, Agency for Toxic substances and Disease Registry (ATSDR), National Institute for Environmental Health Sciences (NIEHS), and affected state, tribal, and local governments. The role of IPSC has been to facilitate and coordinate different accounts and create information transfer links for interagency and intergovernmental activities regarding technological issues (occurrence, health effects, treatability and waste stream handling, analytical detection, and ecological impacts). Although perchlorate does not have a National Primary Drinking Water Regulation (NPDWR), in January 1999, the EPA National Center for Environmental Assessment (NCEA) proposed an oral reference dose (RfD) of 0.0009 mg $kg^{-1}$ $day^{-1}$, which corresponds to an action level of 32 μg $L^{-1}$.[4]

Basic research has been initiated to develop technologies for the clean up of perchlorate-polluted water and soils. Remediation technologies under consideration include adsorption by activated carbon, reverse osmosis, anion exchange, and bioremediation.[5] In a recent study by Nzengung et al.,[6] woody plants were shown to be capable of decontaminating water contaminated with perchlorate. The primary phytoprocesses identified by the latter authors as important in the removal of perchlorate from contaminated water included, uptake and phytodegradation in the tree branches and leaves and rhizodegradation. The authors also reported the temporal accumulation of perchlorate in mainly the leaves and branches of the trees used in their studies. In this study, we investigated possible factors that may influence phytoremediation of perchlorate-contaminated water. The specific objectives were to: (1) show that woody, edible, and wetland plants effectively remove perchlorate from water, (2) identify factors that enhance or inhibit uptake, phytodegradation and rhizodegradation of perchlorate, and (3) isolate and verify the contribution of rhizosphere bacteria in the degradation of perchlorate.

## EXPERIMENTAL

Sodium perchlorate monohydrate $NaClO_4 \cdot H_2O$ was obtained from Aldrich (Milwaukee, WI). Radiolabeled ($^{36}Cl$, 3.993 mCi/mL) ammonium perchlorate was custom prepared by NEN Life Science Products Inc. (Boston, MA). Full strength Hoagland's solution was purchased from Carolina Biological Supply Company, NC. Stern's Miracle-Gro™ (6.8% ammoniacal nitrogen and 8.2% urea nitrogen) in solid form was purchased from a local nursery in Athens, GA. Each nutrient mixture was diluted with deionized water to make the desired strength of growth solution. A 50% w/w NaOH solution obtained from J.T. Baker (Phillipsburg, NJ) was used in preparing the IC eluent and plant extraction solutions. Three woody plants, two edible, one wetland plant and a constructed microbial mat were used in the perchlorate degradation studies (Table 1). Harvested cuttings of cottonwood and willow trees were rooted hydroponically in half-strength Hoagland's solution for three months. A detailed description and schematic of the modified screw cap 2 L culture flasks (bioreactors) used in experiments with woody plants are found in a previous publication by Nzengung et al.[6]

**Table 1.** Different plants and microbial mats used in perchlorate degradation studies

| Plant name | Source | Form | Reactor |
|---|---|---|---|
| Eastern cottonwood *Hybrid populus* | Carswell AFB, Texas | Rooted cuttings [40 cm (L) by 4-mm (D)] | 2 L sand and hydroponic bioreactor |
| Black willow *Salix nigra* | Carswell AFB, Texas | Rooted cuttings 40 cm L × 4 mm D | 2 L sand and hydroponic bioreactor |
| Willow *Salix caroliniana* | Tampa, Florida | Rooted cuttings 40 cm L × 4 mm D | 2 L hydroponic bioreactor |
| *Eucalyptus cineria* | Local nursery, Athens, Georgia | One year old plants | 2 L sand bioreactor |
| Parrot feather *Myriophyllum aquaticum* | Local wetland, Athens, Georgia | Whole plant | 10 gal glass aquarium |
| French tarragon *Artemisia dracunculus* | Local nursery, Athens, Georgia | Minced plant and fresh water extract | 20 mL serum bottles |
| Spinach *Spinacia oleracea* | Local grocery store, Athens, Georgia | Minced plant and fresh water extract | 20 mL serum bottles |
| Mixed-species microbial mats | Constructed using a patented technique* | Whole organism | 60 mL serum bottles |

* Source of the microbial mats was O'Niell et al., 1999. L = length. D – diameter.

Perchlorate degradation studies by woody plants were conducted in 2 L screw cap hydroponic bioreactors with diluted Hoagland's solution or Miracle-Gro™ as the nutrient media. Note that Hoagland's solution is a nitrate N-source and Miracle-Gro is an ammonium and urea N-source. The choice of nutrient solution was based on the desired N-source for that experiment. The initial perchlorate concentrations, as $ClO_4^-$, of 10–100 mg $L^{-1}$ were used in these experiments. Accompanying each set of experiments were dosed blanks (no plant) and undosed-planted controls. A daily record of the volume of water taken up by each tree was maintained over the duration of each study. A 1 mL aliquot of the solution was withdrawn for perchlorate, chlorate, and chlorite analysis once every 24 h. The samples were diluted with deionized water to the ion chromatography (IC) measurement range for the perchlorate ion.

Degradation of perchlorate by the leaves of two edible plants (French tarragon and spinach) was investigated in 20 mL serum bottles. Each plant was minced in a food blender. One gram of the minced plant was added to the vials, filled with deionized water and dosed with perchlorate to obtain an initial solution concentration of 7–8 mg $L^{-1}$. For the microbial mats, 1 g of the wet mat was weighed into 60 mL vials and the headspace filled with 50 mL of deionized water, dosed with perchlorate to obtain an initial solution concentration of 10 mg $L^{-1}$. The vials were mixed continuously on a rotary shaker and sacrificed for analysis at predetermined intervals. A total of 9 sample vials were used in each of the replicate experiments. Controls contained only deionized water dosed with the same concentration of perchlorate as the samples. The pellets were separated from solution by centrifugation, and 1 mL of the liquid-phase was diluted and analyzed by IC.

Perchlorate degradation by *Myriophyllum aquaticum* (parrot feather) was investigated in a 108 L (28 gallon) glass tank. The parrot feather made up one-third of the tank volume and the remaining two-thirds (v/v) was 25% of full-strength Hoagland's solution. The reactor was dosed in succession, after the previous dose was completely depleted from solution with four different concentrations of perchlorate (19, 64, 162, and 23 mg $L^{-1}$, respectively). Samples were taken for perchlorate analysis at least once every 48 h.

## Determination of the Effects of Nitrate and Nitrogen Concentration

Experiments were performed in diluted Hoagland's solution supplemented with $NO_3^-$ to achieve the desired concentration ranges of < 200 mg $L^{-1}$ and 300–600 mg $L^{-1}$. The nitrate concentration in these experiments was maintained at the specified concentrations by adding sodium nitrate as needed to achieve the targeted $NO_3^-$ levels. Other evidence that the N-source of the nutrient solution may influence the phytoremediation of perchlorate was obtained by substituting diluted Miracle-Gro™, an ammonium and urea N-source, for diluted Hoagland's solution in some experiments. The Hydroponic experiments were conducted with willows grown on 2 g $L^{-1}$ and 0.5 g $L^{-1}$ solutions of Miracle-Gro. Thus, we were able to use the perchlorate degradation data from the latter experiments to determine the effect of nitrogen concentration on the reaction kinetics. The chloride, nitrate, and acetate concentrations of the growth solution were measured during the course of each experiment.

### Radiolabeled Experiments

Direct evidence of perchlorate degradation to chloride by the woody plants was obtained from experiments in which $^{36}Cl$ labeled ammonium perchlorate was used as a tracer. 1 mL of the 3.993 mCi/mL stock solution was used to dose each 2 L planted bioreactor. The initial perchlorate concentration in each of the reactors was between 25–30 mg $L^{-1}$. The concentration of perchlorate, chloride and $^{36}Cl$ activity in solution was monitored daily for the duration of the study. Perchlorate and chloride were measured by ion chromatography. The $^{36}Cl$ activity was assayed with a Beckman 5801 liquid scintillation counter. At the termination of the study, the plants were sacrificed for mass balance determination. The whole plant was weighed and sectioned into roots, upper and lower stems, branches and leaves. Each fraction was ground and extracted as described below. The extracts were analyzed for perchlorate and $^{36}Cl$ activity.

### Extraction of Perchlorate from Plant Matter

The concentration of perchlorate in the different tree fractions was measured to verify if perchlorate was hyperaccumulated or phytodegraded in the dosed plants. At the termination of some experiments the trees were sacrificed for extraction and analysis to determine the distribution of perchlorate in different plant fractions. Each fraction (3–5 g) was extracted several times by blending for 30–60 min with a 100 mL solution of 1 mM NaOH (pH = 11). The pellets were separated from solution by centrifugation. Not more than three extractions were needed to completely remove all extractable perchlorate from the plant tissue. The extract was diluted with deionized water to minimize interference from extracted plant organic acids. The extracts were analyzed by IC for perchlorate, chlorate, chlorite, and chloride ions.

In other experiments, the concentration of perchlorate in the tree leaves was monitored for many days or weeks following the complete removal of perchlorate added to the reactors.

Leaves harvested from a cross-section of branches on each tree were extracted and analyzed for perchlorate and its metabolic products. About 3 g of leaves were harvested and used in these analyses.

**Determination of Bacterial Role**

The microbial contribution to perchlorate degradation in the rhizosphere was determined in two ways. Subsamples of the previously described rhizosphere solution were boiled for 2 h or filtered with a 0.45 μm membrane filter before dosed with perchlorate. Equal volumes (50 mL) of the autoclaved, filtered and untreated media were each placed in three sterilized serum bottles dosed immediately with perchlorate and sealed. The vials were continuously mixed on a shaker until sacrificed for analysis by IC. Bacteria were then isolated from the root zone of willow trees that showed perchlorate degradation activity. Each isolate was grown in mineral salt media dosed with 200 mg $L^{-1}$ of perchlorate. The carbon (electron source) was 200–400 mg $L^{-1}$ acetate.

**Ion Chromatography**

A Dionex (Sunnyvale, CA) DX-100 Ion Chromatograph (IC) with SRS control was used for all analyses. The IC was equipped with a Dionex AI-450 Chromatography Automation System and the Advanced Computer Interface Module (ACI). An autosampler with a holding capacity of sixty 5 mL vials was used. Sample injection volume was 25 μL for high perchlorate concentrations (ppm) or 500 μL for low concentrations (ppb). Both an IONPAC® AG11 guard column (2 H 50 mm) and an IONPAC® AS11 analytical column (4 H 250 mm) were used. The analytical conditions developed by Dionex for analysis of low concentrations of perchlorate in drinking water and ground water by IC were followed. Flow rate of eluent was 1 mL $min^{-1}$; 10 mM NaOH solution was used as eluent for analysis of chloride, chlorite, chlorate, nitrate, and acetate ions while 100 mM NaOH solution was used for the perchlorate ion measurement in the rhizosphere. For perchlorate measurement in extracted plant tissue the eluent concentration was 50 mM NaOH. The working perchlorate concentration range was 80–1000 ppb and the conductivity was less than 0.3 μS. The detection limit of perchlorate for the above method was 2 ppb. The run time was 15 minutes. Deionized water (resistance of 18 MΩ_cm) was used as a system blank sample to establish the baseline and to confirm the presence of or lack of contamination in the system. Low and/or high concentration calibration curves were determined each day of sample analysis to ensure accurate quantification of perchlorate (Figure 1).

## RESULTS AND DISCUSSION

Hydroponically grown woody and aquatic plants (Table 1) were observed to remove perchlorate from aqueous solution via two mechanisms. An initial slow uptake and transformation of perchlorate in the plant tissues (phytodegradation) was followed by very rapid removal of perchlorate from solution by degradation in the root zone (rhizodegradation). For woody plants initially exposed to perchlorate-dosed medium, the rate of removal of perchlorate from solution was proportional to the water uptake rate of each plant. Figure 1 shows that of the rooted cuttings of three-woody plants (willow [*Salix* spp.], Eastern Cottonwood [*poplar*], *Eucalyptus cineria*) tested under similar experimental conditions, one-year-old *Eucalyptus* trees removed perchlorate from the nutrient medium at the fastest rate. The *Eucalyptus* trees had the largest fraction root and leaf (6.6 and 46%, respectively) and the highest evapotranspiration rates than the cottonwood and willow trees. Also, temporary accumulation of perchlorate in the plant tissue (mostly branches and leaves) was observed. The perchlorate concentration in the unplanted reactor did not change over the course of this study.

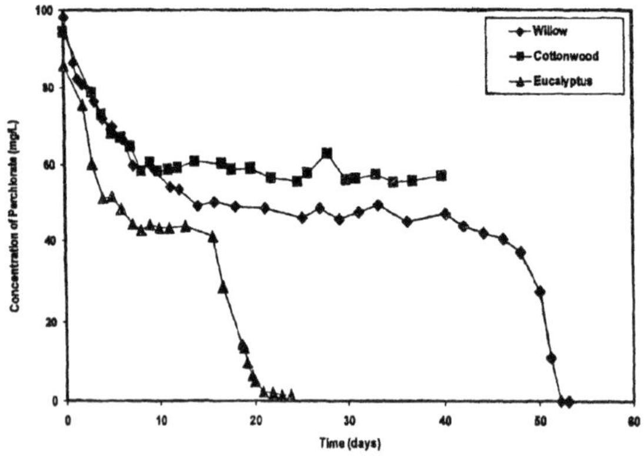


**Figure 1.** Relative rate of perchlorate removal from dosed hydroponic bioreactors by willow, cottonwood, and *Eucalyptus*. *Eucalyptus* trees had the highest fraction root and leaf mass and removed perchlorate from solution at a faster rate than the other trees.

Nzengung et al.[6,7] showed that during the initial phase of perchlorate removal from the bioreactors by uptake and phytodegradation, biostimulation of microbial activity also occurred. Thus, when the population of perchlorate degrading bacteria grew to significant levels in the bioreactor, rhizodegradation became the predominant reaction mechanism. The latter authors obtained zero-order rhizodegradation rate constants of 3.24 mg $L^{-1}$ $h^{-1}$ in studies conducted with three months old willow trees.[6] Generally, a higher fraction root mass corresponded to shorter lag time (biostimulation) and higher rhizodegradation rates. Rhizodegradation has been confirmed in experiments in which willow trees were dosed many times with 100 mg $L^{-1}$ perchlorate and the change in chloride concentration in the rhizosphere was monitored throughout the experiment. Additional evidence from recent studies conducted with radiolabeled perchlorate confirmed the rapid rhizodegradation of perchlorate to chloride in the rhizosphere of willow trees. Specifically, 28 mg $L^{-1}$ of $^{36}Cl$ labeled perchlorate was degraded to undetectable levels in 6 days and the counts per minute (CPM) in solution changed only by about 5%. This indicated that the rapid degradation of perchlorate in the rhizosphere minimized uptake by the plant.

The same two mechanisms observed in studies with woody plants described the removal of perchlorate from solution by parrot feather and microbial mats dosed multiple times with perchlorate. For example, 19 mg $L^{-1}$ of perchlorate initially used to dose the parrot feather was completely removed from solution in 21 days. The lag time preceding the rapid removal of perchlorate was 7 days. Meanwhile, a second dose of 63 mg $L^{-1}$ added to the same reactor was completely depleted from solution in only 14 days with no lag time observed. Subsequent higher doses of perchlorate were also rapidly removed from solution and the zero order kinetic model described the data. The removal of perchlorate from aqueous solution by constructed microbial mats was determined by the state of the mat. Slower kinetics were observed with the living (photosynthesizing) mats than with the dried mats. Specifically, while living mats dosed with 4 mg $L^{-1}$of perchlorate completely degraded the perchlorate to below the IC method detection limit of 2 ppb in about 60 days, the dried mats degraded 4.2 mg $L^{-1}$ to undetectable levels in 16 days. The slow kinetics observed with the living mats was attributed to our experimental design. We speculate that the oxygen formed during photosynthesis by the microbial mats could not escape from the sealed serum bottles and maintained aerobic conditions not favorable for perchlorate degradation. As the mat died from stress and began to decay, anaerobic conditions suitable for very rapid perchlorate degradation were created in the sealed vials. This suggests that a more appropriate design for perchlorate degradation by microbial mats is one in which the mats are allowed to stratify into upper aerobic and lower anaerobic zones, as would occur in natural ecosystems.

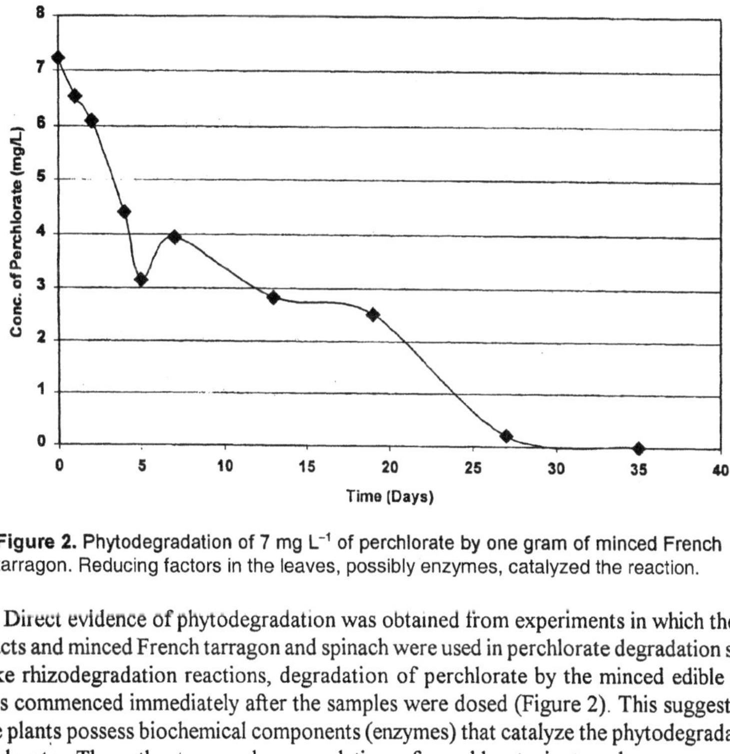


**Figure 2.** Phytodegradation of 7 mg $L^{-1}$ of perchlorate by one gram of minced French tarragon. Reducing factors in the leaves, possibly enzymes, catalyzed the reaction.

Direct evidence of phytodegradation was obtained from experiments in which the crude extracts and minced French tarragon and spinach were used in perchlorate degradation studies. Unlike rhizodegradation reactions, degradation of perchlorate by the minced edible garden plants commenced immediately after the samples were dosed (Figure 2). This suggested that some plants possess biochemical components (enzymes) that catalyze the phytodegradation of perchlorate. Thus, the temporal accumulation of perchlorate in tree leaves reported by Nzengung et al.[7] is due to slow phytodegradation by deoxygenase or reducing plant enzymes.

**Effect of Nitrate and Nitrogen Concentration on Perchlorate Degradation**

Figure 3 shows that the degradation of perchlorate by willow trees was affected by both the concentration of nitrate in the growth media and the N-source. The fastest perchlorate degradation kinetics were observed in planted bioreactors with lower $NO_3^-$ concentrations of < 200 mg $L^{-1}$ and slowest kinetics at higher $NO_3^-$ concentration of 300-600 mg $L^{-1}$. No perchlorate was removed from solution within the first 5 and 10 days by willow trees grown in nutrient solution containing the lower and higher nitrate concentrations, respectively. The influence of $NO_3^-$ on perchlorate reduction was attributed to competing reactions in which rhizosphere bacteria preferentially utilized $NO_3^-$, as TEA. This shortcoming was overcome by replacing the $NO_3^-$ N-source (Hoagland's solution) with an ammonium and urea N-source (i.e.,

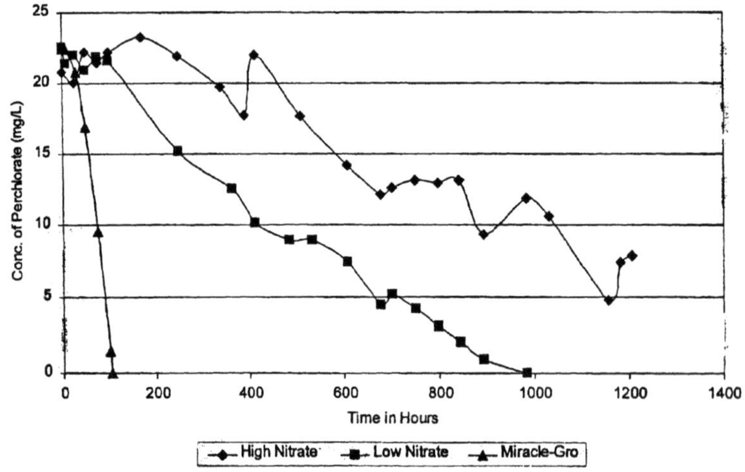


**Figure 3.** Effect of nitrate concentration and N-source on removal of perchlorate from solution by three willow trees. Diluted Hoagland's solution was used as nutrient for trees grown in high ($NO_3^-$ = 300–600 ppm) and low ($NO_3^-$ < 200 ppm) nitrate N-source and diluted Miracle-Gro™ was used for ammoniacal/urea N-source (no $NO_3^-$).

Miracle-Gro). As a result, perchlorate was rapidly reduced in the rhizosphere of the willow trees grown in Miracle-Gro. Uptake of perchlorate by woody plants was minimized in medium with ammonium and urea as the N-sources. In previous work, Nzengung and coworkers[6,7] showed that increasing the concentration of the carbon (electron) source reversed the inhibition of perchlorate degradation under high nitrate concentrations.

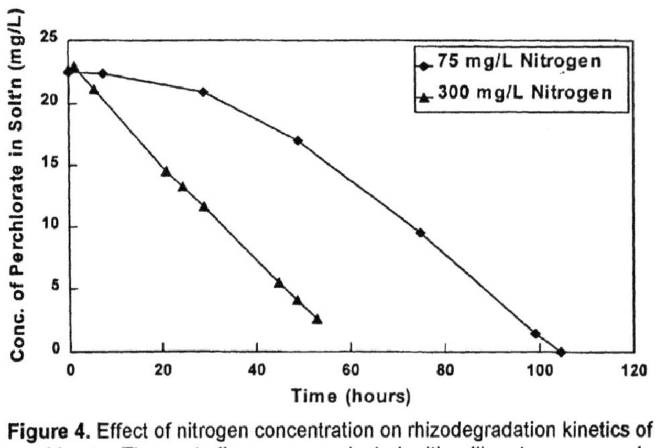


**Figure 4.** Effect of nitrogen concentration on rhizodegradation kinetics of perchlorate. These studies were conducted with willow trees grown in Miracle-Gro™ solution in 2 L hydroponic bioreactors.

Figure 4 shows that the concentration of nitrogen in the growth solution affected the rate of rhizodegradation of perchlorate by willow trees. Faster kinetics were observed in nutrient solutions prepared with Stern's Miracle-Gro containing 300 mg $L^{-1}$ nitrogen. The relatively slower kinetics observed at the lower nitrogen concentration of 75 mg $L^{-1}$ could have been caused by insufficient nitrogen available for optimum tree growth. The yellowing of the willow leaves in the latter study was attributed to nitrogen (nutrient) deficiency or stress. This suggests that woody plants grown under optimum nutrient conditions should be very effective in the decontamination of perchlorate-contaminated soils and water.

Figure 5 shows that one of four pure cultures of bacteria isolated from the rhizosphere of willow trees degraded perchlorate and most likely mediated the rhizodegradation reactions.[9] The perchlorate degrading bacteria utilized acetate as an electron and carbon source. The isolated bacteria have only been tentatively identified and full characterization is in progress. Complete characterization of these bacteria will facilitate biostimulation of rhizodegradation by injection of the perchlorate-degrading isolates into the root zone of local plants growing at impacted sites. Also, the latter approach should minimize unanticipated changes to the natural ecosystem resulting from the introduction of foreign plant species.

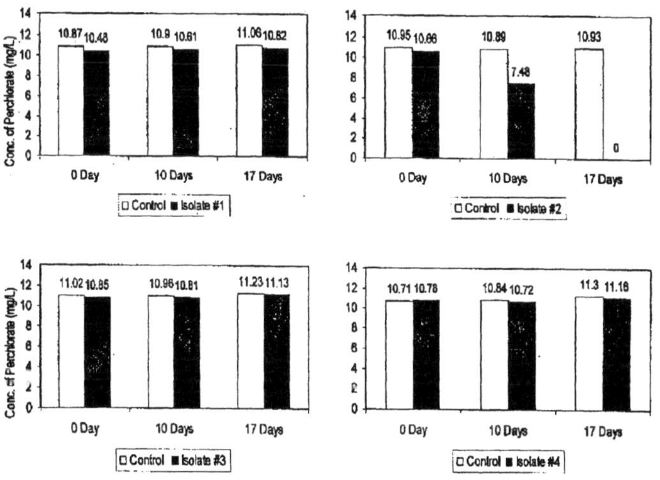


**Figure 5.** Degradation of perchlorate by four bacteria isolated from rhizosphere of willow tree previously used in rhizodegradation of perchlorate. Isolate #2, which completely degraded 11 mg $L^{-1}$ in less than 17 days has been tentatively identified as an *Enterobacter*.

Many factors that enhance or limit the effectiveness of phytoremediation of perchlorate contaminated environments have been identified (Table 2). At any given site some or all of these influences should be important. Identification and management of these factors should ensure optimum performance of phytoremediation of perchlorate polluted sites. Once screen tests have been used to identify the most appropriate plant species for the site clean up; the site geology and environmental conditions should present the only other challenges.

**Table 2.** Summary of specific factors and their influences on phytoremediation of perchlorate contaminated water

| | Specific factors | Influences |
|---|---|---|
| 1 | Plant species | •Willows and *Eucalyptus* show greatest promise for rhizodegradation<br>•Spinach and French tarragon have very high concentrations of perchlorate degrading components (enzymes) in their leaves |
| 2 | Root mass | •Larger fraction root mass corresponds to a shorter lag time (biostimulation) and enhanced rhizodegradation kinetics |
| 3 | Bacteria | •Very rapid degradation predominates in the presence of an isolated root-colonizing bacteria |
| 4 | Nitrate concentration | •High relative concentration of competing terminal electron acceptors is unfavorable<br>•High relative nitrate concentrations causes slow rhizodegradation of perchlorate and increases uptake into tree branches and leaves |
| 5 | N-source | •Ammoniacal and urea N-sources enhance rhizodegradation and minimize uptake into the plant<br>•Nitrate N-source is less favorable |
| 6 | Nitrogen concentration | •Rhizodegradation rates of perchlorate are higher when the plants are grown in media containing optimum ammoniacal and urea nitrogen for the plant species |
| 7 | Concentration of carbon (electron) sources | •Acetate and other carbon and electron sources enhance rhizodegradation kinetics.<br>•High acetate concentration reverses the nitrate effects |

## CONCLUSIONS

The results of this study show that selected woody, edible, and aquatic plants and microbial mats can be used to detoxify environments contaminated with perchlorate. The initial slow uptake and phytodegradation of perchlorate by plants exposed to perchlorate changes to very rapid removal by rhizodegradation after several days depending on the plant physiology and environmental conditions. We believe that initial or prolonged exposure of rooted green plants to perchlorate-dosed media stimulates the growth of perchlorate-degrading microorganisms in the rhizosphere. Exudates secreted by the plant roots supplied nutrients that sustained the growth of rich and diverse consortia of bacteria among which was perchlorate-degrading bacteria. In the presence of isolated and identified root-colonizing bacteria, perchlorate was rapidly degraded to chloride in the rhizosphere with minimum uptake into the tree leaves and branches. Any perchlorate taken up into the green plants was not simply accumulated, but was slowly transformed. Ongoing studies are focused on the design and testing of simulated wetland and hydroponic ebb-and-flow systems to treat perchlorate-contaminated wastewater. Also, the characterization of specific plant enzymes and root zone colonizing bacteria is underway.[9]

## ACKNOWLEDGMENTS

The United States Air Force at Wright Patterson Air Force Base in Dayton, OH, provided initial funding for this work.

## REFERENCES

1. Schilt, A.A. Perchloric Acid and Perchlorates." GFS Chemical: Columbus, OH, 1979; passim.

2. Susarla, S; Bacchus, S.T.; Wolfe, N.L.; McCutcheon, S.C. "Phytotransformation of perchlorate and identification of metabolic products in *Myriophyllum aquaticum*." *International Journal of Phytoremediation* **1999**, *1*, 97–107.

3. Cao, Xi. "Timing of vulnerability of the brain to iodine deficiency in endemic cretinism." *New England Journal of Medicine* **1994**, *331*, 1739–1744.

4. Renner, R. "EPA draft almost doubles safe dose of perchlorate in water." *Environmental Science and Technology* **1999**, *33*, 110A–111A.

5. Herman, D.C., Frankenberger, W.T. Jr. "Microbial-mediated reduction of perchlorate in groundwater." *Journal of Environmental Quality* **1998**, *27*, 750–754.

6. Nzengung, V.A.; Wang, C.; Harvey, G. "Plant-mediated transformation of perchlorate into chloride." *Environmental Science and Technology* **1999**, *33*, 1470–1478.

7. Nzengung, V.A.; Wang, C.; Harvey, G.; McCutcheon, S.C.; Wolfe, N.L. "Phytoremediation of Perchlorate Contaminated water: Laboratory Studies." Symposium Series: Fifth International Symposium. *In Situ and On-Site Bioremediation: Phytoremediation.* A. Leeson; B.C. Alleman, Eds. Battelle: Columbus, OH, 1999; pp. 239–244.

8. O'Niell, L.W.; Nzengung, V.A.; Noakes, J.; Bender, J.; Phillips, P. "Sorption accompanied by biodegradation of PCE and TCE using mixed-species microbial mats." *Journal of Hazardous Waste Research*, in press.

9. Nzengung, V.A. Phytoremediation of perchlorate contaminated water and soils. *Provisional patent application,* 1999.

# Chapter 22

## *IN SITU* BIOREMEDIATION OF PERCHLORATE IN GROUNDWATER[⊗]

Evan E. Cox,[1]* Elizabeth Edwards,[2] and Scott Neville[3]

[1]GeoSyntec Consultants, Inc., 160 Research Lane, Suite 206
Guelph, Ontario N1G 5B2 Canada
[2]University of Toronto, Department of Chemical Engineering
and Applied Chemistry, Toronto, Ontario M5S 1A1 Canada
[3]Aerojet General Corporation, Sacramento, California 95670

## INTRODUCTION

Groundwater contamination related to the production, handling and use of rocket propellants such as ammonium perchlorate has been identified as a widespread problem at United States Department of Defense (DoD), Department of Energy (DOE) and defense contractor facilities. It is estimated that perchlorate has been manufactured and/or used in 44 states, resulting in groundwater contamination in at least 14 of these states.[1] In California, Arizona, and Nevada alone, it is estimated that perchlorate impacts the drinking water supplies of more than 15 million people.[2] The concern surrounding perchlorate in groundwater and drinking water supplies relates to its potential ability to impact thyroid function.[3] While a national regulatory standard has yet to be set, the California Department of Health Services (CDHS) established a provisional action level (PAL) of 18 $\mu g\ L^{-1}$ for perchlorate in drinking water, and this PAL has been adopted as an interim regulatory guideline for groundwater by several regulatory agencies.

Few cost-effective technologies currently exist for the treatment of perchlorate-contaminated groundwater. Of the technologies being developed, bioremediation is among the most promising because it has the potential to destroy perchlorate rather than transferring it to another waste stream (e.g., impacted resin or brine) requiring costly treatment or disposal.

[⊗]This work was presented at the 218th national meeting of the American Chemical Society as part of the Environmental Division symposium Perchlorate in the Environment, held August 22–24, 1999, in New Orleans, Louisiana.

*Author to whom correspondence should be directed. Phone: 519-822-2230. Fax: 519-822-3151. Electronic mail: ecox@geosyntec.com.

Accordingly, significant effort has been directed in recent years to the development of ex situ bioremediation treatment systems that can be integrated into existing groundwater extraction and treatment systems. These ex situ bioreactors have been shown to effectively remove perchlorate from impacted groundwater; however, the systems do not afford relief against costly long-term operations and maintenance (O&M), the Achilles' heel of all ex situ treatment systems. By comparison, passive and semipassive in situ bioremediation may afford a less costly and less O&M-intensive approach to managing and remediating perchlorate-impacted groundwater, and specifically, to treating or controlling the perchlorate source areas that serve as long-term contributors to groundwater, and that drive the expansive perchlorate plumes that threaten our drinking water supplies.

Since 1998, we have been evaluating the applicability of in situ bioremediation for perchlorate-impacted groundwater. Our work has included bench scale microcosm studies using soil and groundwater from multiple impacted sites. Results have been extremely promising, suggesting that bacteria capable of perchlorate reduction are ubiquitous in subsurface environments, and that perchlorate can be rapidly biodegraded over a wide range of concentrations and starting conditions. Field pilot testing is underway to generate design and cost information for technology scale-up and commercialization. This chapter provides a brief description of several current test sites and study methodologies, and presents the results of our technology development efforts for in situ bioremediation of perchlorate in groundwater.

## OVERVIEW OF PERCHLORATE BIODEGRADATION

Perchlorate biodegradation results from microbially mediated redox reactions where perchlorate serves as the electron acceptor and is reduced via chlorate to chlorite. Chlorite then undergoes a biologically mediated dismutation reaction, releasing chloride and oxygen. The oxygen is subsequently reduced, provided sufficient electron donors are available. Figure 1 shows the hypothesized pathway for perchlorate reduction[4].

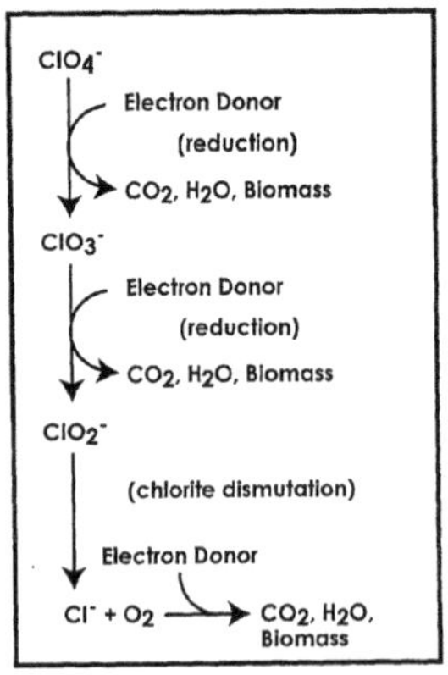


**Figure 1.** Pathway for the reduction of perchlorate.[4]

A variety of electron donors have been used to stimulate perchlorate reduction using pure or mixed cultures, including alcohols (e.g., ethanol, methanol), volatile fatty acids (e.g., acetate) and some sugar mixtures (molasses). To date, a number of microorganisms have been identified as possessing the ability to reduce perchlorate, and this information has been used to develop several ex situ bioreactor processes to remove high perchlorate concentrations from rocket motor wash water[5,6]. A production scale, continuous stirred tank reactor system began treating wastewater from rocket motor production operations in Utah in 1997. Ex situ bioreactor applications have subsequently been modified for groundwater applications. For example, Aerojet has applied anaerobic fluidized bed bioreactors to successfully reduce perchlorate concentrations in groundwater (from over 5000 $\mu g\ L^{-1}$) to the PAL), with ex situ biotreatment being incorporated into the overall groundwater remediation systems at the site. Current directions in perchlorate bioremediation are focusing on the development of in situ bioremediation applications to treat/control perchlorate source areas in situ, and to control groundwater plume migration using passive or semipassive permeable biobarriers.

## SITE DESCRIPTIONS

The first site consists of a former burn area, where perchlorate- and solvent-containing wastes were reportedly burned. The site soil consists of silty-clay; the watertable at the site is shallow. Perchlorate concentrations in the site soil range as high as 4200 $mg\ kg^{-1}$; however average concentrations are typically less than 10 $mg\ kg^{-1}$. Perchlorate has impacted the site groundwater at concentrations of up to 140 $mg\ L^{-1}$, as a result of its high solubility and infiltration. The site groundwater is typically aerobic and oxidizing, with little to no organic carbon available to promote any intrinsic perchlorate reduction reactions. As a result, perchlorate migrates relatively unattenuated in the groundwater at this site.

The second site consists of a former propellant hogout facility, where perchlorate was historically washed out of rocket casings and replaced with fresh perchlorate (due to the limited shelf life of the propellant). Perchlorate is present at concentrations in excess of 1000 $mg\ kg^{-1}$ in less permeable soils at a depth of 30 feet below ground surface (bgs). Perchlorate concentrations in groundwater currently range as high as 1,000 $mg\ L^{-1}$, but have historically ranged as high as 6000 $mg\ L^{-1}$. Interestingly, groundwater redox conditions at this site are anaerobic and reducing, in part due to the intrinsic biodegradation of organic carbon (approximately 20 $mg\ L^{-1}$ as TOC) that appears to result from a former fuel release in the area.

## EXPERIMENTAL

### Site 1: Former Burn Area

Simulated aquifer microcosms were constructed using 1 L stoppered glass jars filled with 1 kg of homogenized soil from the site, and saturated with deionized water (approximately 500 mL). Jars were constructed in duplicate for each test condition. Controls and treatments consisted of:

1. Active Control (AC): no added electron donor or carbon source
2. Simple Electron Donor (SED): ethanol (100 $mg\ L^{-1}$)
3. Complex Electron Donor (CED): manure (5 g)
4. Bioaugmented Treatment (BT): ethanol (100 $mg\ L^{-1}$) + inoculum

Perchlorate concentrations in the soil used to construct the site microcosms ranged between 30 to 40 $mg\ kg^{-1}$, which resulted in starting aqueous concentrations of 90 to 120 $mg\ L^{-1}$ once the soils were saturated. The inoculum for the bioaugmentation treatment consisted of a food waste product known to contain a perchlorate-degrading bacteria.

Microcosms were incubated on bench top at room temperature (approximately 20°C). The water in the microcosms was analyzed on a weekly basis for perchlorate, dissolved oxygen, oxidation-reduction potential and pH by standard electrode methods. The method detection limit for perchlorate by the ion-specific electrode (ISE) method was 0.7 mg $L^{-1}$. Perchlorate samples were collected at selected time points for confirmation analysis by standard ion chromatography (IC) methods (Dionex Protocol; AS11 column). Results of ISE and IC analyses were typically within 15% variability. Treatments were respiked with electron donors/nutrients as required, to maintain availability.

**Site 2: Former Propellant Hogout Facility**

Treatment and control microcosms were constructed using site groundwater. Microcosms consisted of 250 mL (nominal volume) sterile glass bottles filled with approximately 200 mL of groundwater, allowing some headspace for potential gas production from added electron donors. Microcosms were capped with Mininert™ caps to allow repetitive sampling of each microcosm. All microcosms were amended with resazurin to monitor groundwater redox conditions (resazurin is clear under anaerobic-reducing conditions but turns pink if exposed to oxygen). Treatment microcosms were amended with one of the following electron donors/carbon substrates:

- Ethanol (to a target concentration of 250 mg $L^{-1}$);
- Molasses (250 mg $L^{-1}$ chemical oxygen demand equivalent); or
- Ethanol (250 mg $L^{-1}$) + inoculum

Sterile Control (SC) microcosms were amended with ethanol and 1.8 mL of 5% mercuric chloride and 0.5 mL of 5% sodium azide to inhibit microbial activity. Active control (AC) microcosms received no electron donor amendments.

Starting perchlorate concentrations in the microcosm groundwater averaged 100 mg $L^{-1}$. Microcosms were incubated in an anaerobic chamber, in the dark, at room temperature (about 20°C; similar to site groundwater). Water samples were collected on a weekly basis for analysis of perchlorate and chemical oxygen demand (to assess electron donor fate). Perchlorate analyses were conducted through a combination of ISE and IC methods. The effective detection limit for ISE analyses was 1.0 mg $L^{-1}$. The method detection limit for the IC analyses was reported as 1.9 μg $L^{-1}$.

Following reduction of perchlorate in the treatment microcosms, the microcosms were respiked with perchlorate to restore target concentrations of 100 mg $L^{-1}$. Treatment and control microcosms were then spiked with neat TCE (to achieve a target starting TCE concentration of about 750 μg $L^{-1}$) and electron donors. To assist TCE biodegradation, selected control and treatment microcosms were bioaugmented with a known dehalogenating microbial culture referred to as KB-1. The dehalogenation capabilities of KB-1 have been described elsewhere.[7]

## RESULTS AND DISCUSION

**Site 1: Former Burn Area**

Perchlorate concentrations in the AC microcosms (no electron donor addition) did not decline over the 51 day incubation period (Figure 2). In contrast, perchlorate concentrations in all electron donor treatments declined after initial acclimation periods varying from 26 to 40 days (Figure 2). Acclimation appeared to relate to the time required to develop appropriate anaerobic-reducing redox conditions and/or sufficient biomass to actively biodegrade the perchlorate. Alternatively, acclimation may have been related to the time required to exhaust preferential electron acceptors (e.g., nitrate).

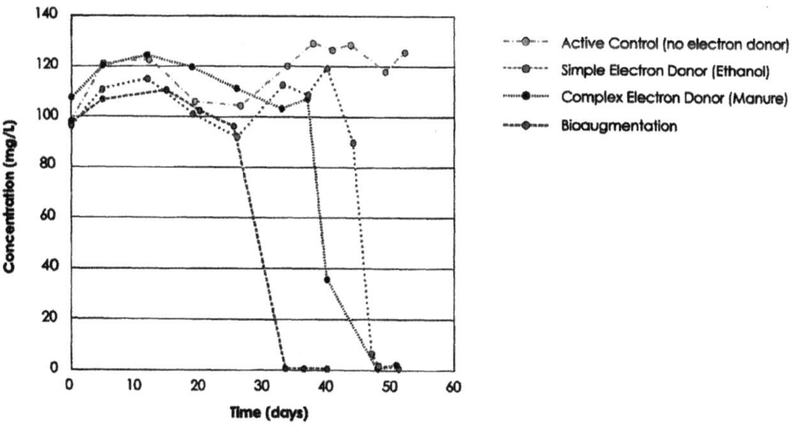


**Figure 2.** Biodegradation of perchlorate in simulated aquifer microcosms, former burn site. Perchlorate biodegradation in treatment microcosms begins after lag periods of up to 40 days. Post acclimation half-lives range between 0.8 to 2 days.

Perchlorate biodegradation in the SED and CED microcosms began after a lag period of about 40 days. Following acclimation, perchlorate concentrations in the SED microcosms declined from 119 mg $L^{-1}$ to < 0.7 mg $L^{-1}$ within 11 days; perchlorate concentrations in the CED microcosms declined from 107 mg $L^{-1}$ to 1.9 mg $L^{-1}$ within 11 days (Figure 2). After a shorter acclimation period (26 days) in the BT microcosms, perchlorate concentrations declined from 89 mg $L^{-1}$ to < 0.7 mg $L^{-1}$ within 11 days (Figure 2). To confirm perchlorate biodegradation, several microcosms were respiked with perchlorate at concentrations up to 351 mg $L^{-1}$ Perchlorate concentrations in the respiked microcosms declined rapidly to <1 mg $L^{-1}$ within 10 days (data not shown). Based on the data from the three treatments, the calculated perchlorate biodegradation half-lives in the acclimated treatment microcosms range from 0.8 to 2 days. While the addition of a perchlorate-degrading inoculum reduced the acclimation period, it did not significantly improve the biodegradation rate or extent compared to the other acclimated treatments. Therefore, it does not appear that bioaugmentation with perchlorate-degraders will be required to stimulate in situ bioremediation of perchlorate in groundwater at this site; appropriate bacteria are already present in the subsurface.

### Site 2: Former Propellant Hogout Facility

Perchlorate biodegradation in all treatment microcosms began immediately following addition of electron donors, with concentrations declining below the ISE detection limit of 1 mg $L^{-1}$ within 8 days (Figure 3). No lag or acclimation period was observed before onset of degradation activity, suggesting that the indigenous bacteria in the groundwater at this site are poised to reduce perchlorate, provided sufficient electron donors are available. By comparison, no perchlorate mass loss was observed in associated sterile or unamended (active) controls. Calculated biodegradation half-lives for perchlorate ranged between 1.2 to 1.8 days, which is similar to post-acclimation rates calculated for Site 1. As observed for Site 1, bioaugmentation did not significantly improve the biodegradation rate or extent compared to electron donor addition only, and therefore bioaugmentation is unlikely to be required to stimulate in situ bioremediation of perchlorate in groundwater at this site.

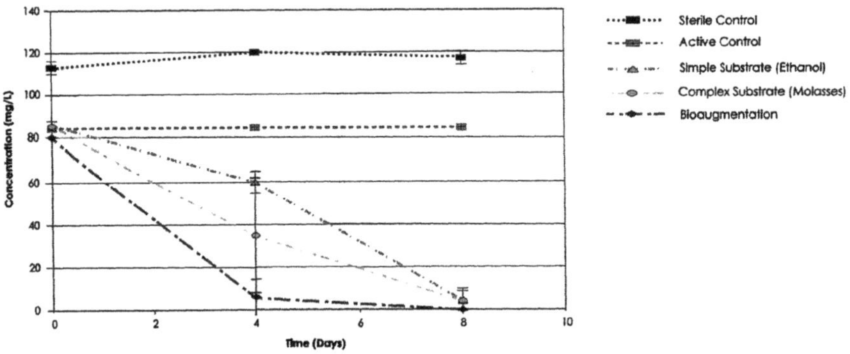


**Figure 3.** Biodegradation of perchlorate in groundwater microcosms, former propellant hogout facility. Perchlorate biodegradation in treatment microcosms begins without acclimation. Half-lives range between 1.2 and 1.8 days.

Following initial biodegradation of perchlorate in the Site 2 treatment microcosms, the microcosms were spiked with both perchlorate and TCE (along with electron donors) to assess the potential to jointly reduce these common groundwater contaminants within the same groundwater system. Groundwater at many DOD and defense contractor sites are impacted by both chemicals as a result of rocket manufacturing and/or handling operations. As expected, perchlorate and TCE concentrations remained relatively stable in the sterile control microcosms over an incubation period of approximately 100 days (Figure 4). By comparison, and somewhat unexpectedly, slow but continuous perchlorate biodegradation was observed in the active (unamended) control. The presence of organic carbon in the groundwater at Site 2 appears to be promoting low level intrinsic biodegradation of perchlorate (Figure 4).

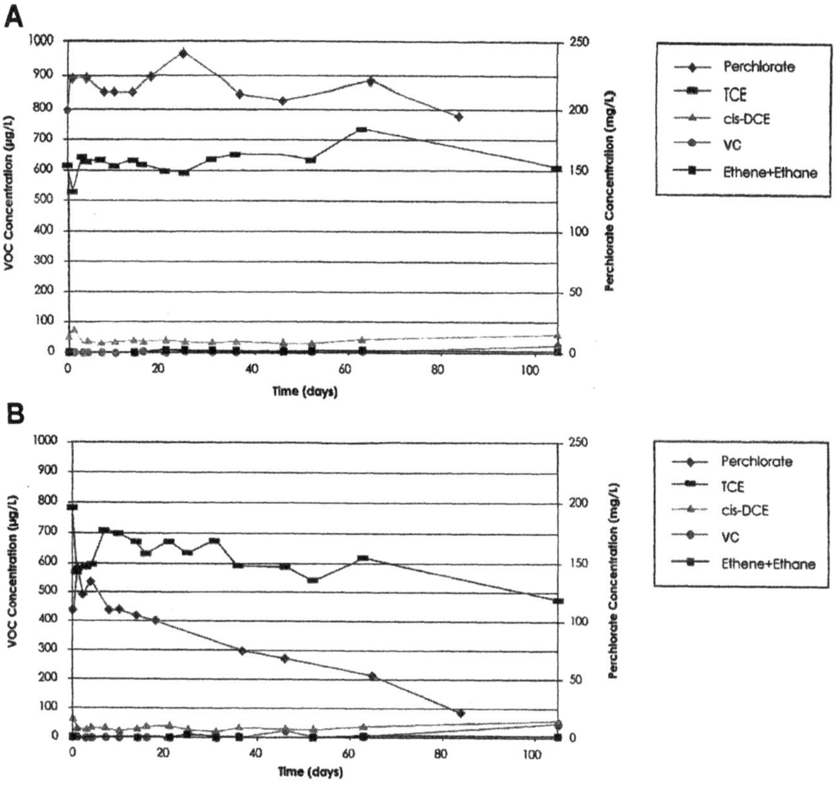


**Figure 4.** Biodegradation of perchlorate in groundwater microcosms, former propellant hogout facility. (A) Sterile control. (B) Active control (no electron donors added). Note the slow, but continuous, perchlorate mass loss in the active controls (half-life of 35 days).

As observed in the initial trial, the addition of molasses promoted rapid perchlorate reduction from starting perchlorate concentrations of nearly 100 mg $L^{-1}$ to less than the PAL of 18 µg $L^{-1}$ (Figure 5). However, simultaneous dechlorination of TCE was only observed in microcosms bioaugmented with KB-1 (Figure 5). In the bioaugmented microcosms, TCE was reductively dechlorinated via *cis*-1,2-dichloroethene (*cis*-1,2-DCE) and vinyl chloride (VC) to ethene and small amounts of ethane. Based on these data, calculated biodegradation half-lives

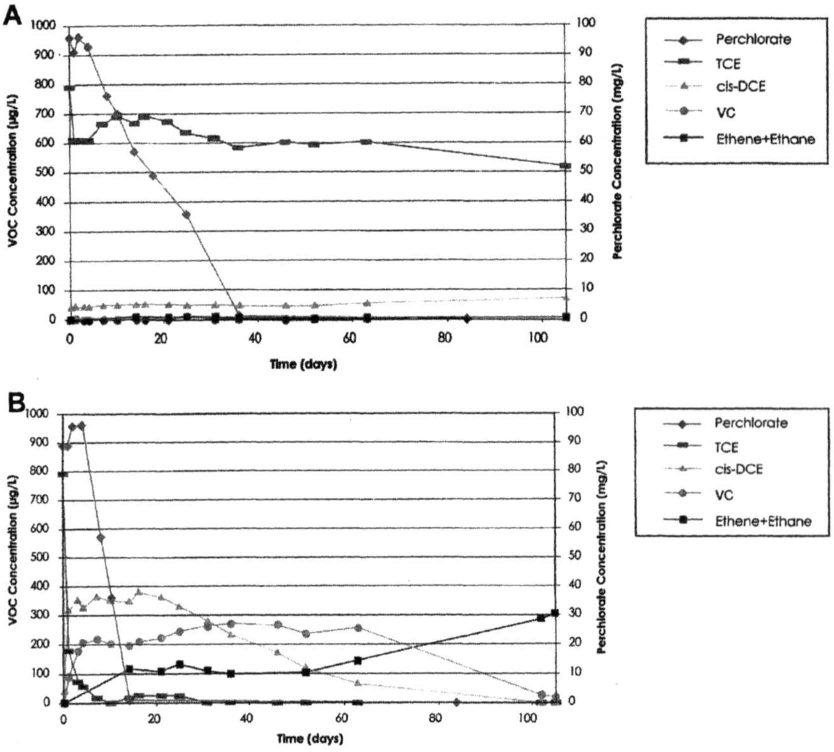


**Figure 5.** Biodegradation of perchlorate in groundwater microcosms, former propellant hogout facility. (A) Molasses treatment. (B) Molasses treatment bioaugmented with halo-respiring culture KB-1.

for perchlorate and TCE dechlorination to ethene in molasses-amended, bioaugmented microcosms are 2.2 and 22 days, respectively. These rates are considered fast by comparison to the groundwater velocity (about 1 foot per day) at this site.

The best rate and extent of joint perchlorate and TCE reduction was observed with addition of the food waste and bioaugmentation with KB-1. In these microcosms, perchlorate was biodegraded within 4 days while complete dechlorination of TCE to ethene was observed within 31 days (Figure 6). Final perchlorate concentrations were less than 7 $\mu g\ L^{-1}$; final concentrations of TCE, *cis*-1,2-DCE and VC were all below 5 $\mu g\ L^{-1}$. Based on these data, the calculated biodegradation half-life for perchlorate was 0.7 days (17 hours), while the half-lives for TCE, *cis*-1,2-DCE and VC were 1, 1.3 and 2.2 days, respectively.

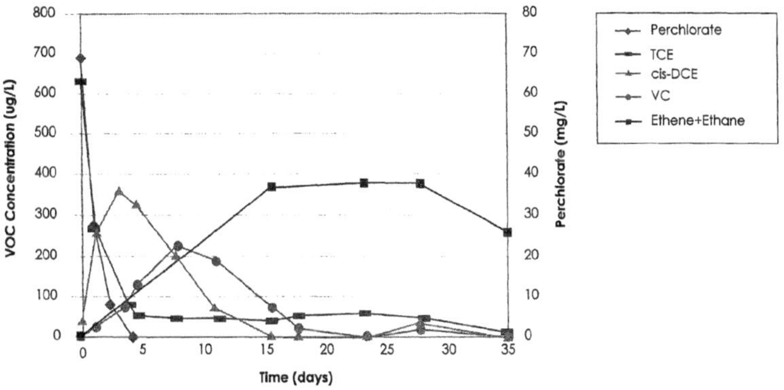


**Figure 6.** Biodegradation of perchlorate in bioaugmented groundwater microcosms amended with food waste, former propellant hogout facility.

## CONCLUSIONS

The results of these studies and our ongoing research suggest that bacteria capable of perchlorate reduction are ubiquitous in subsurface environments, and that perchlorate reduction in soil and groundwater can be stimulated through provision of a wide variety of electron donors. Given this, in situ bioremediation is likely to have a bright future as a low maintenance, cost-effective remedial approach for perchlorate-impacted groundwater in several ways. Firstly, the ability to stimulate rapid biodegradation of perchlorate at starting concentrations ranging between 100 to 1000 mg $L^{-1}$ indicates that in situ bioremediation may be effective in directly treating the perchlorate source areas that are the driving force behind the expansive perchlorate plumes at many sites. Secondly, the use of passive or semipassive permeable biobarriers, as depicted conceptually in Figure 7, should provide a cost-effective means of controlling plume migration and protecting receptors.

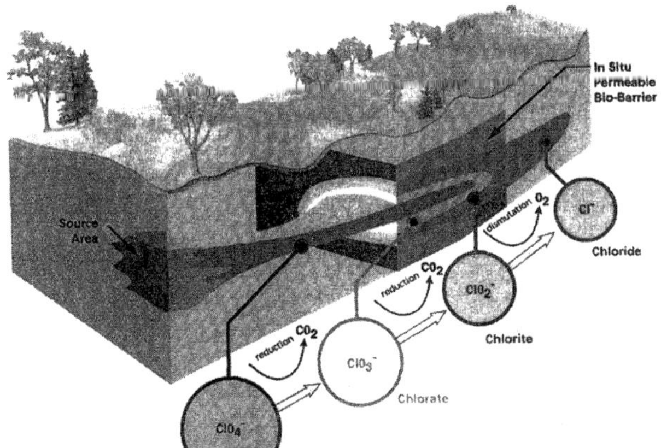


**Figure 7.** Conceptualization of *in situ* bioremediation of perchlorate in groundwater using passive or semipassive permeable biobarriers.

Finally, the potential to jointly biodegrade a variety of common co-contaminants, including chlorinated solvents, petroleum hydrocarbons and byproducts of hydrazine-based liquid rocket fuels (e.g., *N*-nitrosodimethylamine; NDMA) in the same in situ treatment system is of obvious benefit in reducing long-term O&M costs associated with conventional pump and treat based remedies.

**ACKNOWLEDGMENTS**

The authors would like to thank Dr. David Major and Jason Allan at GeoSyntec Consultants, Gerry Swanick and Michael Girard at Aerojet, and Dr. Robert Borch of Applied Environmental for assistance in these and related perchlorate studies.

## REFERENCES

1. Damian, P.; Pontius, F.W. "From rockets to remediation: the perchlorate problem." *Environmental Protection* **1999**, *20*, 24.

2. Environmental Protection Agency. Region 9 Perchlorate Update. May 1999.

3. Urbansky, E.T.. "Perchlorate chemistry: implications for analysis and remediation." *Bioremediation Journal* **1998**, *2*, 79–95.

4. Rikken, G.B.; Kroon, A.G.M.; van Ginkel, C.G. "Transformation of (per)chlorate into chloride by a newly isolated bacterium: reduction and dismutation." *Applied Microbiology and Biotechnology* **1996**, *45*, 420–426.

5. Attaway, H.; Smith, M. "Reduction of perchlorate by an anaerobic enrichment culture." *Journal of Industrial Microbiology* **1993**, *12*, 408–412.

6. Wallace, W.; Ward, T.; Breen, A.; Attaway, H. "Identification of an anaerobic bacterium which reduces perchlorate and chlorate as *Wolinella succinogenes*." *Journal of Industrial Microbiology* **1996**, *16*, 68–72.

7. Major, D.W.; Edwards, E; Cox, E.E. "Bioaugmentation: the future of chlorinated solvent bioremediation." Accepted for presentation at the Partners in Environmental Remediation Symposium, November 1999, Arlington, Virginia.

# Chapter 23

# TREATMENT OF GROUNDWATER CONTAINING PERCHLORATE USING BIOLOGICAL FLUIDIZED BED REACTORS WITH GAC OR SAND MEDIA[⊗]

Mark R. Greene* and Michael P. Pitre

Envirogen, Inc., Princeton Research Center, 4100 Quakerbridge Road, Lawrenceville, New Jersey 08648

## SUMMARY

Ammonium perchlorate has been used by NASA and the U.S. military as a component of solid rocket fuels. The formulation has a limited shelf life and replenishment is done by high pressure water flushing of old fuel followed by replacement with fresh supply as part of regularly scheduled maintenance activities. Past practice allowed the spent flush water to be discharged to the ground. As such, large volumes of ammonium perchlorate have been disposed in Nevada, California, Utah and likely other states since the 1950s.[1] Ammonium perchlorate is very soluble in water (i.e., 20 g $dL^{-1}$ at 25 °C) and dissociates completely to ammonium and perchlorate ions. The perchlorate ion is mobile in aqueous systems and can persist for decades under typical groundwater and surface water conditions due to its lack of reactivity with other compounds.[1]

Some concentrations of perchlorate are believed to be detrimental to human health. At relatively high doses, perchlorate has been found to interfere with the thyroid's ability to produce hormones and regulate metabolism. Biochemical treatment is the technology that offers the most promise for economical cleanup of perchlorate contaminated waters and wastewaters. A laboratory pilot plant study was conducted to compare the perchlorate degradation performance of biological fluidized bed reactors (FBRs) containing either granular activated carbon (GAC) or sand as the fluid bed media. The performance of alternative electron donors (i.e., ethanol, methanol and a mixture of the two alcohols) was

[⊗]This work was presented at the 218th national meeting of the American Chemical Society as part of the Environmental Division symposium Perchlorate in the Environment, held August 22–24, 1999, in New Orleans, Louisiana.

*Author to whom correspondence should be directed. Phone: 609-936-9300. Fax: 609-936-9221. Electronic mail: greene@envirogen.com.

also evaluated. Results from the study showed GAC and sand based FBRs using ethanol as the electron donor both achieved significant perchlorate reduction with GAC media exhibiting superior stability. It was also found that methanol was not as effective as ethanol in promoting perchlorate reduction. However, an ethanol-methanol mixture was found to be as effective an electron donor as ethanol alone for perchlorate removal. FBR perchlorate removal technology has been demonstrated on a full-scale basis in two 4.27 m diameter GAC based FBRs with ethanol added as the electron donor.

## INTRODUCTION

Chlorinated salts such as perchlorate, chlorate, chlorine dioxide and hypochlorite are produced in large quantities in the U.S. and used in a variety of applications.[2] The perchlorate anion enters the environment primarily from the salts of ammonia, potassium and sodium perchlorate. Ninety percent of all perchlorate manufactured in the U.S. is ammonium perchlorate, which has been used as an energetics booster in the formulation for rocket fuel. Most environmental contamination is attributed to the use and disposal of this salt.[3] The majority of perchlorate contaminated sites are associated with current or former propellant and/or rocket manufacturing and testing facilities.[3] Rocket fuel has a limited shelf life and when necessary is removed from the U.S. inventory of missiles and rockets and replaced with fresh material.[4] The replacement process involves a high-pressure washout procedure that generates large volumes of ammonium perchlorate wastewater, which if not properly managed can lead to contamination of surface or groundwaters.

During the next 8–10 years, an estimated 75 million kg of perchlorate may be ready for disposal.[5] Although the presence of perchlorate was first detected in groundwaters in California and Nevada, 44 states have former perchlorate manufacturers or users.[3] Perchlorate compounds are highly energetic, but when dissolved in water, they become quite stable. The perchlorate salt is very soluble in water (i.e., ammonium perchlorate solubility is 20 g $dL^{-1}$ at 25°C) and dissociates completely to ammonia and perchlorate ions. The perchlorate ion is mobile in aqueous systems and can persist for decades under typical groundwater or surface water conditions, due to lack of reactivity with other compounds.[1] Water samples taken near the site of the former Pacific Engineering & Production Company of Nevada in Henderson, which exploded in 1988, contained up to 630 mg $L^{-1}$ of perchlorate.[6]

Large concentrations of perchlorate are believed to be detrimental to human health. At relatively high doses, perchlorate has been found to interfere with the thyroid's ability to produce hormones and regulate metabolism.[4,7] Based on studies completed by the EPA,[6] the California Department of Health Services established a perchlorate provisional action level of 18 μg $L^{-1}$ for drinking water. Below this concentration, perchlorate is not considered to pose a human health risk based on current information.[3]

In 1997, a panel of experts concluded that there was no proven process for the treatment of drinking water containing low concentrations of perchlorate.[8] The lack of volatility and high solubility of perchlorate ions mean that air stripping and micro- or ultrafiltration processes are ineffective means for perchlorate reduction. Nanofiltration and reverse osmosis will remove perchlorate from an aqueous stream, but the costs to implement these technologies at large scale (e.g., treatment of 3.875 $m^3$ $min^{-1}$) are high. Adsorption of perchlorate with granular activated carbon (GAC) is marginal and not attractive. Selective ion exchange is possible for treatment of groundwaters containing low concentrations of perchlorate (e.g., < 200 μg $L^{-1}$), however managing the perchlorate-containing waste brine is an issue. Chemical treatment (i.e., the addition of reducing agents) is not attractive due to the low rate of reduction of perchlorate when present in water. Most researchers agree biochemical treatment is the technology that holds most promise for economical treatment of perchlorate contaminated waters and wastewaters.[3,4,6,9]

In 1998, a chemical manufacturing firm constructed a fluidized bed reactor (FBR) biotreatment system (Figure 1) to remediate groundwater contaminated with perchlorate at a site in California. The system consisted of four FBRs, each 4.27 m in diameter, designed to treat a groundwater flow of 15 $m^3$ $min^{-1}$ containing approximately 8 mg $L^{-1}$ perchlorate. Envirogen supplied the FBRs. Upstream of the FBR treatment, an air stripping step is used to remove volatile organics from the groundwater, which results in a significant dissolved oxygen (DO) concentration (i.e., 4–8 mg $L^{-1}$) in the FBR feed water. The FBRs operating under anoxic conditions (i.e., no air or oxygen addition) are expected to reduce the perchlorate concentration to less than 100 µg $L^{-1}$ and ideally to less than 18 µg $L^{-1}$. The units were designed at a perchlorate volumetric reduction rate of approximately 0.7 kg $m^{-3}$ $day^{-1}$ based on the volume occupied by the fluidized bed (i.e., empty bed volume).

Pilot plant studies conducted at the California site were used as a basis for sizing the full-scale FBRs. The field pilot FBR employed GAC as the bed media and ethanol as the electron donor to promote perchlorate reduction. The pilot plant study provided sufficient information for design of a full-scale FBR system, however a potential operating issue of

**Figure 1.** Full scale FBR units during construction.

uncontrolled GAC loss from the pilot reactor was identified. A low-shear, biofilm-GAC separation device, which is designed to limit the expansion of the bed due to biofilm growth while minimizing GAC attrition, was located near the top of the pilot FBR reactor. This device would not be effective at preventing media loss from the reactor if extensive biofilm growth did occur and/or the nature of biofilm attachment was such that it was very difficult to detach it from the GAC. Normally, silica sand is the preferred bed media for use in FBRs where extensive biofilm growth is anticipated. The use of sand in combination with a high-shear, biofilm-sand separation device allows for more aggressive bed height control and the minimization of media loss from the reactor regardless of the extent of biofilm growth and/or its nature of attachment. FBR experiences with denitrification using ethanol as the electron donor suggest a thick, tenacious biofilm can be expected to form under anoxic conditions. The use of methanol or acetate showed less film growth and different film characteristics. Numerous FBRs have been operated successfully for biological nitrate reduction with methanol without experiencing bed media loss.[10]

The objective of this study was to measure the FBR performance with sand as the fluid bed media versus GAC, and methanol as the electron donor versus ethanol. The work was done in spring 1998 utilizing laboratory FBR pilot plant equipment and facilities provided by Envirogen. The intent of the study was to derive sufficient information to allow a decision to be made as to whether GAC or sand should be used in the full-scale reactors. The purpose of this paper is to report the results from the laboratory pilot study completed in August 1998, and to discuss the results to-date from operation of the full-scale FBR system at the California site. Start-up of two of the four full-scale FBRs began in late 1998.

## LABORATORY PILOT PLANT PROGRAM

### Equipment and Operation

The laboratory pilot plant study involved operation of three FBR pilot units in parallel. One unit contained GAC media with a particle size range of approximately 0.9 to 1.4 mm and used ethanol as the electron donor. The other two units contained sand media with a particle size range of approximately 0.3 to 0.6 mm. One of the sand units used ethanol as the electron donor, and the other used methanol. Each reactor consisted of a 5-cm diameter glass column, approximately 90-cm in length with a spherical glass section mounted on top of the column (Figure 2). The laboratory units were equipped with provisions for measurement of dissolved oxygen above the fluid bed, automatic pH measurement and control, a variable speed effluent recycle fluidization pump, variable speed feed and effluent pumps, feed and effluent collection tanks, and provisions for the addition of the electron donor (i.e., ethanol or methanol) and biomass growth nutrients (i.e., ammonium monohydrogen phosphate). During start-up, hydrochloric acid was manually added to the FBRs on occasion to control the pH near neutral. Subsequently, caustic (i.e., either 0.5 or 1 M NaOH) was used to maintain the pH in the reactors in the range of 7.0 to 7.5. The laboratory units operated at room temperature (i.e., 20–25 °C).

Two sources, synthetic and actual site water, were used to supply the FBR units. The synthetic water was prepared by dissolving ammonium perchlorate in tap water. The site water was shipped to Envirogen in 55-gallon plastic drums. The FBR feed water was dosed with the appropriate amount of biological growth nutrients and combined with the

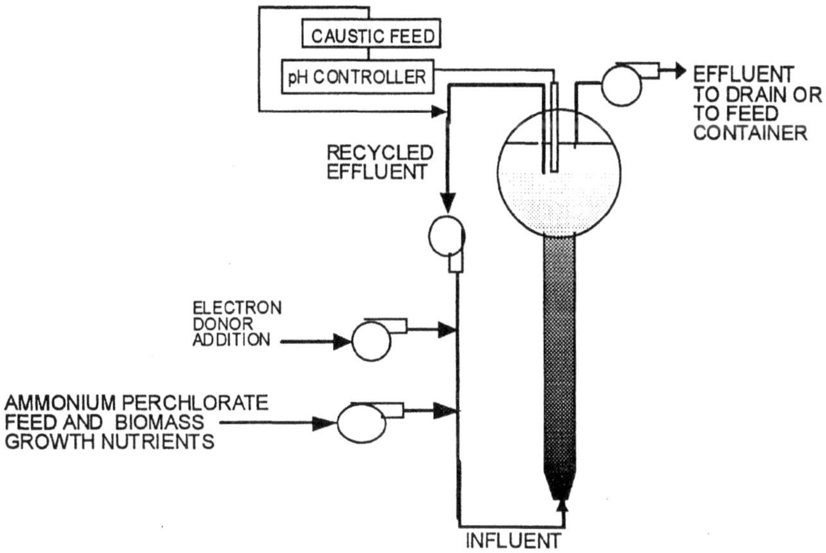


**Figure 2.** Schematic of laboratory pilot FBRs.

recycled effluent to provide sufficient flow to fluidize each FBR column. The effluent was withdrawn from the spherical glass section at the top of the FBR column at a rate equal to the feed rate. Each pilot FBR was fed from a separate 114-l, unmixed container. The effluent was either discharged to drain, to individual effluent collection tanks, or back to the feed container.

### Sampling and Analytical Procedures

The performance of the pilot FBRs was based on the analysis of grab samples collected from various locations. Feed and effluent samples were taken normally 3 times per week for analysis of total organic carbon (TOC) and perchlorate following filtration through 0.45 micron paper. A Dohrmann DC-190 carbon analyzer was used to determine TOC following method 415.1.[11] A Dionex ion chromatograph was used to determine perchlorate following method 300.0[11] with certain modifications. A column specific for separation of the perchlorate ion (Dionex AS–11) was utilized with NaOH and distilled water as the eluent and regenerant solutions, respectively. The detection limits for TOC was 2 mg $L^{-1}$ and for perchlorate was 2 $\mu g\ L^{-1}$. Effluent samples were collected normally one time per week for analysis of ammonia and orthophosphate. Methods 350 and 300[11] were followed to determine respectively ammonia and orthophosphate.

**Experimental Plan**

The pilot study was conducted over a four month period. Before adding the sand media to Reactors B and C, the sand was conditioned for approximately six days by allowing it to sit idle in a pail. The conditioning medium contained ammonium perchlorate, the electron donor specified for operation of Reactor B or C, biomass growth nutrients, and biological solids derived from an anaerobic digester treating municipal wastewater sludge. The carbon media was not pre-conditioned before addition to the FBR column. Following the addition of media to the FBR units and mechanical startup, the test program progressed through five experimental phases (Table 1). Inoculation was done with biological solids derived from an anaerobic digester treating municipal wastewater

**Table 1**. Laboratory Pilot Experimental Program

| Experimental Phase (Days Since Process Startup) | Objective of Experimental Phase and FBR Operating Characteristics |
|---|---|
| Phase 1 (Day 1 Through 34) | • Process start-up and biofilm build-up, utilizing synthetic perchlorate feed<br>• Reactor A: GAC media, ethanol electron donor<br>• Reactor B: silica sand media, ethanol electron donor<br>• Reactor C: silica sand media, methanol electron donor |
| Phase 2 (Day 34 Through 58) | • Quantify effect of fluid bed media (GAC versus sand) and electron donor (ethanol versus methanol) on perchlorate reduction utilizing synthetic perchlorate feed<br>• Reactor A and B characteristics per Phase 1<br>• Reactor C operation discontinued on day 50 |
| Phase 3 (Day 58 Through 66) | • Objective per Phase 2 utilizing perchlorate contaminated pretreated groundwater from the California site<br>• Reactor A and B characteristics per Phase 1 and 2 |
| Phase 4 (Day 66 Through 111) | • Quantify effect of electron donor on perchlorate reduction utilizing synthetic perchlorate feed<br>• Reactor A: GAC media; ethanol, ethanol/methanol, methanol then ethanol/methanol as electron donors<br>• Reactor B: silica sand media; ethanol, ethanol/methanol then methanol as electron donors; operation discontinued on day 85 |
| Phase 5 (Day 111 Through 125) | • Quantify effect of ethanol dose on perchlorate reduction at reduced ethanol to perchlorate mass ratio utilizing synthetic perchlorate feed<br>• Reactor A: GAC media |

sludge. Biofilm growth was initially promoted on the media in all FBRs by operating the reactors without feed addition (i.e., recycle flow only) for approximately 24 hours in the presence of perchlorate and biomass growth nutrients.

Phase 1 involved process startup and biofilm build-up on the GAC or sand media to achieve a perchlorate volumetric reduction rate equivalent to that anticipated according to the full-scale design (i.e., 0.7 kg $m^{-3}$ $day^{-1}$). To maximize the rate of biofilm build-up on the media, the effluent from each FBR was routed back to the individual feed containers. The electron donor and biomass growth nutrients were added directly to the feed containers (i.e., system recycle operation) and replenished as they were depleted. In each new batch of feed, ethanol or methanol was added to achieve a mass ratio of TOC (total organic carbon) to perchlorate of 1.6. TOC was used as an indirect measure of the ethanol or methanol concentration. In experimental Phases 2 through 5, the effluent from each FBR was discharged to drain or to the effluent collection tank (i.e., flow through operation), the electron donor was added directly to the FBR, and biomass nutrients were added to each feed container. In these experimental phases, the ethanol or methanol addition rate was altered as required in an attempt to maintain a residual TOC concentration in the FBR effluent.

The objective of Phases 2 and 3 was to determine the effect of the FBR media (i.e., GAC versus sand) and the electron donor (i.e., ethanol versus methanol) on perchlorate reduction with synthetic and actual site water feeds (Table 1). During these phases, the intent was to operate each FBR at an identical perchlorate feed concentration and volumetric loading conditions. Paired data (i.e., feed and effluent grab samples taken on the same day at the same time) were used to compare performance. At the beginning of Phase 2, the feed perchlorate concentration to each FBR was approximately 25 mg/l. Towards the end of Phase 2, the perchlorate concentration was gradually reduced to the level anticipated in the groundwater from the California site. In Phase 3, pretreated groundwater (i.e., after air stripping) from the California site served as the feed. The feed containers were aerated during Phases 2 and 3 in an attempt to simulate the effect of air stripping pretreatment on the feed dissolved oxygen (DO) concentration. This practice resulted in a feed DO concentration normally in the range from 7.5 to 8 mg/l during these phases versus approximately 1 mg $L^{-1}$ at other times.

In Phases 4 and 5, additional experiments were completed to determine the effect of electron donor mixtures (i.e., mix of ethanol and methanol) and the mass ratio of ethanol to perchlorate on perchlorate reduction performance (Table 1).

## RESULTS AND DISCUSSION

Process start-up (i.e., Phase 1) began on April 15, 1998 with each pilot system operating in a recycle mode as previously described. Prior to start-up, GAC or sand was added to each reactor to a settled depth of 61 cm. Upon start-up, the media was fluidized to a column height of approximately 76-cm. The influent pumping rate required to achieve this level of fluidization in each reactor (i.e., approximately 700 mL $min^{-1}$) was maintained constant throughout the study. The feed container for each reactor was initially charged with approximately 13 mg $L^{-1}$ perchlorate, appropriate concentrations of the electron donor specified for the reactor in question (Table 1), and biomass growth nutrients (i.e., ammonium monohydrogen phosphate). The feed rate to each reactor was in the range of 70 to 80 mL $min^{-1}$. As Phase 1 progressed, the quantity and frequency of new material added to the

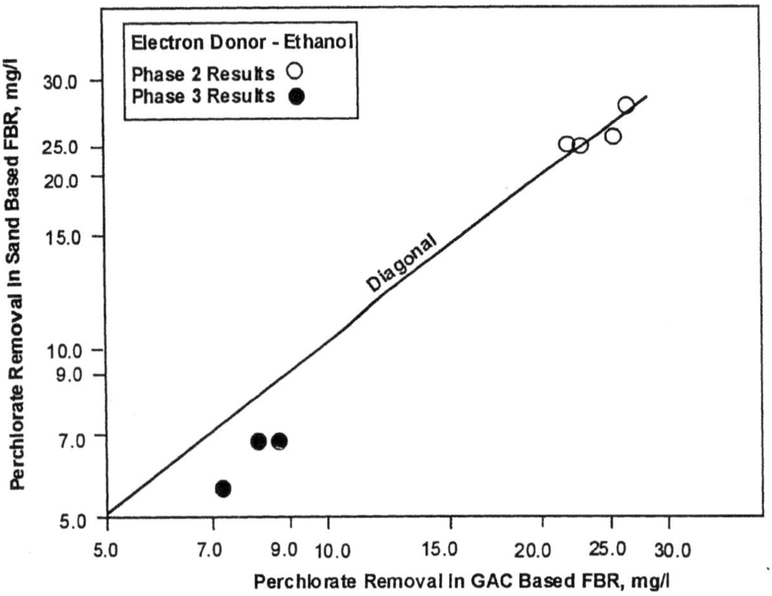


**Figure 3.** Paired data comparison of perchlorate reduction in GAC- and sand-based pilot FBRs.

feed containers was increased as the rate of reduction of perchlorate across each FBR system increased. Phase 1 ended when calculations indicated the perchlorate volumetric reduction rate in all three reactors exceeded 0.7 kg $m^{-3}$ $day^{-1}$ (i.e., after 34 days of operation). Monitoring of the reduction rate in each reactor during Phase 1, showed biomass build-up in Reactor A exceeded that in Reactor B which in turn exceeded that in Reactor C.

During Phases 2 and 3, each FBR was operated at near identical perchlorate feed and volumetric loading conditions, which resulted in paired data, as previously discussed. Figure 3 shows the results for perchlorate removal performance with both FBR media. Equivalent performance between the reactors is identified with the data points scattered around the diagonal line. When the synthetic water was fed (Phase 2 results), a similar removal performance was found. When the pretreated groundwater from the California site was the feed source (Phase 3 results), a greater perchlorate removal was achieved in the FBR containing GAC media (Reactor A) than in the FBR containing sand media (Reactor B). The limited number of data points in Figure 3 reflects the use of paired data for samples when the effluent from both FBRs contained residual ethanol, measured as TOC. The lack of residual electron donor in the effluent might have limited perchlorate reduction and therefore, those paired data were not included. Based on analysis of all data when residual TOC was present in the effluent, the FBRs containing GAC (Reactor A) and

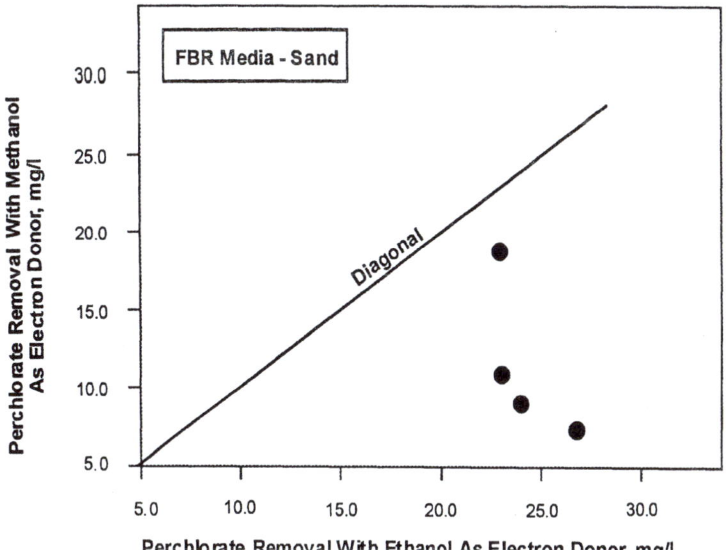


**Figure 4.** Paired data comparison of perchlorate reduction in sand-based pilot FBRs with methanol versus ethanol as electron donor.

sand (Reactor B) both achieved good performance at a perchlorate volumetric loading rate at or exceeding the full-scale design level of 0.7 kg $m^{-3}$ $day^{-1}$.

Figure 4 shows a paired data comparison based on the results from Phase 2 for different electron donors. The results suggest methanol is not nearly as effective as an electron donor as ethanol in promoting perchlorate reduction in sand based FBRs. Only paired data results and results when the electron donor was present in the effluent were considered in this comparison. Operation of the sand based FBR with methanol as the electron donor was discontinued prior to completion of Phase 2 (Table1) due to its relative poorer performance.

In Phase 4, the focus was to quantify the effect on perchlorate reduction of operating the FBRs at various ratios of methanol to ethanol as the electron donor while still attempting to maintain a residual TOC in the effluent. Figure 5 is a chronological plot of the performance of the sand based FBR (Reactor B) as the electron donor mixture was altered. This graph substantiates the advantage of ethanol over methanol in achieving near complete perchlorate reduction. In examining Figure 5, the following should be recognized:

- only results when a residual TOC was measured in the effluent are plotted, and
- Phase 4 began on day 66 at a perchlorate feed concentration to Reactor B of approximately 23 mg $L^{-1}$.

The results in Figure 5 show that when the electron donor consists of equal quantities of ethanol and methanol as TOC and the dosage rate (as TOC) is the same, perchlorate reduction was equivalent to that achieved when ethanol alone was used. Extended operation at this condition is necessary before any full-scale recommendations are justified.

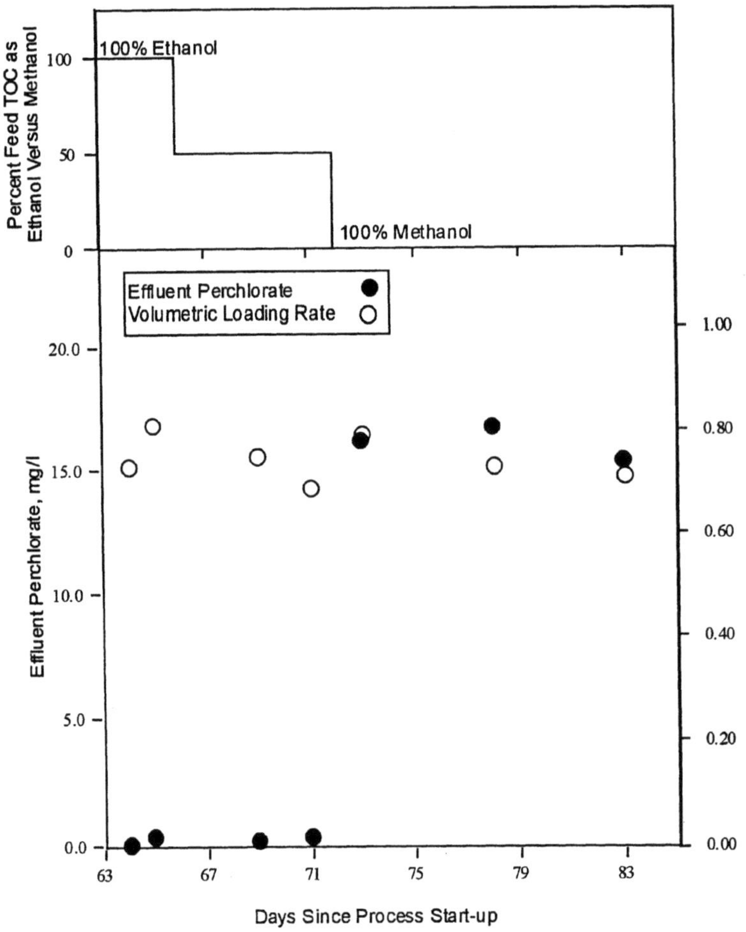


**Figure 5.** Effect of electron donor characteristics (ethanol versus methanol) on effluent perchlorate in sand-based FBR.

Figure 6 is a chronological plot of the performance of the GAC based FBR (Reactor A) as the electron donor mixture was modified. These results confirm the importance of using ethanol to ensure essentially complete perchlorate reduction. The data points plotted in Figure 6 are the results when a residual TOC was measured in the effluent. The perchlorate feed concentration to Reactor A during Phases 4 and 5 ranged from 22–28 mg $L^{-1}$. The Figure 6 results show that when the electron donor consists of equal quantities of ethanol and methanol as TOC, perchlorate reduction is equivalent to that achieved when ethanol alone is used.

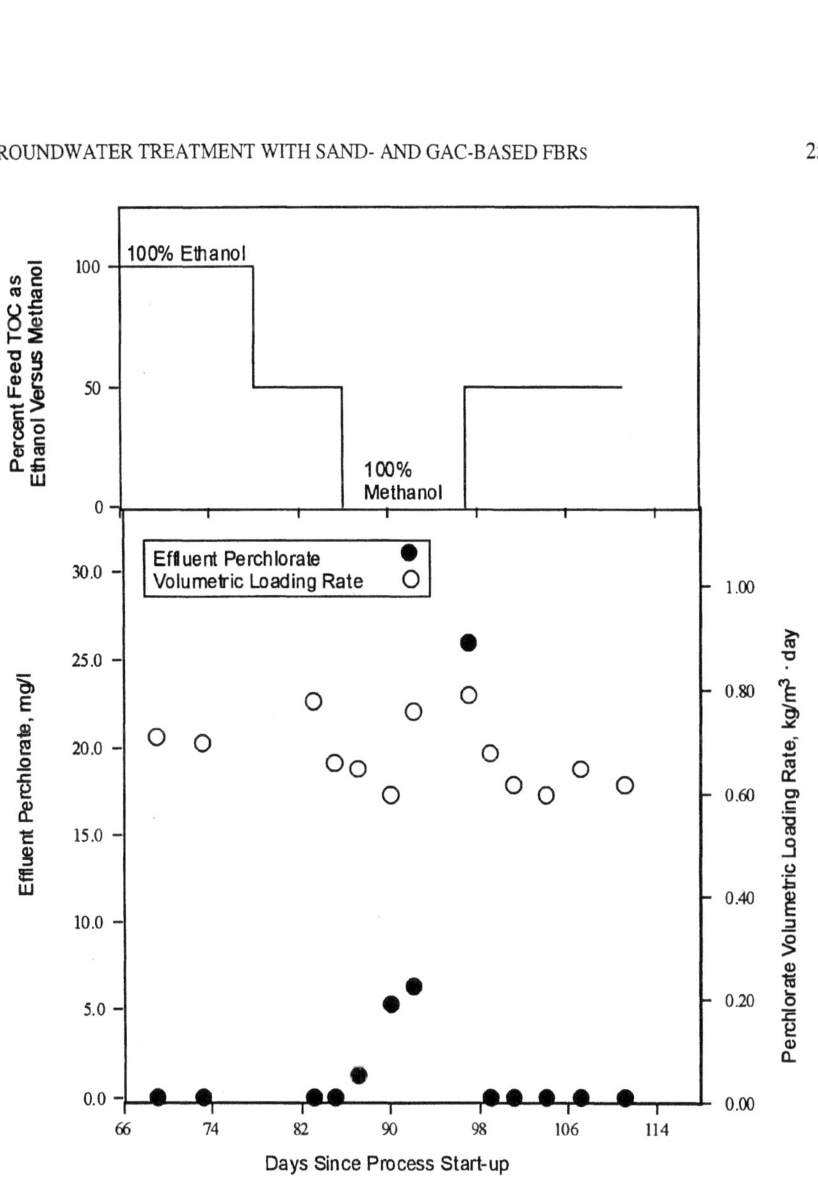


**Figure 6.** Effect of electron donor characteristics (ethanol versus methanol) on effluent perchlorate in GAC-based FBR.

The Phase 4 results indicate that the magnitude of perchlorate removal was unaffected by reducing the feed ethanol to perchlorate mass ratio by 50 %, independent of the addition of methanol. The Phase 5 experimental work was done to determine the performance when just ethanol at a lower mass ratio was used as the electron donor.

Figure 7 is a chronological plot depicting the performance of the GAC based FBR (Reactor A) towards the end of Phase 4 (i.e., day 98 until day 111) and through Phase 5. The figure indicates a deterioration in performance when ethanol alone was used as the electron donor and was added at a rate such that the feed TOC to perchlorate ratio was

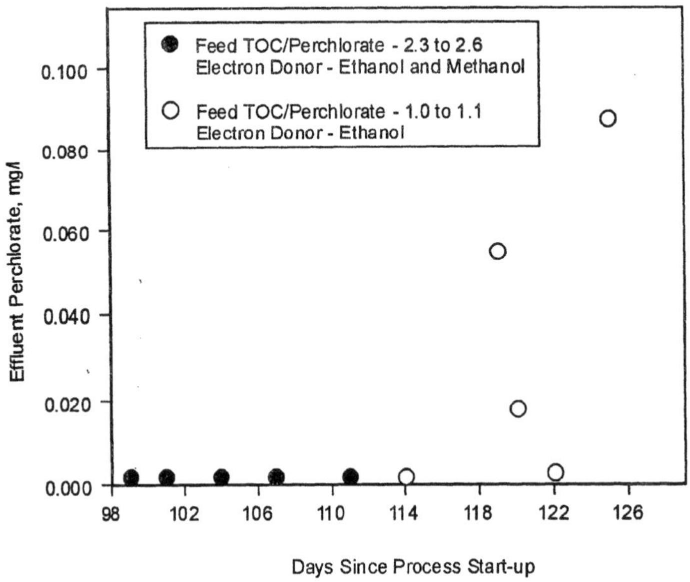


**Figure 7.** Effect of feed TOC:perchlorate concentration ratio versus electron donor characteristics on effluent perchlorate in GAC-based FBR.

reduced by approximately 50 percent. The perchlorate volumetric loading rate throughout Phase 5 was approximately 0.7 kg $m^{-3}$ $day^{-1}$. Phase 5 began on day 111 after sampling and lasted for only 2 weeks. It is likely the adsorptive characteristics of GAC played a role in the results observed in Phase 5. Organic carbon desorption from the GAC may have occurred providing an additional electron donor source. A significant concentration of TOC (i.e., 7 to 45 mg $L^{-1}$) was measured in the FBR effluent during Phase 5 despite the reduction in the TOC fed to the reactor.

## FULL-SCALE FBR SYSTEM OPERATION

Figure 8 is a photograph of the full-scale FBR installation. The four 4.27-m diameter FBR units can be seen on the left-hand side of the photo. Two of the FBRs began process operation in early December 1998 following addition of GAC and subsequent microbial inoculation. The initial pretreated groundwater feed rate was 1635 $m^3$ $day^{-1}$, which was increased to 2453 $m^3$ $day^{-1}$ in mid-December.

Figure 9 shows the FBR effluent concentrations for perchlorate, nitrate and sulfate. Good perchlorate removal was achieved when the dissolved oxygen and nitrate were both depleted from the water. Ethanol was added at approximately 95 dL $m^{-3}$ of feed treated. At this ethanol dosage, the FBR effluent perchlorate concentration was below the analytical detection limit (i.e., 4 μg $L^{-1}$). During operation of the FBRs, a close relationship between the ethanol feed rate and the perchlorate loading rate was identified.

**Figure 8.** Full scale FBR units.

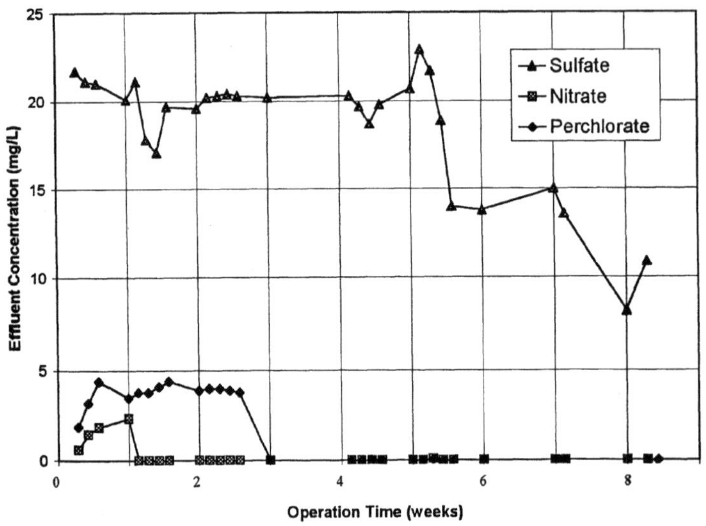


**Figure 9.** Full scale FBR performance with analyte concentrations in effluent.

Figure 10 shows the perchlorate effluent concentration as a function of ethanol dose. Some perchlorate is removed at low ethanol doses, but a dose above the threshold value is required to remove all perchlorate. Ethanol requirements to achieve complete perchlorate removal exceeded expectations based on the laboratory pilot plant results. This is believed to be due to significant quantities of DO, nitrate and sulfate present in the feed.

## CONCLUSIONS

The following conclusions can be drawn from the laboratory pilot study and the full-scale operation regarding GAC and sand based FBRs treating perchlorate contaminated groundwater.

- Paired data analysis shows a GAC based FBR could be expected to achieve a higher level of perchlorate reduction than a sand based FBR when ethanol alone is added as the electron donor.

- GAC and sand based FBRs with only methanol added as the electron donor were not as effective at achieving a high level of perchlorate reduction.

- GAC and sand based FBRs both achieved a high level of a perchlorate reduction when ethanol/methanol mixtures were added as electron donors.

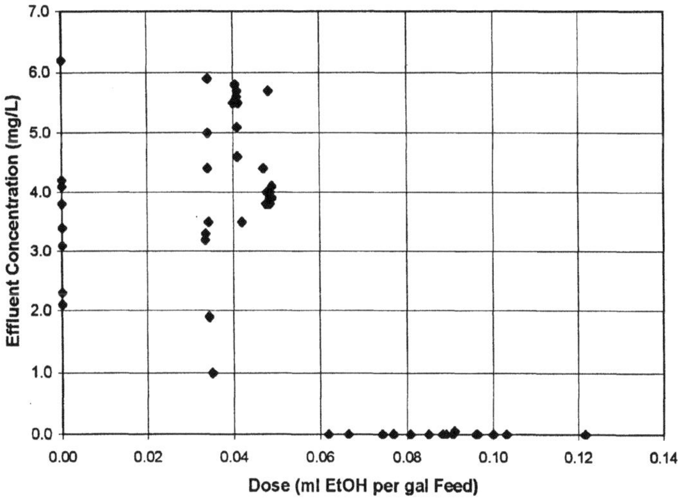


**Figure 10.** Full scale FBR performance, effluent perchlorate concentration versus ethanol dosage.

- The performance achieved in GAC or sand based reactors appears to be unaffected by changing the electron donor from ethanol alone to a mixture consisting of ethanol and methanol in equal quantities on a TOC basis.

- Results from operation of the full-scale, GAC based FBR system have demonstrated complete perchlorate removal can be achieved provided sufficient ethanol is added to the feed.

## ACKNOWLEDGMENTS

Joe Manning and Jim Revak of Envirogen provided support in operation of the pilot FBRs.

## RERERENCES

1. Rogers, D. "Perchlorate Contamination in the Environment." *Proceedings of the Southwest Focused Ground Water Conference: Discussing the Issue of MTBE and Perchlorate in Ground Water*, National Ground Water Association: Anaheim, CA, 1998; pp. 63–84.

2. Van Ginkel, C.G.; Kroon, A.G.M.; Rikken, G.B.; Kengen, S.W.M. "Microbial conversion of perchlorate, chlorate and chlorite." *Proceedings of the Southwest Focused Ground Water Conference: Discussing the Issue of MTBE and Perchlorate in Ground Water*, National Ground Water Association: Anaheim, CA, 1998; pp. 92–95.

3. Damian, P.; Pontius, F.W. "From rockets to remediation: the perchlorate problem." *Environmental Protection* **1999**, *20*, 24–31.

4. Renner, R. "Perchlorate-tainted wells spur government action." *Environmental Science and Technology* **1999**, *33*, 110A–111A.

5. Susarla, S.; Bacchus, S.T.; Wolfe, N.L.; McCutcheon, S.C. "Phytotransformation of perchlorate using parrot-feather." *Soil and Groundwater Cleanup* **1999**, *February-March*, 20–23.

6. Urbansky, E.T. "Perchlorate chemistry: implications for analysis and remediation." *Bioremediation Journal* **1998**, *2*, 81–95.

7. Stanbury, J.B.; Wyngaarden, J.B. "Effect of perchlorate on the human thyroid gland." *Metabolism* **1952**, *1*, 533–539.

8. American Water Works Association Research Foundation. *Final Report of the Perchlorate Research Group Workshop*. September 30 to October 2, 1997.

9. Logan, B.E. "A review of chlorate- and perchlorate-respiring microorganisms." *Bioremediation Journal* **1998**, *2*, 69–79.

10. Environmental Protection Agency. *Nitrogen Control Manual*. EPA/625/R-93/100, Washington, D.C., September 1993, pp. 232–248.

11. Environmental Protection Agency. *Methods for Chemical Analysis of Water and Wastes*. EPA-600/4-91-020, Cincinnati, OH, June 1991.

# Chapter 24

# THE DIVERSE MICROBIOLOGY OF (PER)CHLORATE REDUCTION[⊗]

John D. Coates,[*] Urania Michaelidou, Susan M. O'Connor, Royce A. Bruce, and Laurie A. Achenbach

Department of Microbiology and Center for Systematic Biology, Southern Illinois University, Carbondale, Illinois 62901

## INTRODUCTION

In 1997, following the development of a highly sensitive analytical method for perchlorate, monitoring studies identified perchlorate as a contaminant of major drinking water resources in the southwestern U.S.[1,2,3] The only known natural source of perchlorate is associated with mineral deposits found in Chile where the perchlorate may represent as much as 6–7% of the total mass.[4] Although these Chilean deposits have been extensively mined as a mineral and nitrate source for fertilizer manufacture, this is not thought to represent a significant source of perchlorate in the environment. The primary industrial use of perchlorate is as an oxidant in munitions manufacture and the unregulated release of perchlorate-containing wastestreams from munition manufacturing and handling facilities has been identified as the predominant source of this contaminant in the U.S.[1,3]

Perchlorate has been shown to affect iodide accumulation in the thyroid gland.[5] In 1992, the Environmental Protection Agency reviewed the health effects of perchlorate administered to patients with hyperthyroidism and identified that doses of 6 mg $kg^{-1}$ $day^{-1}$ or more over a two month period resulted in fatal bone marrow changes.[2] Using these data, the California Department of Health Services calculated an action level in 1998 of 18 $\mu g\ L^{-1}$ for drinking water supplies which—if exceeded—would require stopping water usage and initiation of remediation efforts.[2,6] This action level has now been adopted by several states throughout the U.S. In a recent release of a draft toxicology assessment, EPA suggested that the action level might be raised to 32 $\mu g\ L^{-1}$.[7] The presence of perchlorate has been confirmed in surface and ground waters in Texas, Arkansas, Maryland, New York,

[⊗]This work was presented at the 218th national meeting of the American Chemical Society as part of the Environmental Division symposium Perchlorate in the Environment, held August 22–24, 1999, in New Orleans, Lousiana

[*]Author to whom correspondance should be directed. Phone: 618-453-6132. Fax: 618-453-8036. Electronic mail: jcoates@micro.siu.edu.

*Perchlorate in the Environment*, edited by Urbansky.
Kluwer Academic/Plenum Publishers, New York, 2000.

California, Utah, and Nevada[1] and is particularly prevalent in the southwestern states where it has been identified in the Las Vegas Wash, Lake Mead, and the Colorado River.

Although it has been recognized for over half a century that microbial reduction of chlorine oxyanions under anaerobic conditions is possible, [8–20] there is still relatively little known about the microorganisms involved in this respiratory metabolism. Generally, these organisms are assumed to use either chlorate or perchlorate as terminal electron acceptors[21] although this has only been demonstrated in a few isolated cases.[15,17,20] The recent discovery of an organism which is capable of chlorate reduction but not perchlorate reduction is demonstrative that this assumption is incorrect.[46] The endproduct of (per)chlorate reduction is inocuous chloride.[16,17,20] Early studies indicated that microorganisms rapidly reduced chlorate as a competitive reaction for the nitrate reductase pathway[10–12] although similar studies with perchlorate have not been performed. Chlorite was produced as the end product and growth was not associated with this reaction.[22,23] Until recently only six microorganisms were described which can grow by dissimilatory (per)chlorate reduction.[14–17,20,24] Of these six isolates, only four, strain CKB,[20] strain GR-1,[16] *Ideonella dechloratans,*[14] and *Wolinella succinogenes* strain HAP-1,[17] have been studied in any detail; however, only three of these have been characterized both phenotypically and genotypically.[14,15,20] Thus, the true ubiquity and diversity of (per)chlorate-reducing bacteria is still unknown.

In order to determine the ubiquity and diversity of organisms capable of dissimilatory (per)chlorate reduction, we enumerated (per)chlorate-reducing bacteria (PCRB) populations in a broad spectrum of environments and isolated more than twenty new (per)chlorate-reducing bacteria from many of diverse environments. Several of the isolates obtained represent new genera in the *Proteobacteria* and demonstrate that dissimilatory reduction of (per)chlorate is a much more ubiquitous and diverse metabolism than was previously considered.

## EXPERIMENTAL

### Sources of Soils and Sediments

Soil samples were collected from the top 6 cm of an uncontaminated soil in Thompson Woods on the Carbondale campus of Southern Illinois University and also from a hydrocarbon-contaminated soil at Tulsa Tape Incorporated, Carbondale, IL. In addition, sediment samples were collected from campus lake and farm swine lagoons, Southern Illinois University, Carbondale, IL; Potomac River, Pohic Bay, VA; Mississippi River, Chester, IL; South Dakota gold mine drainage sediment, Hotsprings, SD; and swamp lands, Reston, FL. All samples were freshly collected and transported directly back to the lab where they were immediately assayed for (per)chlorate-reducing bacteria.

### Medium and Culturing Conditions

Standard anaerobic culturing techniques were used throughout.[25–27] The medium was boiled under $N_2/CO_2$ (80/20) to remove dissolved $O_2$ and then dispensed into anaerobic pressure tubes or serum bottles under $N_2/CO_2$, capped with thick butyl rubber stoppers, and sterilized by autoclaving. The basal medium was the bicarbonate-buffered freshwater medium that had previously been used for culturing strain CKB.[20] Unless otherwise noted, sodium salts of acetate and chlorate (10 mM each) were used as the electron donor and acceptor, respectively, which were added from sterile anoxic stocks.

Alternative electron donors were added from sterile anoxic aqueous stocks. Pure aromatic hydrocarbons (benzene, hexadecane, and toluene) were added directly (1 μl to 10 mL of medium). Electron acceptors were also added from anoxic aqueous stocks. Soluble Fe(III) was supplied as Fe(III) chelated with nitrilotriacetic acid (Fe(III)-NTA) (10 mM).[28]

Mn(IV) was supplied as synthetic $MnO_2$ that was prepared as previously described [29] to give a final concentration of 10–30 mM . Sulfur was supplied as a polysulfide solution prepared as previously outlined.[30] All other electron acceptors were prepared as anoxic aqueous stocks of the sodium salts to give final concentrations of 10 mM.

### Isolation of (Per)chlorate-Reducing Bacteria (PCRB)

(Per)chlorate-reducing enrichments were established by transferring 1 g subsamples from each of the freshly collected soil and sediment samples into 9 mL of prepared anoxic medium under a gas stream of $N_2/CO_2$. Acetate (10 mM) was the electron donor and chlorate (10 mM) was the electron acceptor. Incubations were done at 30 °C in the dark. Positive enrichments were identified by visual increase in optical density and by microscopic examination. Once a positive enrichment was established the (per)chlorate-reducing culture was transferred (10% inoculum) into 9 mL of fresh anoxic medium. Isolated colonies were obtained from transfers of positive enrichments by the standard agar shake-tube technique outlined previously[20,31] with acetate as the sole electron donor and $ClO_3^-$ (10 mM) as the sole electron acceptor.

### Most Probable Number Counts

Numbers of dissimilatory (per)chlorate-reducing bacteria were determined by three-tube most probable number counts (MPN) with 10 mM acetate as the electron donor. The medium contained (in g $L^{-1}$): $NH_4Cl$ (0.25); $NaClO_3$ (1.03); $CH_3COONa$ (1.36); $NaH_2PO_4$ (0.60); KCl (0.1); $NaHCO_3$ (2.5). Vitamins and trace metals were added (10 ml/L) from stock solutions prepared as previously described.[20] MPN series were incubated at room temperature in the dark for 60 days prior to analysis. Positives in the MPN series were identified visually by increase in optical density and also by microscopic examination.

### Chlorite Dismutase Activity Determination

Washed cell suspensions of each of the PCRB isolates were analyzed for chlorite dismutase activity using a Clark $O_2$ electrode as previously described.[20,32]

### 16S rRNA Gene Sequencing and Analysis

Cells from 2 mL cultures of PCRB were harvested by centrifugation, resuspended in 40 µl sterile water, and lysed by the addition of 5 µl chloroform with a 10 min incubation at 95 °C. Primers specific to bacterial 16S rDNA (8F: 5′–AGAGTTTGATCCTGGCTCAG–3′; 1525R: 5′–AAGGAGGTGATCCAGCC–3′) were used in a polymerase chain reaction (PCR) that consisted of 10 mM Tris-HCl (pH 9.0), 50 mM KCl, 0.1% (w/v) Triton X-100, 1.2 mM $MgCl_2$, 0.2 mM each dNTP, 75 ng of each primer, 0.5 µL Taq polymerase (Gibco/BRL), and 1 µL of lysed cells in a 50 µL reaction. Amplifications were performed at these parameters: 94 °C for 3 min, followed by 30 cycles of 94 °C for 1 min, 55 °C for 1 min, and 72 °C for 2 min with a final incubation of 10 min at 72 °C. The amplification products were gel-purified (GeneClean II, BIO101 Inc., Carlsbad, CA) and cycle sequenced (ThermoSequenase, Amersham Pharmacia Biotech, NJ) using internal primers. Sequence entry and manipulation was performed with the MacVector 6.1 sequence analysis software program for the Macintosh. Sequences of select 16S rRNAs were downloaded from the Ribosomal Database Project[37] and Genbank[38] into the computer program SeqApp.[39] PCRB 16S rDNA sequences were manually added to the alignment using secondary structure information for proper alignment. Distance, parsimony, and maximum likelihood analysis of the aligned sequences was performed on a Power Macintosh G3 using PAUP* 4.0d65.[40] Bootstrap analysis was conducted on 100 replications using a heuristic search strategy to assess the confidence level of various clades.

### Analytical Techniques

Acetate concentrations were analyzed by HPLC with UV detection (Shimadzu SPD-10A, Shimadzu Scientific Instruments, Columbia, MD) using a HL-75H$^+$ cation exchange column (Hamilton #79476, Hamilton Company, Reno, NA). The eluent was 0.016N $H_2SO_4$ at a flow rate of 0.4 mL per min. Perchlorate, chlorate and chloride concentrations were analyzed by HPLC with conductivity detection (Shimadzu CDD-6A, Shimadzu Scientific Instruments, Columbia, MD) using a PRP-X100 anion exchange column (Hamilton #79434, Hamilton Company, Reno, NA). The eluent was 4 mM *p*-hydroxybenzoic acid in 2.5% methanol with pH adjusted to 8.5, and a flow rate of 2.0 ml per min. Growth of cultures on soluble electron acceptors was measured by increase in optical density at 600 nm. Oxygen concentrations from chlorite dismutation were determined by an $O_2$ electrode (YSI, model 5300, Yellow Springs, OH).

## RESULTS AND DISCUSSION

### Most Probable Number Studies

Most probable number counts with chlorate as the electron acceptor indicated that acetate-oxidizing, (per)chlorate-reducing bacteria are present in many diverse environments. The (per)chlorate-reducing microbial community represented a significant population in all environments tested, even those which have had no prior contact with oxyanions of chlorine (Table 1). The PCRB numbers ranged from $(2.31 \pm 1.33) \times 10^3$ to as high as $(2.40 \pm 1.74) \times 10^6$. The highest MPN counts observed were in swine waste lagoons.

**Table 1.** Most probable number counts of acetate-oxidizing, (per)chlorate-reducing organisms in different environments

| Environment | Counts (cells per gram) |
|---|---|
| Swine waste lagoon | $(2.40 \pm 1.74) \times 10^6$ |
| Pristine aquatic sediment | $(4.62 \pm 1.75) \times 10^3$ |
| Pristine soil | $(2.31 \pm 1.33) \times 10^3$ |
| Petroleum contaminated soil | $(9.33 \pm 4.17) \times 10^3$ |
| Aquatic sediments | $(1.49 \pm 0.60) \times 10^4$ |
| Swamp sediments | $(2.31 \pm 1.33) \times 10^4$ |

The fact that that microbial (per)chlorate reduction is prevalent in all of the environments tested in this study supports and further expands the observations of a previous investigation[41] in which it was shown that chlorate reduction was prevalent in several diverse environments. This is, however, unexpected as the only known natural sources of these compounds are salt deposits in Chile[4] where perchlorate may represent as much as 6–7% of the total mass and many of the environments screened were pristine. Although early studies suggested that microbial (per)chlorate reduction may simply be a competitive reaction for the nitrate reductase system of denitrifying bacteria in the environment,[10-12] this does not explain the presence of chlorate reductase enzymes in organisms which can only use chlorate as a substrate such as the chlorate reductase C purified from *Proteus mirabilis*.[42]

### PCRB Isolates

To date, there are only six dissimilatory (per)chlorate-reducing organisms that have been identified and only four of these have been studied in detail. Thus, the true diversity of the microorganisms capable of (per)chlorate reduction is still unknown. After two weeks incubation of the primary enrichments from all of the environments sampled good growth was observed. Enrichments were transferred into fresh basal medium (10% inoculum) and good growth was observed in the transfer after 24 hours as determined by increase in optical density and microscopic examination. Highly enriched (per)chlorate-reducing cultures were obtained by sequential transfer over the following week prior to serial dilution into agar tubes. Small colonies of consistent morphology were apparent in the higher dilution agar tubes from each enrichment after one week of incubation. Colonies were generally pink, wet, domed, entire, smooth, and small, being 1–4 mm in diameter. Several of these colonies were selected from each of the enrichment series and (per)chlorate-reducing isolates were obtained from all environments sampled.

## Phylogeny of the PCRB Isolates

Analyses of the 16S rDNA sequences indicated that all isolates were members of the Proteobacteria (Figure 1). The PCRB isolates belonged to three subgroups (alpha, beta, and gamma) of the Proteobacteria, demonstrating that this metabolism is widespread throughout the phylum). Some of these isolates were closely related to previously described genera not recognized for the potential to grow by dissimilatory (per)chlorate reduction while others had no close relatives and represented novel genera in the Proteobacteria (Figure 1).

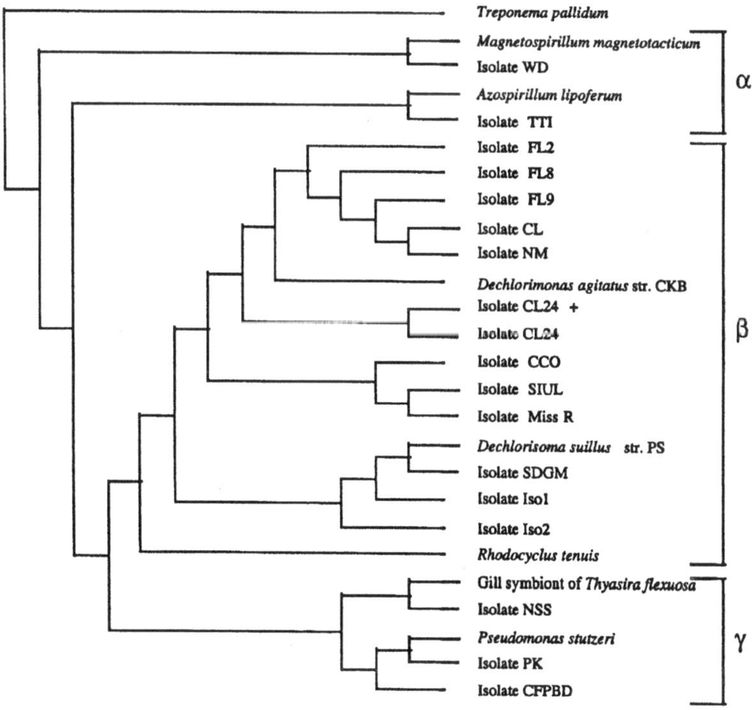


**Figure 1.** Phylogenetic tree of newly isolated (per)chlorate-reducing bacteria and their closest relatives. A heuristic search from distance data with Jukes-Cantor correction was performed in the computer program PAUP* 4.0.[40]

Of the known dissimilatory (per)chlorate-reducers only strain GR-1,[16] *Ideonella dechloratans,*[14] *Wolinella succinogenes* strain HAP-1,[17] and strain CKB[20] have been well characterized. Strains CKB [20], GR-1 [16], and *I. dechloratans* [14] are members of the beta subclass of the *Proteobacteria*. *I. dechloratans* is phylogenetically distinct from any of the PCRB isolates obtained during this study. Similar comparisons with strain GR-1 could not be made as the 16S rDNA sequence for this isolate is not available. *W. succinogenes* strain HAP-1 is a member of the epsilon subclass of the Proteobacteria and as such is very distantly related to any of the PCRB isolates obtained.[17] The broad phylogenetic diversity of organisms observed in this study which are capable of this metabolism has some interesting evolutionary implications due to the relatively short time in which (per)chlorate reduction could have evolved.

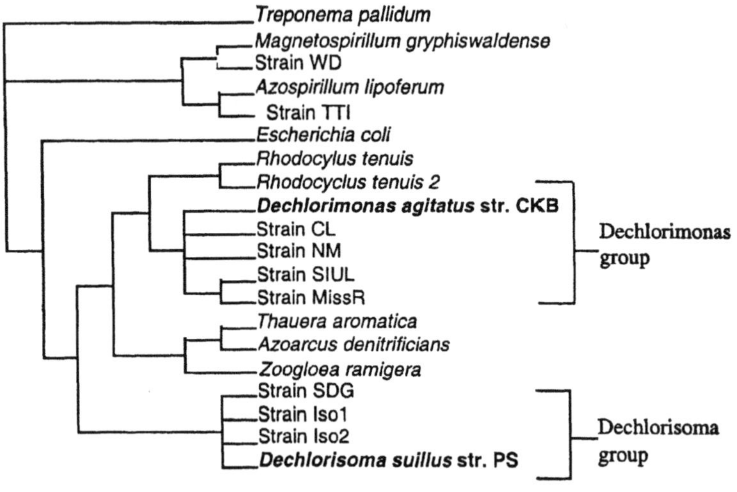


**Figure 2.** Phylogenetic tree inferred from 16S rRNA sequences showing the tight cluster of PCRB closely related to the phototrophic *Rhodocyclus* species in the beta subclass of the Proteobacteria.

The majority of the (per)chlorate-reducing isolates obtained in this study were closely related to one another and to the phototrophic *Rhodocyclus* species (Figure 2). This assemblage of (per)chlorate-reducing bacteria appears to have branched evolutionarily to form two related subgroups denoted *Dechlorimonas* and *Dechlorisoma* respectively (Figure 2). These groups were named after the typ strains, *Dechlorimonas agitatus* strain CKB and *Dechlorisoma suillus* strain PS which were isolated from a paper mill waste sludge and a swine waste lagoon respectively. Members of these groups have been isolated from all environments tested in this study suggesting that members of these two groups may be the predominant (per)chlorate-reducing bacteria in the environment. In addition, detailed inspection of the 16S rRNA sequences revealed the presence of nucleotides characteristic of either of these two groups [43]. Previous studies using specific molecular probes based on these characteristic nucleotide sequences, have identified the presence of members of both the *Dechlorimonas*group and *Dechlorisoma* group in other diverse environments[32] which further supports the hypothesis that these groups represent the dominant (per)chlorate-reducing bacteria in the environment.

**Phenotypic Characteristics**

Similarly to previously described PCRB,[14,16,17,20] all of the PCRB isolates were gram-negative, non-fermentative bacteria. Morphologically, most of the isolates were short motile rods 0.5 µm in diameter by 2 µm length. Spores were not visible in wet-mounts of any of the isolates by phase contrast microscopy and no growth was observed in fresh acetate-chlorate medium after pasteurization at 80 °C for 3 min. All of the isolates could grow aerobically on L-broth and colonies on L-broth agar plates were generally white, smooth, and approximately 0.5 mm in diameter.

All of the PCRB isolates were strict respirers and could not grow on anoxic basal media amended with glucose (10 mM), yeast extract (10 $g.L^{-1}$) and casamino acids (10 $g.L^{-1}$). All of the PCRB isolates could couple the complete oxidation of acetate to the reduction of chlorate in defined basal medium (Figure 3). The increase in cell numbers coincided with the oxidation of acetate and the production of chloride (Figure 3).

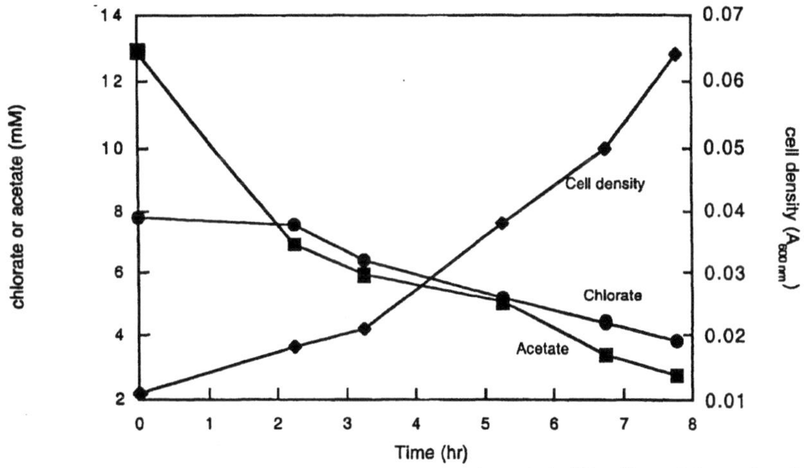


**Figure 3.** Growth curve of (per)chlorate-reducing isolate strain PS with acetate as the electron donor and chlorate (10 mM) as the sole electron acceptor.

As previously observed with other dissimilatory (per)chlorate-reducers,[14,16,17,20] chlorate and perchlorate were completely reduced to inocuous chloride for all isolates tested. In all tested cases the ratio of acetate oxidized to chlorate reduced gave a stoichiometry range of 1.10–1.30 which, when assimilation into biomass is considered, is in close agreement with the theoretical value according to eq 1:

$$3CH_3COO^- + 4ClO_3^- + 3H^+ \rightarrow 6CO_2 + 4Cl^- + 6H_2O \quad (1)$$

Chlorite, the potential intermediate of chlorate reduction, was not detected in the culture broth. In addition to acetate, the PCRB isolates tested used short chain volatile fatty acids and simple dicarboxylic acids as alternative electron donors. In contrast to previous described (per)chlorate-reducers,[14–17] the new PCRB isolates were relatively limited in the range of electron donors or acceptors used. None of the isolates utilized carbohydrates which are used by *I. dechloratans*.[14] In addition, none of the new PCRB isolates could oxidize $H_2$, an important endproduct of fermentation which does serve as an electron donor for (per)chlorate reduction by *W. succinogenes* strain HAP-1.[17] The only electron acceptors of those tested utilized by the PCRB isolates were $O_2$, perchlorate, chlorate, and in some cases nitrate. The ability of these isolates to grow aerobically or microaerophilically is similar to the previously described (per)chlorate-reducing bacteria and suggests that all (per)chlorate-reducing bacteria are facultative anaerobes. Although it was originally suggested that *W. succinogenes* strain HAP-1 was a strict anaerobe,[17] a recent study indicated that it is in fact a microaerophile.[44] The fact that not all of the PCRB isolates can use nitrate further supports the hypothesis that chlorate reduction and nitrate reduction are two unrelated pathways and is in contrast to the suggestions of earlier studies.[10–12]

A broad range of alternative electron acceptors were not used by the PCRB isolates including sulfate, fumarate, selenate, malate, manganese(IV), iron(III), and the humic substances analog 2,6-anthraquinone disulfonate.

The (per)chlorate-reducing isolates tested could grow over a broad range of environmental conditions. In the case of *Dechlorimonas agitatus* strain CKB, good growth was observed over a temperature range of 25 to 40 °C, a pH of 6.5 to 8.5, and a salinity range of 0 to 2% NaCl (Figure 4) . Optimum growth was observed at 35 °C, 1% NaCl, and pH 7.5 (Figure 4).

**Chlorite Dismutase**

Similarly to strain GR-1[47] and *D. agitatus* strain CKB[19,20,32,45] washed whole cell suspensions of all of the PCRB isolates could dismutate chlorite to chloride and molecular oxygen. No $O_2$ production was observed in the absence of cells or if the PCRB being tested was heat killed. $O_2$ evolution was extensive and formed copious quantities of froth at higher chlorite concentrations (10 mM). Previous studies in our laboratory have demonstrated that this unique metabolic activity of (per)chlorate-reducing bacteria can be used to stimulate microbial degradation of contaminating hydrocarbons in anoxic environments, including notoriously recalcitrant and toxic compounds such as benzene or naphthalene.[19,32]

A single enzyme with chlorite dismutase activity has previously been purified to homogeneity from two (per)chlorate-reducing bacteria, strain GR-1[47] and *D. agitatus* strain CKB.[45] In strain GR-1 this enzyme was found exclusively in the soluble fraction of cell lysates while, in contrast, the chlorite dismutase of *D. agitatus* was present in both the soluble and membrane fractions.[20] Although chlorite dismutation has not yet been demonstrated for all the known dissimilatory (per)chlorate reducing bacteria, the fact that all of the PCRB tested to date, including the isolates obtained in this study, contain chlorite dismutase activity implies that it is a common metabolism to (per)chlorate-reducing bacteria which is in contrast to recently published suggestions.[21] As such this has significant implications from both a thermodynamic and physiological standpoint.

The Gibbs free energy for the reduction of perchlorate to chloride coupled to the oxidation of acetate indicates that microbial perchlorate reduction should be more energetically favorable than aerobic respiration (Table 2). However, if all PCRB only gain energy for the partial reduction of perchlorate to chlorite, which is then dismutated in a non-energy yielding reaction,[45,47] the thermodynamics of microbial perchlorate reduction are significantly lower than $O_2$ respiration and fall in the same range as nitrate respiration (Table 2). This has important implications for the microbial attenuation of environments contaminated with perchlorate and suggests that for a reliable bioremediative strategy to be implemented, the environment should have low partial pressures of $O_2$ and low concentrations of nitrate relative to the perchlorate concentration.

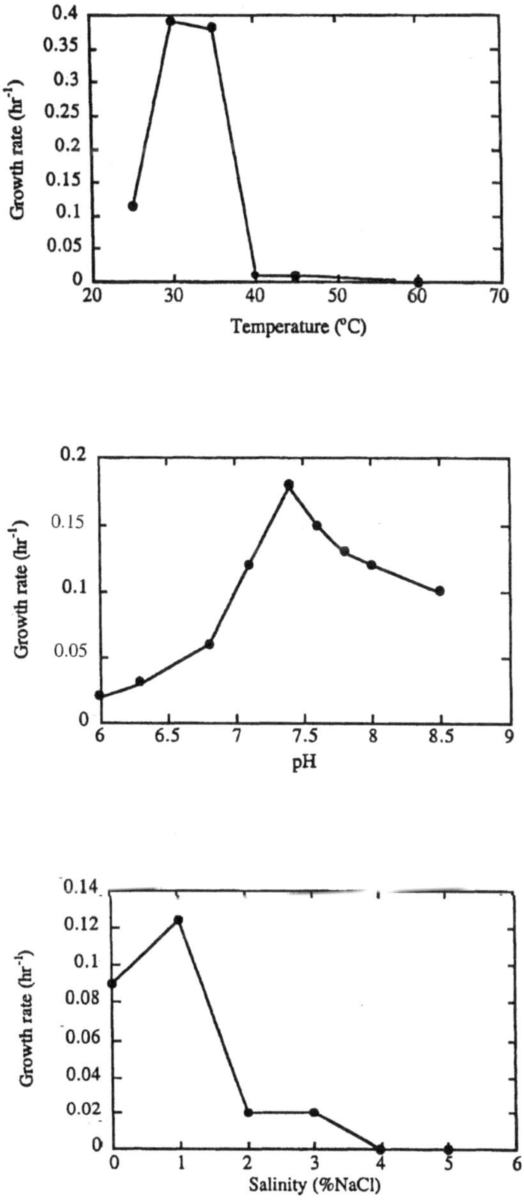


**Figure 4.** Good growth of the (per)chlorate-reducing isolates was observed over a broad range of environmental conditions as shown here for *D. agitatus* strain CKB.

**Table 2.** Gibbs free energy changes for acetate oxidation coupled to $O_2$ reduction, nitrate reduction, and both complete and incomplete perchlorate reduction.

| Stoichiometric Reaction | $\Delta G°$ [kJ (mol acetate)$^{-1}$] |
|---|---|
| $CH_3COO^- + 2\ O_2 \rightarrow 2\ HCO_3^- + H^+$ | –844 |
| $CH_3COO^- + ClO_4^- \rightarrow 2\ HCO_3^- + Cl^- + H^+$ | –966 |
| $CH_3COO^- + 2\ ClO_4^- \rightarrow 2\ HCO_3^- + 2\ ClO_2^- + H^+$ | –801 |
| $5\ CH_3COO^- + 8\ NO_3^- + 3\ H^+ \rightarrow 10\ HCO_3^- + 4\ H_2O + 4\ N_2$ | –792 |

In addition, the fact that all perchlorate-reducing bacteria produce $O_2$ as a result of chlorite dismutation during the reductive metabolism of perchlorate suggests that $O_2$ is not toxic to these organisms as would be the case with strict anaerobes and explains why all (per)chlorate reducing bacteria are either facultatively anaerobic or microaerophilic.

**Environmental Significance and Bioremediative Potential**

The results of this study together with our previously published reports[20,43,45] demonstrate the hitherto unrecognized ubiquity of microbial (per)chlorate reduction and the broad phylogenetic diversity of the organisms capable of this metabolism. Contamination of drinking water, ground water, and surface water by oxyanions of chlorine, especially chlorate and perchlorate has only recently been recognized as a potentially serious health risk.[1–3] Although microbial reduction of (per)chlorate has been recognized for the last 50 years and was identified as a potentially important metabolism for the treatment of perchlorate and chlorate contamination in the environment,[1,2,44] there is still very little known about the microorganisms involved in (per)chlorate reduction. Several organisms have been shown to be capable of the reduction of chlorate to chlorite, including *Proteus mirabilis,*[22] *Rhodobacter capsulatus* and *Rhodobacter sphaeroides,*[23] however, no growth is associated with this metabolism and the chlorite endproduct is generally toxic to these organisms.

The role of (per)chlorate-reducing bacteria in environments that have no previous exposure to chlorine oxyanions has yet to be determined. Although a few dissimilatory (per)chlorate-reducers have now been described,[14–17,20] all of these isolates were obtained from contaminated sediments or wastewater treatment sludges. This study demonstrates that organisms with (per)chlorate-reducing capability can be readily isolated from pristine environments.

Although (per)chlorate reduction has been recognized for more than fifty years, the presence of oxyanions of chlorine in the environment is thought to be the result of human activities over the last hundred years. As such, the evolution of such a phylogenetically diverse group of organisms with the ability to couple growth to the reduction of (per)chlorate is unexpected. This metabolic capability appears to be centered around the unique ability of these organisms to dismutate chlorite into chloride and oxygen. Although chlorite dismutation has not yet been demonstrated for all the known dissimilatory (per)chlorate reducing bacteria, it was shown previously for strain GR-1,[16] strain CKB[19,20,32,45] and in this study for all of the PCRB isolates obtained thus suggesting that it is common to all PCRB. The fact that the purified chlorite dismutase enzymes from strains

GR-1 and CKB were found to be similar in general structure, molecular mass, and specific activity[45] suggests that a single gene encoding for this enzyme may be conserved amongst these (per)chlorate-reducing organisms and further suggests that (per)chlorate reduction may be the result of horizontal gene transfer events. The pathway used for this reductive metabolism has some important implications with regard to the development of bioremediative strategies for the attenuation of perchlorate contaminated environments as thermodynamic considerations demonstrate that for effective cleanup, such environments should have low partial pressures of $O_2$ and low concentrations of nitrate relative to the perchlorate. At higher concentrations, the metabolic reduction of these compounds will simply outcompete microbial perchlorate reduction and inhibit its removal.

## ACKNOWLEDGMENTS

Support for this research was in part from grant DE-FG02-98ER62689 from the Department of Energy to J.C. and L.A. and from the 1998 Oak Ridge Associated Universities Junior Faculty award to J.C.

## REFERENCES

1. U.S. Environmental Protection Agency. **1999** URL: http://www.epa.gov/ncea /perch.htm.

2. Urbansky, E.T. "Perchlorate chemistry: implications for analysis and remediation." *Bioremediation Journal* **1998,** *2,* 81–95.

3. Renner, R. "Perchlorate-tainted wells spur government action." *Environmental Science and Technology* **1998,** *32,* 210A.

4. Ericksen, G.E. "The Chilean nitrate deposits." *American Scientist* **1983,** *71,* 366–374.

5. Stanbury, J.B.; Wyngaarden, J.B. "Effect of perchlorate on the human thyroid gland." *Metabolism* **1952,** *1,* 533–539.

6. California Department of Health Services website "Perchlorate in California drinking water." 1997 URL: http://www.dhs.cahwnet.gov/ps/ddwem/chemicals/perchl/perchl index.htm

7. Renner, R. "EPA draft almost doubles safe dose of perchlorate in water." *Environmental Science and Technology* **1999,** *33,* 110A–111A.

8. Bryan, E.H.; Rohlich, G.A. "Biological reduction of sodium chlorate as applied to measurement of sewage BOD." *Sewage and Industrial Wastes* **1954,** *26,* 1315–1324.

9. Bryan, E.H. "Application of the chlorate BOD procedure to routine measurement of wastewater strength." *Journal of Water Pollution Control Federation* **1966,** *38,* 1350–1362.

10. Hackenthal, E.; Mannheim, W.; Hackenthal, R.; Becher, R. "Die reduktion von perchlorat durch bakterien. I. untersuchungen an intakten zellen." *Biochemical Pharmacology* **1964,** *13,* 195–206.

11. Hackenthal, E. "Die reduktion von perchlorat durch bacterien II. Die identitat der nitratreduktase und des perchlorat reduzierenden enzyms aus *B. cereus*." *Biochemical Pharmacology* **1965,** *14,* 1313–1324.

12. Stouthamer, A. "Nitrate reduction in *Aerobacter aerogenes*. I. Isolation properties of mutant strains blocked in nitrate assimilation and resistant against chlorate." *Archives of Microbiology* **1967,** *56,* 68–75.

13. Korenkov, V.; Romanenko, V.; Kuznetsov, S.; Voronov, J. "Process for purification of industrial waste waters from perchlorates and chlorates." U.S. Patent No. 3,943,055, 1976.

14. Malmqvist, A.; Welander, T.; Moore, E.; Ternstrom, A.; Molin, G. Stenstrom, I.-M. "*Ideonella dechloratans* gen. nov., sp. nov., a new bacterium capable of growing anarobically with chlorate as an electron acceptor." *Systematic and Applied Microbiology* **1994,** *17,* 58–64.

15. Stepanyuk, V.; Smirnova, G.; Klyushnikova, T.; Kanyuk, N.; Panchenko, L.; Nogina, T.; Prima, V. "New species of the *Acinetobacter* genus *Acinetobacter thermotoleranticus* sp. nov." *Mikrobiologiya* **1992,** *61,* 347–356.

16. Rikken, G.; Kroon, A.; van Ginkel, C. "Transformation of (per)chlorate into chloride by a newly isolated bacterium: reduction and dismutation." *Applied Microbiology and Biotechnology* **1996,** *45,* 420–426.

17. Wallace, W.; Ward, T.; Breen, A.; Attaway, H. "Identification of an anaerobic bacterium which reduces perchlorate and chlorate as *Wolinella succinogenes*." *Journal of Industrial Microbiology* **1996,** *16,* 68–72.

18. Malmqvist, A.; Welander, T.; Gunnarsson, L. "Anaerobic growth of microorganisms with chlorate as an electron acceptor." *Applied and Environmental Microbiology* **1991,** *57,* 2229–2232.

19. Coates, J.D.; Bruce, R.A. Haddock, J.D. "Anoxic bioremediation of hydrocarbons." *Nature* **1998,** *396,* 730.

20. Bruce, R.A.; Achenbach, L.A.; Coates, J.D. "Reduction of (per)chlorate by a novel organism isolated from a paper mill waste." *Environmental Microbiology* **1999,** *1,* 319–331.

21. Logan, B. "A review of chlorate- and perchlorate-respiring microorganisms." *Bioremediation Journal* **1998,** *2,* 69–79.

22. De Groot, G.N.; Stouthamer, A.H. "Regulation of reductase formation in *Proteus mirabilis*. I. Formation of reductases and enzymes of the formic hydrogenlyase complex in the wild type and in chlorate resistant mutants." *Archives of Microbiology* **1969,** *66,* 220–233.

23. Roldan, M.D.; Reyes, F.; Moreno-Vivian, C.; Castillo, F. "Chlorate and nitrate reduction in phototrophic bacteria *Rhodobacter capsulatus* and *Rhodobacter sphaeroides*." *Current Microbiology* **1994,** *29,* 241–245.

24. Romanenko, V.I.; Korenkov, V.N.; Kuznetsov, S.I. "Bacterial decomposition of ammonium perchlorate." *Mikrobiologiya* **1976,** *45*, 204–209.

25. Balch, W.E.; Fox, G.E.; Magrum, L.J.; Woese, C.R.; Wolfe, R.S. "Methanogens: reevaluation of a unique biological group." *Microbiological Reviews* **1979,** *43,* 260–296.

26. Hungate, R.E. "A roll tube method for cultivation of strict anaerobes." *Methods in Microbiology* **1969,** *3B,* 117–132.

27. Miller, T.L.; Wolin, M.J. "A serum bottle modification of the Hungate technique for cultivating obligate anaerobes." *Applied Microbiology* **1974,** *27,* 985–987.

28. Roden, E.E.; Lovley, D.R. "Dissimilatory Fe(III) reduction by the marine microorganism *Desulfuromonas acetoxidans*." *Applied and Environmental Microbiology* **1993,** *59,* 734–742.

29. Lovley, D.R.; Phillips, E.J.P. "Novel mode of microbial energy metabolism: organic carbon oxidation coupled to dissimilatory reduction of iron or manganese." *Applied and Environmental Microbiology* **1988,** *54,* 1472–1480.

30. Widdel, F.; Bak, F. "Gram-negative mesophilic sulfate-reducing bacteria." *Environmental Science and Technology* **1992,** *26,* 725–733.

31. Widdel, F.; Hansen, T.A. "The Dissimilatory Sulfate- and Sulfur-Reducing Bacteria." In *The Prokaryotes*. Springer-Verlag: New York, 1992; pp. 583–624.

32. Coates, J.D.; Bruce, R.A.; Patrick, J.A.; Achenbach, L.A. "Hydrocarbon bioremediative potential of (per)chlorate-reducing bacteria." *Bioremediation Journal* **1999,** *3,* in press.

33. Lovley, D.R.; Giovannoni, S.J.; White, D.C.; Champine, J.E.; Phillips, E.J.P.; Gorby, Y.A.; Goodwin, S. "*Geobacter metallireducens* gen. nov. sp. nov., a microorganism capable of coupling the complete oxidation of organic compounds to the reduction of iron and other metals." *Archives of Microbiology* **1993,** *159,* 336–344.

34. Coates, J.D.; Lonergan, D.J.; Lovley, D.R. "*Desulfuromonas palmitatis* sp. nov., a long-chain fatty acid oxidizing Fe(III) reducer from marine sediments." *Archives of Microbiology* **1995,** *164,* 406–413.

35. Coates, J.D.; Lovley, D.R.; Blunt-Harris, E. "Anaerobic microbial metabolism with humic substances." *Symposium on refractory organic substances in the environment – ROSE* **1997,** 29–30.

36. Coates, J.D.; Phillips, E.J.P.; Lonergan, D.J.; Jenter, H.; Lovley, D.R. "Isolation of *Geobacter* species from a variety of sedimentary environments." *Applied and Environmental Microbiology* **1996,** *62,* 1531–1536.

37. Maidak, B.L.; Olsen, G.J.; Larsen, N.; Overbeek, R.; McCaughey, M.J.; Woese, C.R.; "The RDP (Ribosomal Database Project)." *Nucleic Acids Research* **1997,** *25,* 109–111.

38. Benson, D.A.; Boguski, M.S.; Lipman, D.J.; Ostell, J.; Ouellette, B.F. "GenBank." *Nucleic Acids Research* **1998,** *26,* 1–7.

39. Gilbert, D.G. "SeqApp. version 1.9a157" Biocomputing Office, Biology Department, Indiana University: Bloomington, IN, 1993.

40. Swofford, D.L. "PAUP*: Phylogenetic Analysis Using Parsimony (and other methods)." 1999.

41. Van Ginkel, C.G.; Plugge, C.M.; Stroo, C.A. "Reduction of chlorate with various energy substrates and inocula under anaerobic conditions." *Chemosphere* **1995,** *31,* 4057–4066.

42. Oltmann, L.F.; Reifnders, W.N.M.; Stouthamer, A.H. "Characterization of purified nitrate reductase A and chlorate reductase C from *Proteus mirabilis*." *Archives of Microbiology* **1976,** *111,* 25–35.

43. Achenbach, L.A.; Bruce, R.A.; Michaelidou, U.; Coates, J.D. "Phylogenetic analysis of dissimilatory (per)chlorate-reducing bacteria and development of molecular probes." *Applied and Environmental. Microbiology*, in press.

44. Wallace, W.; Beshear, S.; Williams, D.; Hospadar, S.; Owens, M. "Perchlorate reduction by a mixed culture in an up-flow anaerobic fixed bed reactor." *Journal of Industrial Microbiology and Biotechnology* **1998,** *20,* 126–131.

45. Coates, J.D.; Michaelidou, U.; Bruce, R.A.; O'Connor, S.M.; Crespi, J.N.; Achenbach, L.A. "The ubiquity and diversity of dissimilatory (per)chlorate-reducing bacteria." *Applied and Environmental Microbiology* **1999,** *65,* 5234–5241.

46. Bruce, R.A.; Achenbach, L.A.; Coates, J.D. "*Dechlorimarinus chloriphilus* gen nov., sp. nov., a novel chlorate-reducing organism isolated from marine sediments." *Applied and Environmental Microbiology*, submitted.

47. Van Ginkel, C.; Rikken, G.; Kroon, A.; Kengen, S. "Purification and characterization of chlorite dismutase: a novel oxygen-generating enzyme." *Archives of Microbiology* **1996,** *166,* 321–326.

# Chapter 25

# ISOLATION AND CHARACTERIZATION OF TWO NOVEL (PER)CHLORATE-REDUCING BACTERIA FROM SWINE WASTE LAGOONS[⊗]

Urania Michaelidou, Laurie A. Achenbach, and John D. Coates*

Department of Microbiology and Center for Systematic Biology,
Southern Illinois University, Carbondale, Illinois 62901

## INTRODUCTION

Microbial chlorate ($ClO_3^-$) and perchlorate ($ClO_4^-$) reduction has recently been recognized as an important form of microbial metabolism for the removal of chlorine oxyanion contamination in the environment.[1] Chlorine oxyanions have many industrial applications including use as bleaching agents by the paper and pulp industry,[2,3] as disinfectants and defoliants by the agricultural industry,[4] and as components of explosives and rocket propellants by the aerospace and defense industries.[5] Chlorates can also be formed as a result of ozonation of drinking waters which have been treated with chlorine[6] or photodecomposition of chlorite or chlorine dioxide[7] which are used in addition to chlorine for water disinfection.

Perchlorate has been shown to affect iodide accumulation in the thyroid gland[8] and at concentrations greater than 6 mg (kg body weight)$^{-1}$ day$^{-1}$ perchlorate can cause fatal bone marrow disease.[1] In addition, acute haemolysis in animals[9,10] and enzyme damage to human erythrocytes[11] has been associated with both chlorate and chlorite ($ClO_2^-$). These pollutants also have an effect on aquatic plants and invertebrates,[3,12,13] and are assumed to be the cause of the disappearance of the brown algae, *Fucus vesiculosus,* from the Baltic sea.[14]

It was originally assumed that (per)chlorate reduction in the environment was the result of the activity of nitrate-respirers which coincidentally used chlorine oxyanions in place of nitrate. In support of this, many nitrate respiring bacteria have been shown to be

[⊗]This work was presented at the 218th national meeting of the American Chemical Society as part of the Environmental Division symposium Perchlorate in the Environment, held August 22–24 1999, in New Orleans, Louisiana.

*Author to whom correspondence should be directed. Phone: 618-453-6132. Fax: 618-453-8036. Electronic mail: jcoates@micro.siu.edu.

*Perchlorate in the Environment*, edited by Urbansky.
Kluwer Academic/Plenum Publishers, New York, 2000.

capable of the reduction of chlorate to chlorite, including *Proteus mirabilis*,[15] *Rhodobacter capsulatus* and *Rhodobacter sphaeroides*,[16] however growth was not associated with this metabolism and the chlorite formed as the endproduct was toxic to these organisms.

The first organism which was shown to couple growth to the reduction of chlorate or perchlorate was recovered as part of an investigation into the biological attenuation of (per)chlorate contamination in the environment.[17] Chloride was produced as the innocuous endproduct of this reductive metabolism. This organism was putatively identified as a *Vibrio* species based on physiology and morphology and was named *Vibrio dechloraticans*. In the last decade, four more dissimilatory (per)chlorate-reducing strains including *Ideonella dechloratans*, *Wolinella succinogenes* strain HAP-1, *Acinetobacter thermotoleranticus*, and strain GR-1 have been isolated from contaminated sediments or wastewater treatment sludge.[18–21] Of these (per)chlorate-reducing isolates, only two, *Ideonella dechloratans* and *Wolinella succinogenes* strain HAP-1, have been characterized both phenotypically and genotypically.[18,19] More recently *Dechlorimonas agitatus* strain CKB was isolated from a paper mill waste sludge as part of a study on the bioremediative potential of (per)chlorate-reducing bacteria.[22–28] Similarly to most of the other known (per)chlorate-reducing isolates, this organism was a facultative anaerobe which coupled chlorate or perchlorate reduction to the oxidation of simple organic acids or alcohols.[22,23] In contrast to the other known isolates, this organism could not grow by nitrate reduction. This further supports the assumption that chlorate reduction and nitrate reduction are two unrelated pathways as suggested recently[18,21] and is in contrast to the suggestions of earlier studies.[29–31]

As part of a study of the ubiquity and diversity of organisms involved in microbial (per)chlorate reduction[27,28,32] we isolated several novel (per)chlorate-reducing organisms from a broad diversity of environments. Here we report on the phenotypic and genotypic characteristics of two of these isolates, *Dechlorisoma suillus* strain PS and *Dechlorospirillum anomalous* strain WD. Both of these organisms were isolated from samples collected from animal waste lagoons on the Southern Illinois University Agricultural Research Center.

## EXPERIMENTAL

### Source of Organisms and Sediments

*Dechlorisoma suillus* strain PS and *Dechlorospirillum anomolous* strain WD were both isolated from sediment samples freshly collected from the swine waste primary lagoons on the Southern Illinois University campus Agricultural Research Center. Samples were collected in small glass jars (100 mL) which were filled to capacity to reduce exposure to atmospheric $O_2$ to a minimum. Samples were immediately transported back to the laboratory where they were used within two hours. Enrichment cultures were established by inoculating 1.0 g sediment subsamples into 9 mL of anoxic, defined freshwater medium described below with acetate (10 mM) as the sole electron donor and chlorate (10 mM) as the sole electron acceptor.

*Dechlorimonas agitatus* was previously isolated in our laboratory[22] and stored as a frozen stock culture. *Rhodocyclus tenuis* was kindly provided from the laboratory culture collection of Dr. Michael T. Madigan. *R. tenuis* was grown phototrophically on a modification of the RCVB medium as described previously.[33] *Duganella zoogloeoides* was purchased from the American Type Culture Collection, Manassas, VA, ATCC# 19544 and was grown aerobically on the recommended yeast-peptone medium (ATCC# 1858), unless otherwise noted. The two magnetotactic bacteria, *Magnetospirillum magnetotacticum* strain MS-1, and *Magnetospirillum* species strain AMB-1 were provided by the laboratory of Dr. Dennis Bazylinski. Both strains were grown under microaerobic conditions (1 kPa

$O_2$) in basal media outlined below with acetate as the electron donor and nitrate as the electron acceptor unless otherwise noted.

**Media and Culturing Techniques**

Standard anaerobic culturing techniques were used throughout[34] unless otherwise specified. Anoxic medium was prepared by boiling under $N_2/CO_2$ (80/20 v/v) to remove dissolved $O_2$ and dispensed under $N_2/CO_2$ into anaerobic pressure tubes or serum bottles which were then capped with thick butyl stoppers. The freshwater medium contained (in grams per liter): $NH_4Cl$ (0.25); $NaClO_3$ (1.03); $CH_3CO_2Na$ (1.36); $NaH_2PO_4$ (0.60); KCl (0.1); $NaHCO_3$ (2.5). Vitamins and trace metals were added (10 mL $L^{-1}$ respectively) from stock solutions. The vitamin stock contained (mg $L^{-1}$): biotin (2), folic acid (2), pyridoxine HCl (10), riboflavin (5), thiamin (5), nicotinic acid (5), pantothenic acid (5), vitamin $B_{12}$ (0.1), *p*-aminobenzoic acid (5), thioctic acid (5). The trace metal stock contained (g $L^{-1}$): nitrilotriacetic acid (1.5), $MgSO_4$ (3.0), $MnSO_4{\bullet}H_2O$ (0.5), NaCl (1.0), $FeSO_4{\bullet}7H_2O$ (0.1), $CaCl_2{\bullet}2H_2O$ (0.1), $CoCl_2{\bullet}6H_2O$ (0.1), ZnCl (0.13), $CuSO_4$ (0.01), $AlK(SO_4)_2{\bullet}12H_2O$ (0.01), $H_3BO_3$ (0.01), $Na_2MoO_4$ (0.025), $NiCl_2{\bullet}6H_2O$ (0.024), $Na_2WO_4{\bullet}2H_2O$ (0.025). The pH of the final medium was 6.8–7.0. Electron donors and acceptors were added to the sterile medium from sterile anoxic stocks of the sodium salts. Incubation was done at 30 °C unless otherwise stated.

Solidified media was prepared using the same media components and amending with 2% w/v noble agar (Difco) as previously outlined.[22] The bicarbonate buffer was replaced with 10 mM TES [*N*-tris(hydroxymethyl)methyl-2-aminoethanesulfonic acid) buffer, pH 7.0 and the agar plates were poured, streaked and incubated in an anaerobic glove bag with a gas phase of $N_2/H_2$ (95/5 v/v).

**Cell Suspension Preparation**

Cells of the various respiratory organisms used in this study were grown in 500 mL volumes anaerobically with acetate as the electron donor and chlorate or nitrate as the electron acceptor as necessary. In the case of *Rhodocyclus tenuis*, cells were grown on RCVB medium phototrophically in the presence of 10 mM sodium chlorate. After dense growth of all cultures, cells were harvested by centrifugation. Cell pellets were washed twice and resuspended in 1 mL volumes of anoxic bicarbonate buffer (2.5 g $L^{-1}$) undcr a headspace of $N_2/CO_2$ or, in the case of *R. tenuis*, anoxic phosphate buffer (10 mM) under a headspace of $N_2$.

**Analytical techniques**

Acetate concentrations were analyzed by HPLC with UV detection (Shimadzu SPD-10A, Shimadzu Scientific Instruments, Columbia, MD) using a HL-75H$^+$ cation exchange column (Hamilton #79476, Hamilton Company, Reno, NA). The eluent was 8 mM $H_2SO_4$ at a flow rate of 0.4 mL $min^{-1}$. Chlorate, chloride, nitrate and nitrite concentrations were analyzed by ion chromatography with conductivity detection (Shimadzu CDD-6A, Shimadzu Scientific Instruments, Columbia, MD) using a PRP-X100 anion exchange column (Hamilton #79434, Hamilton Company, Reno, NA). Elution was done at 2.0 mL $min^{-1}$ with 4 mM *p*-hydroxybenzoic acid in 2.5% methanol at pH 8.5. Concentrations of $N_2$ and $CO_2$ in headspace samples was followed by gas chromatography (Shimadzu GC-8A, Shimadzu Scientific Instruments, Columbia, MD), with thermal conductivity detection. Growth of cultures on soluble electron acceptors was measured by increase in optical density at 600 nm. Molecular oxygen production from chlorite dismutation was detected by an $O_2$ electrode (YSI model 5300, Yellow Springs, OH). Concentrations of HCl-extractable Fe(II) were determined colorimetrically by the ferrozine assay.

**16S rRNA Gene Sequencing and Phylogenetic Analysis**

Cells from 1.5 mL of pure cultures of strain PS and WD were harvested and resuspended in 20 μL sterile water. Chloroform (2 μL) was added and the suspension was boiled for 15 minutes to lyse the cells. Polymerase chain reaction (PCR) was performed to amplify the 16S rDNA sequence with bacterial-specific primers (8F: 5′-AGAGTTTGAT-CCTGGCTCAG-3′; 1525R: 5′-AAGGAGGTGATCCAGCC-3′). Reagents consisted of 10 mM Tris-HCl (pH 9.0), 50 mM KCl, 0.1% Triton X-100, 1.2 mM $MgCl_2$, 0.2 mM each dNTP, 75 ng of each primer, 0.5 μlTaq polymerase (Gibco/BRL), and 1 μL of lysed cells in a 50 μlreaction. Amplification parameters were: 94 °C for 3 min, followed by 30 cycles of 94 °C for 1 min, 55 °C for 1 min, and 72 °C for 2 min with a 2-s auto-extension. The PCR product was electrophoresed on a 1% agarose gel to ensure correct amplification. The DNA was gel-purified on DEAE membrane and extracted as previously described[35] and cycle sequenced (Thermosequenase, Amersham). Sequence entry and analysis was performed with the MacVector 6.0 sequence analysis software program for the Macintosh (Oxford Molecular). Aligned sequences of various 16S rDNAs were downloaded from the Ribosomal Database Project[36] into the computer program SeqApp.[37] Strain PS and strain WD 16S rDNA sequences were manually entered and properly aligned using secondary structure information. Parsimony and bootstrap analysis was performed on a Macintosh G3 using PAUP* 4.0.[38] Bootstrap analysis was conducted on 100 replicates using a heuristic search strategy to assess the confidence level of various clades. 16S rDNA sequences are from the following GenBank accession numbers: *Magnetospirillum gryphiswaldense* (Y10109); *Dechlorospirillum anomolous* str. *WD* (AF170352); *Magnetospirillum magnetotacticum* (M58171); *Magnetospirillum* sp. str. *AMB-1* (D17514); *Phaeospirillum fulvum* (*Rhodospirillum fulvum*, M59065); *Phaeospirillum molischianum* (*Rhodospirillum molischianum*, M59067); *Rhodocyclus tenuis* (D16209); *Rhodocyclus purpureus* (M34132); *Dechlorimonas agitatus* str. *CKB* (AF047462); *Dechlorisoma suillus* str. *PS* (AF170348); *Azoarcus evansii* (X77679); *Azoarcus denitrificians* (L33689); *Azoarcus indigens* (L15531); *Thauera selenatis* (X68491); *Duganella zoogloeoides* (*Zoogloea ramigera*, X74913).

## RESULTS

**Enrichment and Isolation**

After seven days incubation, microscopic observation revealed the presence of a variety of rod-shaped bacteria in primary enrichment cultures with acetate as the sole electron donor and chlorate as the sole electron acceptor. These were transferred into fresh, sterile anoxic medium and good growth resulted within 72 h. After several more transfers into fresh anoxic media, samples from each enrichment culture were streaked onto anaerobic plates with chlorate and acetate as the electron acceptor and donor respectively. Growth on plates, as indicated by the presence of individual visible pink colonies of a consistent morphology, occurred within 10–15 days. Several colonies were picked and inoculated into liquid media. From these active cultures strains PS and WD were isolated.

**Cell and Colony Morphology**

Strains PS and WD are both gram-negative, complete-oxidizing, non-fermentative facultative anaerobes, although strain WD grows optimally under microaerophilic conditions. Cells of strain PS are rod-shaped, 0.5 μm by 1–2 μm, while strain WD is a spirillum, 0.2 μm × 7 μm. Cells of both strains are motile and non-spore forming. Both organisms grow slowly on solid media either aerobically or anaerobically. When grown anaerobically with chlorate as the electron acceptor colonies appear small (0.1–0.5 mm diam.), round, smooth and pink colored. In contrast, when grown aerobically, colonies of both organisms appear white.

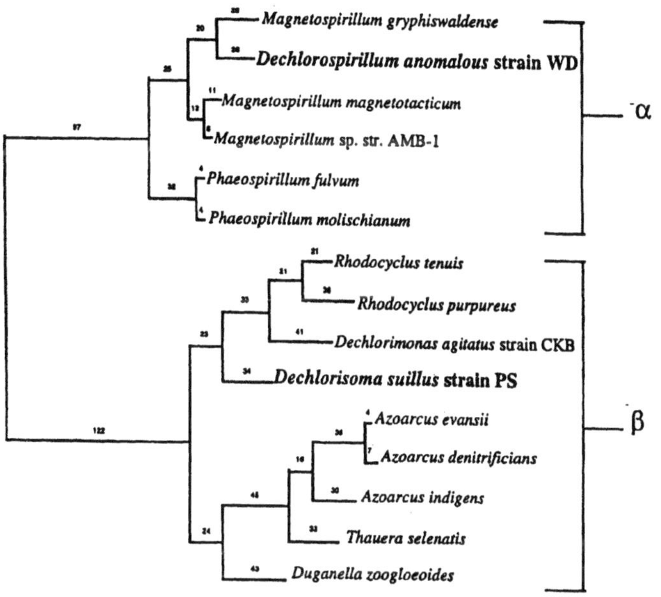


**Figure 1.** Phylogenetic tree of the 16S rDNA sequences of strains WD and PS and their closest relatives resulting from a heuristic search using parsimony analysis. Branch length values are indicated. The same topology was obtained using either distance or maximum likelihood and was supported by bootstrap analysis.

## Phylogenetic Analysis

Comparative analysis of the 16S rDNA sequence from both strains placed strain PS in the beta subdivision of the Proteobacteria closely related to the phototrophic *Rhodocyclus* species and our previously isolated (per)chlorate-reducer, *Dechlorimonas agitatus* strain CKB (Figure 1).[22]

*Rhodocyclus tenuis* is the closest relative (94.4% similarity). Although it is phylogenetically distinct (93.1% similarity), strain PS is physiologically more similar to *Dechlorimonas agitatus* strain CKB which can also grow by (per)chlorate respiration coupled to the oxidation of acetate.In contrast to strain PS, phylogenetic analysis of strain WD placed it in the alpha subdivision of the Proteobacteria, closely related to the magnetotactic *Magnetospirillum* species (Figure 1). *Magnetospirillium gryphiswaldense* is the closest relative. Similarly to strain WD, the *Magnetospirillium* species are facultative microaerophilic spirilla. However, none of the *Magnetospirillium* species tested (*M. gryphiswaldense, M. magnetotacticum* and *Magnetospirillium* strain AMB-1) could couple growth to the reduction of perchlorate or chlorate.

## Optimum Growth Conditions

Both strains WD and PS grew over a broad range of environmental conditions. Strain PS grew over a temperature range of 25–42 °C and a pH range of pH 5.0–8.0. Optimum growth was observed at 35 °C and pH 7.2 respectively. Strain WD grew over a temperature range of 25–37 °C and a pH range of pH 6.5–7.5. Optimum growth for strain WD was observed at similar conditions to strain PS, i.e., 35 °C and pH 7.2. Neither strain tolerated high concentrations of NaCl and concentrations above 1% were completely inhibitory. Both strains WD and PS grew preferentially in freshwater medium with a sodium chloride concentration of 0%.

**Electron Donors and Acceptors**

Strains PS and WD have similar metabolic capabilities and both organisms grow by the complete oxidation of acetate coupled to the reduction of chlorate or perchlorate (Fig. 2). In addition to acetate, strains PS and WD also grow on propanoate, butanoate, *iso*-butanoate, pentanoate (valerate), ethyl alcohol, 2-oxopropanoate (pyruvate), lactate (2-hydroxypropanoate), butanedioate (succinate), hydroxybutanedioate (malate), *trans*-butenedioate (fumarate) and acid-hydrolyzed casein (casamino acids). Neither strain grew or reduced (per)chlorate with $H_2$ as the electron donor. Strains PS and WD also did not grow by fermentation in complex, organic rich medium in which the basal medium was amended with yeast extract (5 g $L^{-1}$), casamino acids (10 g $L^{-1}$) and glucose (1.8 g $L^{-1}$). The organisms grew without the addition of vitamins, however, no growth was apparent if the trace element solution was omitted from the media. A wide variety of other potential electron donors did not support growth or (per)chlorate reduction (data not shown).

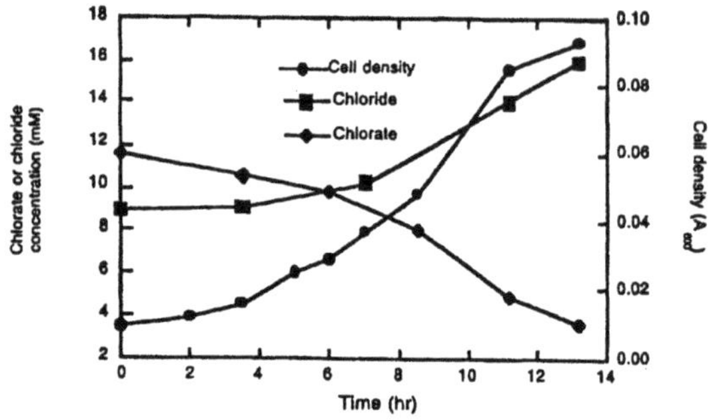


**Figure 2.** Cell growth and chlorate reduction by strain PS with acetate as the electron donor. Chloride and $CO_2$ are formed as the innocuous endproducts of this metabolism.

With acetate (10 mM) as the electron donor, strain PS grew preferentially with chlorate as the electron acceptor (Figure 3). Strain PS grew in a perchlorate or chlorate concentration range of 1 to 40 mM (Figure 3). Optimum growth was observed at concentrations of 10 mM for both perchlorate and chlorate respectively.

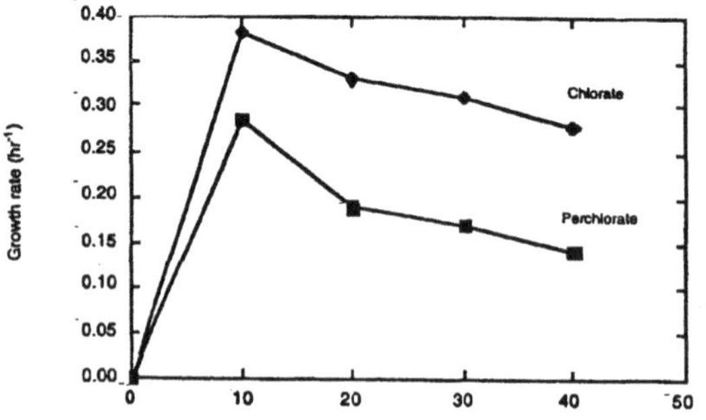


**Figure 3.** Growth rate of strain PS on different concentrations of chlorate or perchlorate. Chlorate is the preferentially utilized electron acceptor by this organism.

Similarly, strain WD grew preferentially with chlorate as the electron acceptor, although, the growth rate difference between chlorate and perchlorate was not as significant as observed for strain PS (Figure 4). Strain WD grew optimally in chlorate or perchlorate concentrations of 10 mM respectively. However, in contrast to strain PS, strain WD could grow and actively reduce perchlorate concentrations as high as 80 mM. For both strains perchlorate and chlorate were completely reduced to innocuous chloride. The reduction of 8.00 ± 0.64 mM (mean + standard deviation, $n = 3$) chlorate resulted in the production of 7.14 ± 0.71 mM (mean + standard deviation, $n = 3$) chloride which is in close agreement with the expected values. Chlorite was not detected in the medium during growth.

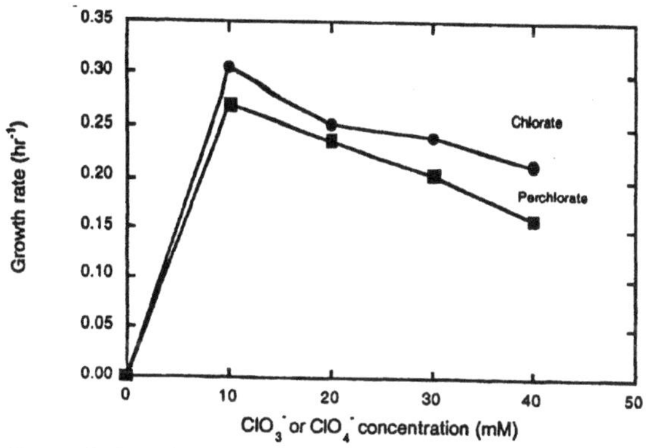


**Figure 4.** Growth rate of strain WD on different concentrations of chlorate or perchlorate. Although chlorate is the preferentially utilized electron acceptor by this organism, the difference in growth rate on chlorate and perchlorate is not as significant as that observed for strain PS.

In addition to chlorate and perchlorate, both strains can also utilize nitrate as an alternative electron acceptor. In the case of strain PS, $N_2$ gas is the major endproduct of nitrate reduction and this organism could couple the oxidation of Fe(II) to this reductive metabolism (Figure 5). The endproducts of nitrate respiration were not determined for strain WD.

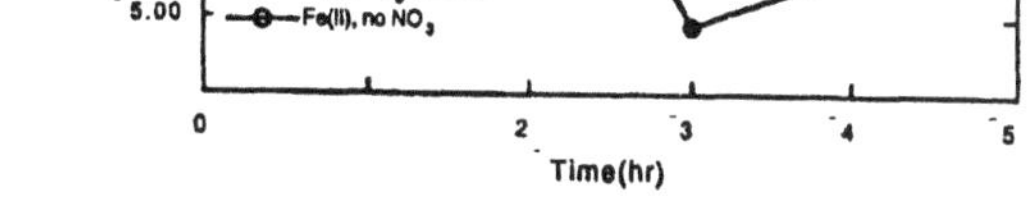


**Figure 5.** Oxidation of ferrous iron by strain PS coupled to the reduction of nitrate. Insoluble amorphous ferric oxyhydroxide and $N_2$ are the endproducts of this metabolism.

**Chlorite Dismutase Activity**

In the presence of chlorite, washed cell suspensions of both strains PS and WD produced $O_2$. Chlorite was not dismutated in controls in which the cells were omitted or which were heat-treated (Figure 6). $O_2$ production was rapid and proportional to the initial concentration of chlorite and $O_2$ production was stoichiometric with initial chlorite concentrations suggesting that chlorite was completely dismutated into $Cl^-$ and $O_2$ (Figure 6).

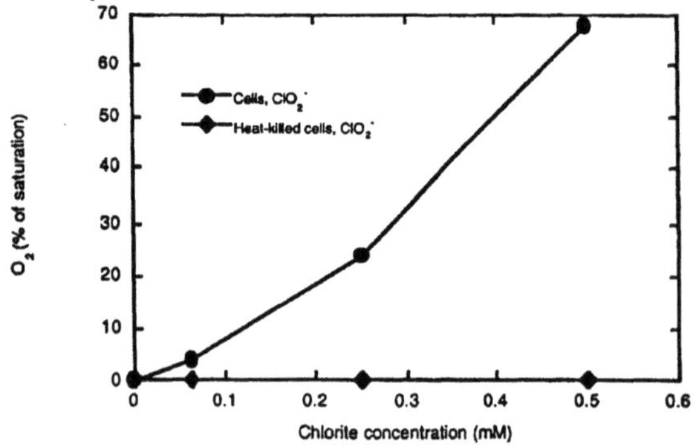


**Figure 6.** $O_2$ production from the dismutation of increasing concentrations of chlorite by washed whole cell suspensions of strain PS.

Other anions such as nitrite ($NO_2^-$), hypophosphite ($H_2PO_2^-$), dithionite (hydrosulfite, $S_2O_4^{2-}$), or arsenite ($AsO_3^{3-}$) were not dismutated even after 15 min incubation. If chlorite was added after 15 min., rapid $O_2$ production and chlorite dismutation was observed in all cases demonstrating that these compounds were not inhibiting the chlorite dismutase enzyme (data not shown). The non-(per)chlorate-reducing, close relatives of strain PS, (*Rhodocyclus tenuis* and *Duganella zoogloeoides*) did not dismutate chlorite in similar experiments. Similarly, the close relatives of strain WD (*M. magnetotacticum, M. gryphiswaldense* and *Magnetospirillium* strain AMB-1) also did not dismutate chlorite.

## DISCUSSION

Two new (per)chlorate-reducing isolates, strains PS and WD, were obtained from sediments collected from swine waste lagoons at the Southern Illinois University Agricultural Research Center. The two organisms differ from each other and from previously described ClRB and strain WD is the first example of a dissimilatory (per)chlorate-reducing organism in the alpha subclass of the Proteobacteria.

To date there are only six other dissimilatory perchlorate-reducing bacteria that have been described in the literature.[18–22,39] Interestingly, these organisms are quite diverse in their physiological characteristics.[18–22,39] Although phylogenetic comparisons can not be made with most of these isolates because the 16S rDNA sequence data is not available, the three isolates that have been genotypically characterized to date are all members of the Proteobacteria. Similarly to strain PS, two of these, *D. agitatus* strain CKB[22] and *Ideonella dechloratans*,[18] belong to the beta subclass of the Proteobacteria. In contrast, the third, *Wollinella succinogenes* strain HAP-1,[19] is a member of the epsilon subclass of the Proteobacteria. In addition, recent studies in our laboratory have resulted in the isolation of (per)chlorate-reducing bacteria that are members of the gamma subclass of the Proteobacteria.[27] These findings demonstrate that microbial (per)chlorate reduction is spread throughout the Proteobacteria and further emphasizes the diversity of microorganisms capable of this novel metabolism. This diversity is unexpected as the only known natural source of perchlorate that has been identified are mineral deposits in Chile and it is assumed that the presence of oxyanions of chlorine in the environment is due to anthropogenic intervention over the last 100 years which represents a relatively short timeframe for such a diverse group of organisms to have evolved this metabolic capability.

Strain PS shows little physiological resemblance to its closest relative, *R. tenuis*. *R. tenuis* is a purple, non-sulfur photosynthetic organism that cannot respire (per)chlorate or reduce (per)chlorate in cell suspension.[22] In contrast, strain PS is a strictly respiratory, heterotrophic, dissimilatory (per)chlorate-reducer that cannot grow phototrophically. In addition, *R. tenuis* cannot dismutate chlorite in cell suspension. Although *D. agitatus* strain CKB is a more distant relative than *R. tenuis* , there is much greater physiological similarity between this organism and strain PS. Both are capable of coupling the complete oxidation of acetate or other simple organic acids and alcohols to the reduction of chlorate or perchlorate under a broad range of environmental conditions. However, in contrast to strain PS and the previously described (per)chlorate-reducing bacteria, *D. agitatus* strain CKB cannot couple growth to the reduction of nitrate.

In contrast to strain PS, strain WD is a member of the alpha subclass of the Proteobacteria and its closest relatives are the *Magnetospirillum* species. Strain WD is morphologically identical to the known species of this genus, however, there are marked physiological differences between strain WD and its closest relatives. All members of the *Magnetospirillum* genus that have been isolated to date form magnetosomes—an intracellular form of magnetite—when growing microaerophilically on iron-based media which gives these organisms a unique magnetotactic characteristic. In contrast, strain WD does not produce the fine grained magnetite characteristic of magnetotactic bacteria when grown microaerophilically on iron-based medium. In addition, none of the *Magnetospirillum* species tested could grow by dissimilatory (per)chlorate reduction or could dismutate chlorite into chloride and $O_2$.

Strain WD and strain PS were both isolated from swine waste lagoons and represent the first described examples of (per)chlorate-reducing bacteria that were isolated from an environment not known to be contaminated with oxyanions of chlorine. Even though the two strains differ morphologically, their physiologies are very similar (temperature, pH, salinity optima) and optimum growth for both organisms is observed on short chain fatty acids and dicarboxylic acids while respiring on perchlorate, chlorate, nitrate or oxygen. In contrast to most other known (per)chlorate-reducing isolates and similarly to its closest relatives, strain WD is a microaerophile. Of the previously described (per)chlorate-reducing bacteria, only *Wolinella succinogenes* strain HAP-1 is known to be microaerophilic.[40] However, in contrast to strain WD, *W. succinogenes* is a member of the epsilon subclass of the Proteobacteria and incompletely oxidizes organic compounds while growing with oxyanions of chlorine.[19] *W. succinogenes* can also couple the oxidation of $H_2$, an important endproduct of microbial fermentation processes, to the reduction of (per)chlorate which is unique among the described (per)chlorate-reducing bacteria.

Growth rates on either perchlorate or chlorate for both strains PS and WD are comparable over a broad range of (per)chlorate concentrations. Strains WD and PS grow best at a concentration of 10 mM of either electron acceptor respectively which is similar to previous observations made for *D. agitatus* strain CKB.[22] Growth rates of both strains PS and WD are not significantly decreased even at concentrations of 40 mM chlorate/perchlorate. Strain WD is remarkably tolerant to perchlorate and is capable of growth in concentrations as high as 80 mM or approximately 7000 ppm. Similarly to strain GR-1[41] and *D. agitatus* strain CKB,[22,24,26,27] both strains PS and WD demonstrate the ability to dismutate chlorite into $O_2$ and chloride in washed whole-cell suspensions. Although this metabolism was until recently assumed to be unique to the (per)chlorate-reducing organism strain GR-1, studies in our laboratory have demonstrated that this metabolism is also found in more than twenty other perchlorate-reducing isolates obtained from a broad diversity of environments and suggest that this is probably the central enzyme in the reductive pathway for chlorate and perchlorate used (per)chlorate-reducing microorganisms.[22,27]

The distinct phenotypic and genotypic differences of strains WD and PS from each other, from their closest relatives, and from the previously described (per)chlorate-reducing bacteria indicate that these organisms represent two new genera of perchlorate-reducers in the Proteobacteria. The names *Dechlorisoma suillus* and *Dechlorospirillum anomalous* are proposed for strains PS and WD, respectively.

## ACKNOWLEDGMENTS

Support for this research was in part from grant DE-FG02-98ER62689from the Department of Energy to J.C. and L.A. and from the 1998 Oak Ridge Associated Universities Junior Faculty award to J.C.

## REFERENCES

1. Urbansky, E.T. "Perchlorate chemistry: implications for analysis and remediation." *Bioremediation Journal* **1998**, *2,* 81–95.

2. Germgard, U.; Teder, A.; Tormund, D. "Chlorate formation during chlorine dioxide bleaching of softwood kraft pulp." *Paperi ja Puu* **1981**, *3,* 127–133.

3. Rosemarin, A.; Lehtinen, K.; Notini, M. "Effects of treated and untreated softwood pulp mill effluents on Baltic sea algae and invertebrates in model ecosystems." *Nordweigen Pulp and Paper Research Journal* **1990**, *2,* 83–87.

4. Agaev, R.; Danilov, V.; Khachaturov, V.; Kasymov, B.; Tishabaev, B. "The toxicity to warm-blooded animals and fish of new defoliants based on sodium and magnesium chlorates." *Uzbekistan Biologiya Zhurnal* **1986**, *1,* 40–43.

5. Urbanski, T. *Composite propellants*. Pergamon: Oxford, England (UK), 1988; pp. 602–620.

6. Siddiqui, M. "Chlorine-ozone interactions: formation of chlorate." *Water Research* **1996**, *30,* 2160–2170.

7. Cosson, H.; Ernst, W.R. "Photodecomposition of chlorine dioxide and sodium chlorite in aqueous solution by irradiation with ultraviolet light." *Industrial Engineering Chemistry Research* **1994**, *33,* 1468–1475.

8. Stanbury, J.B.; Wyngaarden, J.B. "Effect of perchlorate on the human thyroid gland." *Metabolism* **1952**, *1,* 533–539.

9. Calabrese, E.J.; Moore, G.; Brown, R. "Effects of environmental oxidant stressors on individuals with a G-6-PD deficiency with particular reference to an animal model." *Environmental Health Perspectives* **1979**, *29,* 49–55.

10. Heffernan, W.P.; Guion, C.; Bull, R.J. "Oxidative damage to erythrocyte induced by sodium chlorite in vivo." *Journal of Environmental Pathology and Toxicology* **1979**, *2,* 1487–1499.

11. Singelmann, E.; Wetzel, E.; Adler, G.; Steffen, C. "Erythrocyte membrane alterations as the basis of chlorate toxicity." *Toxicology* **1984**, *30,* 135–147.

12. Van Wijk, D.J.; Hutchinson, T.H. "The ecotoxicity of chlorate to aquatic organisms: a critical review." *Ecotoxicology and Environmental Safety* **1995**, *32,* 244–253.

13. van Wijk, D.J.; Croon, S.G.M. Garttener-Arends, I.C.M. "Toxicity of chlorate and chlorite to selected species of algae, bacteria, and fungi." *Ecotoxicology and Environmental Safety* **1998**, *40,* 206–211.

14. Rosemarin, A.; Mattsson, J.; Lehtinen, K.; Notini, M. Nylen, E. "Effects of pulp mill chlorate ($ClO_3^-$) on *Fucus vesiculosus*—a summary of projects." *Ophelia* **1986**, *4,* 219–224.

15. De Groot, G.N.; Stouthamer, A.H. "Regulation of reductase formation in *Proteus mirabilis*. I. Formation of reductases and enzymes of the formic hydrogenlyase complex in the wild type and in chlorate resistant mutants." *Archives of Microbiology* **1969**, *66,* 220–233.

16. Roldan, M.D.; Reyes, F.; Moreno-Vivian, C. Castillo, F. "Chlorate and nitrate reduction in phototrophic bacteria *Rhodobacter capsulatus* and *Rhodobacter sphaeroides*." *Current Microbiology,* 1994, *29,* 241–245.

17. Korenkov, V.; Romanenko, V.; Kuznetsov, S. Voronov, J. "Process for purification of industrial waste waters from perchlorates and chlorates." U.S. Patent 3,943,055, 1976.

18. Malmqvist, A.; Welander, T.; Moore, E.; Ternstrom, A.; Molin, G. Stenstrom, I.-M. "*Ideonella dechloratans* gen. nov., sp. nov., a new bacterium capable of growing anaerobically with chlorate as an electron acceptor." *Systematic and Applied Microbiology* **1994**, *17,* 58–64.

19. Wallace, W.; Ward, T.; Breen, A.; Attaway, H. "Identification of an anaerobic bacterium which reduces perchlorate and chlorate as *Wolinella succinogenes*." *Journal of Industrial Microbiology* **1996**, *16,* 68–72.

20. Stepanyuk, V.; Smirnova, G.; Klyushnikova, T.; Kanyuk, N.; Panchenko, L.; Nogina, T.; Prima, V. "New species of the *Acinetobacter* genus *Acinetobacter thermotoleranticus* sp. nov." *Mikrobiologiya* **1992**, *61,* 347–356.

21. Rikken, G.; Kroon, A.; Van Ginkel, C. "Transformation of (per)chlorate into chloride by a newly isolated bacterium: reduction and dismutation." *Applied Microbiology and Biotechnology* **1996**, *45,* 420–426.

22. Bruce, R.A.; Achenbach, L.A.; Coates, J.D. "Reduction of (per)chlorate by a novel organism isolated from a paper mill waste." *Environmental Microbiology* **1999**, *1,* 319–331.

23. Bruce, R.A. *The Microbiology and Bioremediative Potential of (Per)chlorate-Reducing Bacteria.* Master's Thesis. Southern Illinois University: Carbondale, IL, 1999.

24. Coates, J.D.; Bruce, R.A.; Patrick, J.A.; Achenbach, L.A. "Hydrocarbon bioremediative potential of (per)chlorate-reducing bacteria." *Bioremediation Journal,* **1999**, *3,* in press.

25. Coates, J.D.; Bruce, R.A.; Patrick, J.; Achenbach, L.A. "Stimulation of benzene degradation in anaerobic sediments by microbial chlorite dismutation." *Proceedings of the 5th International Petroleum Environmental Conference,* 1999.

26. Coates, J.D.; Bruce, R.A.; Haddock, J.D. "Anoxic bioremediation of hydrocarbons." *Nature* **1998**, *396,* 730.

27. Coates, J.D.; Michaelidou, U.; Bruce, R.A.; O'Connor, S.M.; Crespi, J.N.; Achenbach, L.A. "The ubiquity and diversity of dissimilatory (per)chlorate-reducing bacteria." *Applied and Environmental. Microbiology* **1999**, *65,* 5234–5241.

28. Achenbach, L.A.; Bruce, R.A.; Michaelidou, U.; Coates, J.D. "Phylogenetic analysis of dissimilatory (per)chlorate-reducing bacteria and development of molecular probes." *Applied and Environmental Microbiology* **1999**, in press.

29. Hackenthal, E.; Mannheim, W.; Hackenthal, R.; Becher, R. "Die reduktion von perchlorat durch bakterien I. Untersuchungen an intakten zellen." *Biochemical Pharmacology* **1964**, *13,* 195–206.

30. Hackenthal, E. "Die reduktion von perchlorat durch bacterien II. Die identitat der nitratreduktase und des perchlorat reduzierenden enzyms aus *B. cereus.*" *Biochemical Pharmacology* **1965**, *14,* 1313–1324.

31. Stouthamer, A. "Nitrate reduction in *Aerobacter aerogenes.* I. Isolation properties of mutant strains blocked in nitrate assimilation and resistant against chlorate." *Archives of Microbiology* **1967**, *56,* 68–75.

32. Coates, J.D.; Michaelidou, U.; O'Connor, S.M.; Bruce, R.A.; Achenbach, L.A. "The diverse microbiology of (per)chlorate reduction." In THIS VOLUME: *Perchlorate in the Environment,* E.T. Urbansky, Ed. Kluwer/Plenum: New York, NY, 2000; Ch. 24.

33. Tayeh, M.A.; Madigan, M.T. "Malate dehydrogenase in phototrophic purple bacteria: purification, molecular weight, and quaternary structure." *Journal of Bacteriology* **1987**, *169,* 4196–4202.

34. Hungate, R.E. "A roll tube method for cultivation of strict anaerobes." *Methods in Microbiology* **1969**, *3B,* 117–132.

35. Nickrent, D.L. "From field to film: rapid sequencing methods for field-collected plant species." *BioTechniques* **1994**, *16,* 470–475.

36. Maidak, B.L.; Olsen, G.J.; Larsen, N.; Overbeek, R.; McCaughey, M.J.; Woese, C.R. "The RDP (Ribosomal Database Project)." *Nucleic Acids Research* **1997**, *25,* 109–111.

37. Gilbert, D.G. "SeqApp. version 1.9a157" Biocomputing Office, Biology Department, Indiana University: Bloomington, IN, 1993.

38. Swofford, D.L. "PAUP*: Phylogenetic Analysis Using Parsimony (and other methods)." 1999, Version 4.0. Sinauer Associates: Sunderland, MA. Smithsonian Institution: Washington, D.C.

39. Romanenko, V.I.; Korenkov, V.N.; Kuznetsov, S.I. "Bacterial decomposition of ammonium perchlorate." *Mikrobiologiya* **1976**, *45*, 204–209.

40. Wallace, W.; Beshear, S.; Williams, D.; Hospadar, S.; Owens, M. "Perchlorate reduction by a mixed culture in an up-flow anaerobic fixed bed reactor." *Journal of Industrial Microbiology and Biotechnology* **1998**, *20*, 126–131.

41. Van Ginkel, C.; Rikken, G.; Kroon, A.; Kengen, S. "Purification and characterization of chlorite dismutase: a novel oxygen-generating enzyme." *Archives of Microbiology* **1996**, *166*, 321–326.

# Contributors

**Achenbach, Laurie A.**, Chapters 24,[®] 25[®]
Department of Microbiology and Center for Systematic Biology
Southern Illinois University
Carbondale, Illinois 62901

**Alexandratos, Spiro D.**, Chapter 15[®], 16[®]
Department of Chemistry
University of Tennessee
Knoxville, Tennessee 37996

**Amadei, Giulio A.**, Chapter 9[®]
Department of Chemistry
Georgetown University
Washington, DC 20057

**Batista, Jacimaria R.**,[*+] Chapter 13[®]
Department of Civil and Environmental Engineering
University of Nevada–Las Vegas
4505 Maryland Parkway
Las Vegas, Nevada 89154-4015
Email: Jaci@ce.UNLV.edu

**Bonnesen, Peter V.**, Chapter 15[®]
Chemical and Analytical Sciences Division
Oak Ridge National Laboratory
Oak Ridge, Tennessee 37831

**Brown, Gilbert M.**,[*+] Chapter 15[®], 16[®]
Chemical and Analytical Sciences Division
Oak Ridge National Laboratory
Oak Ridge, Tennessee 37831
Email: BrownGM1@ORNL.gov

**Bruce, Royce A.**, Chapter 24[®]
Department of Microbiology and Center for Systematic Biology
Southern Illinois University
Carbondale, Illinois 62901

**Cassady, Rachael S.**, Chapter 6[®]
Utah Department of Environmental Quality
288 North 1460 West
Salt Lake City, Utah 84114-2102

**Chaudhuri, Sanwat**, Chapter 6[®]
Utah Department of Health Laboratory
46 North Medical Drive
Salt Lake City, Utah 84113

[*]Indicates corresponding author; email addresses given for corresponding authors only.
[®]Indicates material was presented at the 218th national ACS meeting.
[+]Indicates author who presented paper at the 218th national ACS meeting.

**Clark, James J.J.,*⁺** Chapter 3
Komex•H2O Science, Inc.
11040 Santa Monica Boulevard, Suite 335
Los Angeles, California 90025
Email: JClark@KomexH2O.com

**Clewell, Rebecca A.,⁺** Chapters 6,® 7
U.S. Air Force Research Laboratory
Human Effectiveness Directorate
Toxicology Branch
Building 79, 2856 G Street
Wright-Patterson AFB, Ohio 45433-7400

**Clifford, Dennis A.,*** Chapter 12®
Civil and Environmental Engineering Department
University of Houston
4800 Calhoun Road
Houston, Texas 77204
Email: DAClifford@UH.edu

**Coates, John D.,*⁺** Chapters 24,® 25®
Department of Microbiology and Center for Systematic Biology
Southern Illinois University
Carbondale, Illinois 62901
Email: JCoates@micro.SIU.edu

**Cox, Evan E.,*⁺** Chapter 21®
GeoSyntec Consultants, Inc.
160 Research Lane, Suite 206
Guelph, Ontario N1G 5B2
Canada
Email: ECox@Geosyntec.com

**Dale, James A.**, Chapter 16®
Purolite International
Cowbridge Road
Pontyclun, Wales CF72 8YL (UK)

**Damian, Paul**, Chapter 4®
Earth Tech
Englewood, Colorado 80111

**Dickson, Steve**, Chapter 6
Utah Department of Health Laboratory
46 North Medical Drive
Salt Lake City, Utah 84113

**Earley, Joseph E., Sr.,*⁺** Chapter 9®
Department of Chemistry
Georgetown University
Washington, DC 20057
Email: EarleyJ@GUsun.Georgetown.edu

**Eaton, Andrew D.**, Chapter 4
Montgomery Watson Laboratories
Pasadena, California 91101

**Edwards, Elizabeth**, Chapter 21[⊗]
University of Toronto
Department of Chemical Engineering and Applied Chemistry
Toronto, Ontario M5S 1A1
Canada

**Eldridge, J. Eric**, Chapters 6,[⊗] 7
U.S. Air Force Research Laboratory
Human Effectiveness Directorate
Toxicology Branch
Building 79, 2856 G Street
Wright-Patterson AFB, Ohio 45433-7400

**Espenson, James H.**,* Chapter 1
Ames Laboratory and Department of Chemistry
Iowa State University of Science and Technology
Ames, Iowa 50011
Email: Espenson@AmesLab.gov

**Flowers, Tracey C.**,[+] Chapter 17[⊗]
Department of Civil and Environmental Engineering
University of California–Berkeley
Berkeley, California 94720-1710

**Frankenberger, William T., Jr.**,* Chapter 19[⊗]
Department of Soil and Environmental Sciences
University of California–Riverside
Riverside, California 92521
Email: WilliamF@orange.UCR.edu

**Giblin, Tara L.**,[+] Chapter 19[⊗]
Department of Soil and Environmental Sciences
University of California–Riverside
Riverside, California 92521

**Gokhale, Swati**, Chapter 5[⊗]
Dionex Corporation
1228 Titan Way
Sunnyvale, California 94088-3606

**Greene, Mark R.**,*[+] Chapter 23[⊗]
Envirogen, Inc.
Princeton Research Center
4100 Quakerbridge Road
Lawrenceville, New Jersey 08648
Email: Greene@Envirogen.com

**Gu, Baohua,***+ Chapter 15®, 16®
Environmental Sciences Division
Oak Ridge National Laboratory
Oak Ridge, Tennessee 37831
Email: GuB1@ORNL.gov

**Gurol, Mirat D.,*** Chapter 10®
Department of Civil and Environmental Engineering
San Diego State University
San Diego, California 92182-1324
Email: MGurol@mail.SDSU.edu

**Guter, Gerald A.,***+ Chapter 11®
Guter Consulting
215 Monte Vista
San Clemente, California 92672
Email: ionexchange@earthlink.net

**Herman, David C.,** Chapter 19®
Department of Soil and Environmental Sciences
University of California–Riverside
Riverside, California 92521

**Hunt, James R.,*** Chapter 17®
Department of Civil and Environmental Engineering
University of California–Berkeley
Berkeley, California 94720-1710
Email: Hunt@CE.Berkeley.edu

**Jackson, Peter E.,***+ Chapter 5®
Dionex Corporation
1228 Titan Way
Sunnyvale, California 94088-3606
Email: Peter_Jackson@Dionex.com

**Jennings, Dale L.,** Chapter 14
Calgon Carbon Corporation
500 Calgon Carbon Drive
Pittsburgh, Pennsylvania 15205

**Kim, Kyehee,**+ Chapter 10®
Department of Civil and Environmental Engineering
San Diego State University
San Diego, California 92182-1324

**Klara, Scott M.,** Chapter 14
Calgon Carbon Corporation
500 Calgon Carbon Drive
Pittsburgh, Pennsylvania 15205

**Logan, Bruce E.,***+ Chapter 18®
Department of Civil and Environmental Engineering
The Pennsylvania State University
University Park, Pennsylvania 16802
Email: BLogan@PSU.edu

**Long, John R.**,[*+] Chapter 2[⊗]
GFS Chemicals, Inc.
Columbus, Ohio 43222
Email: JRLong@GFSchemicals.com

**Magnuson, Matthew L.**, Chapter 8[⊗]
United States Environmental Protection Agency
Office of Research and Development
National Risk Management Research Laboratory
Water Supply and Water Resources Division
26 West Martin Luther King Drive
Cincinnati, Ohio 45268

**Mattie, David R.**, Chapter 7[⊗]
U.S. Air Force Research Laboratory
Human Effectiveness Directorate
Toxicology Branch
Building 79, 2856 G Street
Wright-Patterson AFB, Ohio 45433-7400

**McGarvey, Frank X.**, Chapter 13[⊗]
39 Woodbury Drive
Cherry Hill, New Jersey 08003

**Michaelidou, Urania**,[+] Chapters 24,[⊗] 25[⊗]
Department of Microbiology and Center for Systematic Biology
Southern Illinois University
Carbondale, Illinois 62901

**Moyer, Bruce A.**, Chapter 15[⊗]
Chemical and Analytical Sciences Division
Oak Ridge National Laboratory
Oak Ridge, Tennessee 37831

**Neville, Scott**, Chapter 22[⊗]
Aerojet General Corporation
Sacramento, California 95670

**Nzengung, Valentine A.**,[*+] Chapter 21[⊗]
Department of Geology
University of Georgia
Athens, Georgia 30602
Email: VNzengun@UGA.edu

**Ober, Robert**, Chapter 15[⊗], 16[⊗]
Department of Chemistry
University of Tennessee
Knoxville, Tennessee 37996

**O'Connor, Susan M.**, Chapter 24[⊗]
Department of Microbiology and Center for Systematic Biology
Southern Illinois University
Carbondale, Illinois 62901

**Patel, Vijay**, Chapter 15[®]
Department of Chemistry
University of Tennessee
Knoxville, Tennessee 37996

**Pitre, Michael P.**, Chapter 23[®]
formerly with Envirogen, Inc.
Princeton Research Center
4100 Quakerbridge Road
Lawrenceville, New Jersey 08648

**Plant, Steven**, Chapter 16[®]
Purolite International
Cowbridge Road
Pontyclun, Wales CF72 8YL (UK)

**Pontius, Frederick W.**,[*+] Chapter 4[®]
Regulatory Affairs Consultant
Lakewood, Colorado 80226
Email: Pontius@polnow.net
Formerly with American Water Works Association

**Robles, Heriberto**,[*+] Chapter 20[®]
LFR Levine Fricke, Inc.
1920 Main Street, Suite 750
Irvine, California 92614-7211
Email: Heriberto.Robles@LFR.com

**Rohrer, Jeff S.**, Chapter 5[®]
Dionex Corporation
1228 Titan Way
Sunnyvale, California 94088-3606

**Tofan, Daniel C.**, Chapter 9[®]
Department of Chemistry
Georgetown University
Washington, DC 20057

**Tripp, Anthony R.**,[+] Chapter 12[®]
Civil and Environmental Engineering Department
University of Houston
4800 Calhoun Road
Houston, Texas 77204

**Tsui, David T.**,[*] Chapters 6,[®] 7
U.S. Air Force Research Laboratory
Human Effectiveness Directorate
Toxicology Branch
Building 79, 2856 G Street
Wright-Patterson AFB, Ohio 45433-7400
Email: David.Tsui@he.WPAFB.AF.mil

**Urbansky, Edward T., Editor,***+ Chapter 8®
United States Environmental Protection Agency
Office of Research and Development
National Risk Management Research Laboratory
Water Supply and Water Resources Division
26 West Martin Luther King Drive
Cincinnati, Ohio 45268
Email: Urbansky.Edward@EPA.gov

**Venkatesh, K. Raman,** Chapter 14
Calgon Carbon Corporation
500 Calgon Carbon Drive
Pittsburgh, Pennsylvania 15205

**Vieira, Adriano R.**, Chapter 13®
Department of Civil and Environmental Engineering
University of Nevada–Las Vegas
4505 Maryland Parkway
Las Vegas, Nevada 89154-4015

**Wagner, Norman J.**, Chapter 14
Calgon Carbon Corporation
500 Calgon Carbon Drive
Pittsburgh, Pennsylvania 15205

**Wallner, William M.**, Chapter 6®
Utah Department of Environmental Quality
288 North 1460 West
Salt Lake City, Utah 84114-2102

**Wang, Chuhua**, Chapter 21®
Technic, Inc.
1 Spectacle Street
Cranston, Rhode Island 02910

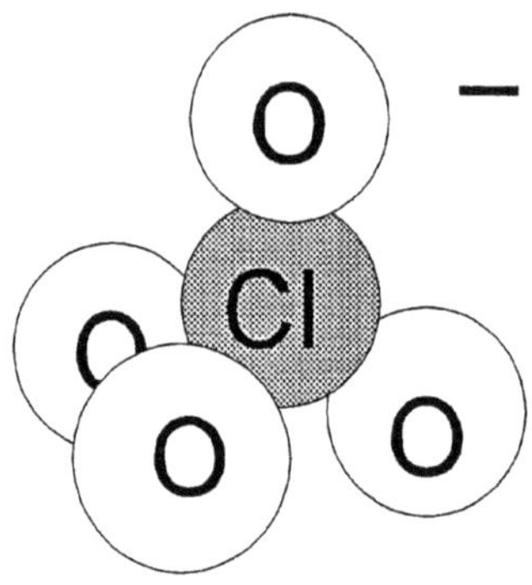
O
−
Cl
O
O
O

# Index

* The asterisk indicates a structure, photograph, drawing, or other illustration.

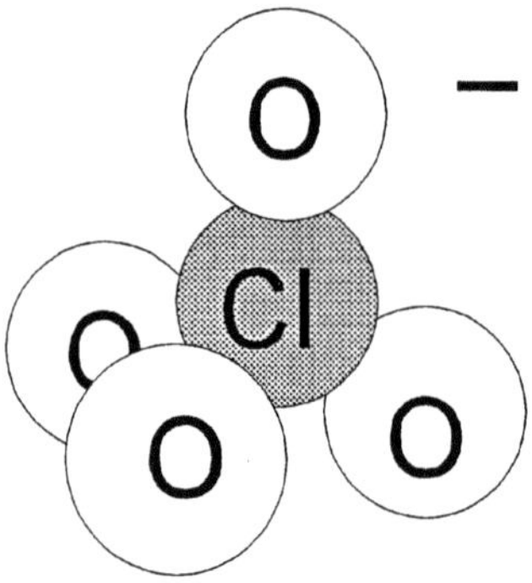
O
Cl
O
O
O
−

# Common abbreviations and initials

**AFRL** U.S. Air Force Research Laboratories. Perchlorate group is at Wright-Patterson AFB, Dayton, Ohio.

**CDHS** California Department of Health Services, state agency that oversees drinking water, responsible for many of the initial ion chromatography developments that led to California's actions on wells.

**DoD** U.S. Department of Defense, the federal executive cabinet agency that oversees the military.

**DOE** U.S. Department of Energy.

**EPA** U.S. Environmental Protection Agency, the federal executive agency that has much of the responsibility for environmental regulation, including drinking water. Groundwater concerns are shared with the U.S. Department of Agriculture. Bottled water is under the Food and Drug Administration.

**IPSC** Interagency Perchlorate Steering Committee. IPSC is composed of employees from multiple federal agencies. Not to be confused with PSG.

**NCEA** National Center for Environmental Assessment, part of EPA's ORD, responsible for evaluating ecological and toxicological data and establishing levels appropriate for health, such as NOAELs and LOAELs, perchlorate group is located in Research Triangle Park, NC.

**NCERQA** National Center for Environmental Research and Quality Assurance, part of EPA's ORD, EPA's main center for sponsored research, NCERQA sponsors external studies through grants and other funding vehicles.

**NERL** National Exposure Research Laboratory, part of part of EPA's ORD, responsible for assessing materials for possible exposure routes and the extent of exposure in the environment. Perchlorate group is in Athens, GA.

**NHEERL** National Health and Environmental Effects Research Laboratory, part of EPA's ORD, responsible for performing fundamental ecological and toxicological studies. Perchlorate group is located in Research Triangle Park, NC.

**NRMRL** National Risk Management Research Laboratory, part of EPA's ORD, responsible for research on strategies and technologies to meet health and regulatory objectives, the Editor's duty station. Perchlorate group is in Cincinnati, OH.

**ORD** EPA's Office of Research and Development.. It is composed of NERL, NRMRL, NCEA, NHEERL, and NCERQA.

**OSWER** EPA's Office of Solid Waste and Emergency Response, one of the program offices responsible for establishing regulations, in this case dealing mostly with contaminated soil, waste disposal sites, and Superfund/CERCLA sites through its Office of Emergency and Remedial Response, OERR.

**OW** EPA's Office of Water, one of the program offices responsible for establishing regulations, in this case National Primary Drinking Water Regulations (**NPDWRs**), the Unregulated Contaminants Monitoring Rule (**UCMR**), and the Contaminant Candidate List (**CCL**). OW's **OGWDW** (Office of Groundwater and Drinking Water) carries out this work.

**PSG** Perchlorate Study Group, composed of industry (producers and consumers of ammonium perchlorate) and other parties potentially responsible for perchlorate clean-up. This is not a federal government agency.

**TSC** Technical Support Center, usually refers to OW's TSC in Cincinnati, OH, which does research on test method development and other issues directly of concern to OW.

MIX
Papier aus verantwortungsvollen Quellen
Paper from responsible sources
FSC® C105338

If you have any concerns about our products,
you can contact us on
**ProductSafety@springernature.com**

In case Publisher is established outside the EU,
the EU authorized representative is:
**Springer Nature Customer Service Center GmbH**
**Europaplatz 3, 69115 Heidelberg, Germany**

Printed by Libri Plureos GmbH
in Hamburg, Germany